Developments
in
Structural Form

Rowland J. Mainstone

Developments
in
Structural Form

Architectural Press

OXFORD BOSTON JOHANNESBURG MELBOURNE NEW DELHI SINGAPORE

Architectural Press
An imprint of Butterworth-Heinemann
Linacre House, Jordan Hill, Oxford OX2 8DP
225 Wildwood Avenue, Woburn, MA 01801-2041
A division of Reed Educational and Professional Publishing Ltd

℞ A member of the Reed Elsevier plc group

First published by Allen Lane in association with RIBA Publications Ltd 1975
Reprinted by Allen Lane and first published in Penguin Books 1983
Second edition 1998

British Library Cataloguing in Publication Data
A catalogue record for this book is available from the British Library

Library of Congress Cataloguing in Publication Data
A catalogue record for this book is available from the Library of Congress

ISBN 0 7506 2893 6

Printed and bound in Great Britain

Chapter 3: Structural materials   47

Chapter 4: Construction and form   65

Chapter 5: Structure and form   83

Part 2: Structural elements

Part 3: Complete structures

Part 4: Design

Preface to the first edition

Professionally I have been chiefly concerned with structural forms of today – with research into the design and structural behaviour of some of those that we now use for our buildings and bridges. But I have long been almost equally interested in their earlier counterparts and the ways in which these have been conceived and built, and have never found it easy to regard this interest as something apart.

All the forms, new and old, have at least one thing in common: none was conceived in isolation. All those of which we now know anything have owed much to what had been built before, even if the debt has varied greatly both in extent and in the way in which it has been incurred. The chief difference in this respect is simply that in the past it was, of necessity, incurred more directly. Today, thanks to a large body of abstract theory about possible structural actions and related almost equally abstract principles of design, much of it may be incurred only indirectly. But, for various reasons, I still do not see it as desirable or even truly possible that such abstract theory and principles should ever become the sole basis of otherwise completely *ab initio* design.

Underlying this similarity has always been the same basic structural requirement that the form shall stand more or less immobile under whatever loads it is called upon to bear, and the same limited range of structural actions has, in principle, been available to meet it. The chief differences between the forms themselves have stemmed from the sometimes very different non-structural requirements, from the often less onerous loadings in the past, and from the more restricted and simpler range of actions that was usually exploited then on account partly of more limited technical resources and partly of a more limited understanding of the actions. By virtue of this characteristically greater simplicity, the actions of the earlier forms are usually more easily legible than those of the forms built today if both are viewed in terms of present-day understanding. Quite apart from their importance in the actual sequences of development, some of the earlier forms may thus serve as useful introductory illustrations of these actions.

In striving to present a widely intelligible and wholly non-mathematical account of the development of structural forms up to the present day that will not be a mere historical listing but will give some insight, equally relevant to the immediate future, into the real processes of development and the constraints surrounding them, I have therefore adopted a dual approach. On the one hand, I have used some of the forms themselves, particularly some of the earlier ones just referred to, to introduce and illustrate in a concrete way the main possibilities of structural action as we now understand them. On the other, I have made full use of our present understanding to look critically at both our present successes and failures and those of the past. As far as I am aware, this has not been attempted before in recent years. It is probably necessary to look back as far as Viollet-le-Duc and Choisy for a rough parallel, and their subject matter was of course much more limited. I hope at least that I shall not, as a result of a deliberately almost exclusive emphasis on the structural aspects of design, be accused of the same sort of extreme rationalism as they were.

For two reasons I have avoided a simple chronological framework. One is that this has made it easier to concentrate on those developments that seemed structurally most significant. The other and more fundamental one is that the approach just outlined seemed to demand considerable freedom to range backwards and forwards in time. The chief demerit of so doing is that those whose main interest is in the straightforward history of what-was-done-when will, to some extent, have to look elsewhere. What they will find here is, from the purely historical point of view, more of a commentary, though some additional information on dates and designers (where known) will be found in the Index and the figure captions. But, on the credit side, it is hoped that even the historian will in the end benefit, because he should be much better equipped after reading the book to interpret for himself forms and situations to which no specific reference has been made.

The more analytical framework actually adopted has indeed made it possible to cater almost equally effectively for several distinct types of reader. I have had in mind four groups in particular: architects and architectural students, engineers and engineering students, architectural historians and archaeologists, and the general reader with an intelligent and enquiring interest in the man-made world around him. Material that will be of varying interest to these different groups has been arranged in five introductory chapters supplemented by a Glossary, and in the concluding chapter. Some suggestions about the orders in which these chapters might be read by, for instance, those with a purely historical interest will be found at the end of Chapter 1. The book has however been conceived as a whole and I feel that the greatest benefit will come from reading it, eventually, as a whole.

Finally a few words may be appropriate about the selection of examples for discussion. In comparison with most histories, the book is I think unusual in that, while giving most attention to the developments of the past two hundred years or so, it gives almost as much space to earlier ones, rather than concentrating almost exclusively on one set or the other. In part this is simply a natural consequence of the approach that has already been outlined. But it does also reflect my assessment of relative long-term significance from a structural point of view. Within each broad period I have tried to adopt similar criteria of long-term significance, except that I have preferred always to discuss a few examples in some depth rather than do little more than list a larger number. Where, as often, this has still left a choice, I have always preferred to discuss examples that I knew from personal inspection and was able to illustrate fairly adequately. But I have tried to avoid any undue emphasis on those that I happened to know best.

The need for a book of this kind was first put to me by the Professional Literature Committee of the Royal Institute of British Architects, and it was initially commissioned, at their suggestion, as part of the 'gap filling' programme of RIBA Publications Limited but with the welcome encouragement to write with the needs of a considerably wider readership in mind. I was doubly fortunate in having also, from the start, the interest of Allen Lane under whose imprint the book now appears and, in particular, of John Fleming and Hugh Honour as their architectural editors. I have been most grateful for their continued support and constructive criticism. I am also grateful to the Director of the Building Research Establishment not only for permission to write the book (in a private capacity) but also for the grant of two periods of special leave without which it would have been impossible to give to it the necessary continued attention. The opinions expressed are, of course, entirely my own.

For early stimulus and encouragement in pursuing the studies on which it is based, I should like to thank Sir Frederick Lea, Director of Building Research from the time when I joined the Station until 1965, William Allen, George Atkinson, Dr Norman Davey, Graham Denyer, and Dr Stanley Hamilton among my past and present colleagues there, and Sir Hubert Shirley Smith. The last two have also been among those who have been kind enough to read the manuscript in whole or in part and to comment helpfully on it. I owe a further debt of this kind to those Schools of Architecture, Engineering, and Art History, particularly the Institute of Advanced Architectural Studies at York, that have from time to time provided me with a forum for exploring some of the basic ideas.

My largest debt is unfortunately impossible to acknowledge adequately. It is to all those, whether known to me personally or not, on whose earlier researches I have drawn, and especially to those who have been kind enough to send me copies of their published work or answer queries or discuss particular points with me; to the libraries of which I have also made extensive use in this connection; to those with whom I have studied in detail a number of structures both old and new; and to the many more with whom I have visited and examined others, who have otherwise assisted me in seeking them out or gaining access to them, or whose hospitality has made my travels more enjoyable and rewarding than they might otherwise have been. The first part of this debt is acknowledged to some extent in the Notes and Bibliography. Of libraries I have made the greatest use of the British Museum. But I should also like to mention the considerable help that I have received from the librarians and library staffs of the Building Research Station, the Institution of Civil Engineers, the Royal Institute of British Architects, the Victoria and Albert Museum, and the Warburg Institute, and the special courtesies of the Bibliothèque Nationale in allowing me to examine the priceless sketchbook of Villard de Honnecourt and of the Houghton Library at Harvard in loaning another rare volume. As for the more personal debts, I regret that I cannot list more than a few names, running thereby the risk of seeming ungrateful to those whose names are omitted. But it would seem even more ungrateful to make no personal acknowledgements of this kind, so I hope that the names that follow will be read as standing for a much larger number: Professor David Billington, Blair Birdsall, Miss Edith Clay, Professor George Collins, Professor Kenneth Conant, Dr Yolande Crowe, Bay Feridun Dirimtekin, Professor Sedad Eldem, Professor John Fitchen, Professor James Fitzgibbon, Neal FitzSimons, Professor George Forsyth, Nomer Gray, Professor Guido Guerra, Professor Martin Harrison, Ernest Hawkins, Dr Basil Hennessy, Professor Jacques Heyman, Richard Huber, Professor Tine Kurent, Dr Caro Minasian, Dr Gamal ed-Din Mukhtar, Frank Prager, Professor Howard Saalman, Professor Piero Sanpaolesi, Professor Alec Skempton, David Stronach, Dr John Thacher (through whom the Dumbarton Oaks Center for Byzantine Studies, Harvard University, sponsored the largest single study that I have made of any early structure), Dr

Ahmed Tutenk, Robert Van Nice, Robert Vogel, and John Ward-Perkins.

Acknowledgement for photographs and drawings not my own is made in the list of illustrations, but I am particularly grateful to Professors Conant and Forsyth for the unique prints which they have allowed me to reproduce. Though most of the photographs are my own they were nearly all taken as colour transparencies. For the skill and painstaking care that he has brought to the task of making black-and-white negatives and prints of these for reproduction I am deeply indebted to John Russell. For their equal skill, care, and patience in preparing the manuscript for publication, I am similarly indebted to my publishers, and especially to Miss Sylvia Bruce and Gerald Cinamon. Mrs Joanna Sharp prepared the first draft of the place and name indexes.

The text was largely completed in December 1971. References in the notes cover most of the relevant literature up to that date, albeit selectively. References to a few works published subsequently have been added while the book was in the press, but these additions have necessarily been much more selective.

Rowland Mainstone

January 1974

Preface to this edition

When this book was first reissued in 1982, a major revision did not seem justified. Much has happened subsequently to change this situation. For too long, other commitments nevertheless made it difficult to find the time for extensive revision. I am grateful for patience of all those who continued to press for a new edition and am happy to have been able finally to undertake it.

Most obviously, there have been major developments in actual construction and – through the availability of ever more powerful computers – in the ways in which designers work. At the same time, there has been far more research into past achievements. To give due weight both to all the new developments, and to the large number of publications stemming from historical research, has not been easy without an excessive increase in length. Some abbreviation of the earlier text, and particularly of the Notes, has been unavoidable. So has the omission of a considerable number of earlier illustrations. But the same overall balance has been sought – based primarily on assessments of long-term significance and relevance to the situation in which we now find ourselves. The chief additions will be naturally found in the discussions of recent developments in Chapters 6 to 10 and 13 to 15. The treatment of design in Chapter 16 has also been extensively revised. As before, wherever a choice between examples of equal relevance occurred, I have preferred to consider those of which I had personal knowledge and could illustrate best.

The aim remains that of giving the reader, whatever his prime interest, a rounded and creative understanding of the pattern and processes of development and, thereby, also of future possibilities. In doing this, chronology cannot be ignored. Time and place have always partly determined objectives, precedents, and available means. But, given some knowledge of these, it is more rewarding to look at the challenges that designers have faced, and at the ways in which they have been met and the resulting achievements.

The main text should, again, be intelligible to the general reader without specialist knowledge, either historical or structural. But the interests of those with more specialist knowledge and/or a desire to enquire more deeply into some aspect of the development are served, as before, by full Notes and References at the end, although the references to the literature have had to be more selective than before and the previous Bibliography of suggested further reading has been omitted on account of the frequency with which new texts now appear, only to go out of print a few years later.

Some acknowledgement of my debt to others was made in the Preface to the first edition. That debt is now even greater than it already was 25 years ago, making adequate acknowledgement yet more difficult. Among those to whom I should now like to add my thanks are Dr Bill Addis, Michel Bancon, Yves Boiret, Dr Robert Berger, Julia Elton, Professor Rainer Graefe, Dr Bruce Boucher, Professor Geoffrey Broadbent, Michael Chrimes, Professor Gorgio Croci, Dr Sergei Federov, Sir Bernard Feilden, Professor Wolfram Jäger, Dr Manolis Korres, the late Professor Raymond Lemaire, Professor Fritz Leonhardt, Frank Newby, Hugh Pagan, Professor George Penelis, Professor Andrei Punin, Stephen Rustow, James Sutherland, Ettore Vio, Professor Fritz Wenzel, Professor Mufit Yorulmaz, and Dr Costas Zambas.

Acknowledgements for all photographs and drawings not my own are now made in the captions. But I should like to add special thanks for some of the new ones to David Brown, Anthony Hunt, and Professors Michael Barnes, Horst Berger, Klaus Linkwitz, and Michel Virolgeux. Also to Peter Adler of Kall Kwik, St Albans, for his great help with the copying of early engravings and the like and of a further selection of my colour transparencies as well as with other printing, and to Michael Caldwell for his care in printing a selection of my negatives.

Finally I must express my warmest thanks to my wife Rhoda, to whom with my late first wife, this edition is dedicated. Without her patience, good humour and encouragement during the long processes of gestation and production, this edition could not have appeared.

Rowland Mainstone

June 1998

To Madeleine and Rhoda

Part 1: Introductory

Chapter 1: Introduction

The structural forms with which we shall be concerned are those built to provide shelter, protection, a way over an obstacle, or to serve other closely related but increasingly diverse ends by enclosing space and bridging voids. They cover a wide range of shape and size and manner of construction, and their development spans a period of at least 10 000 years. The simple hut seen under construction in Figure 1.1 looks back towards its beginnings. Its manner of construction has more in common with that of many birds' nests and bowers than with most construction today. The smaller of Nervi's Roman Sports Palaces, built in 1957 to accommodate up to 5 000 spectators, is one modern counterpart [1.2]. It is a very different structure in all but its basic domical form; and it could be built as it was only by drawing on many of the skills and much of the experience acquired in the course of those 10 000 years in building countless other structures to meet more and more challenging requirements. Not all these other structures are of interest today beyond the light that they throw on their own times. Some of them are of great interest though, both structurally and architecturally. On the one hand, they illustrate in various ways the kinds of problems to be overcome in all structural design and the means available for overcoming them. On the other, they mark significant steps forward in the facing of these problems and illustrate the growing freedom of choice of form that

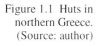

Figure 1.1 Huts in northern Greece. (Source: author)

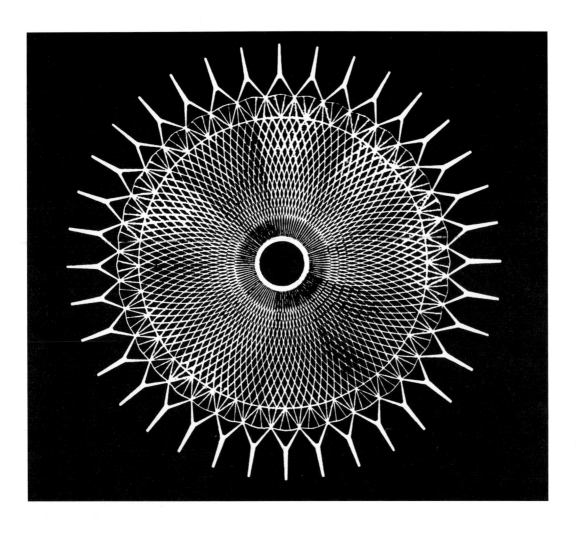

Figure 1.2 Plan of the
dome and its supporting
struts, Small Sports
Palace, Rome (Nervi).
(Source: Cement and
Concrete Association)

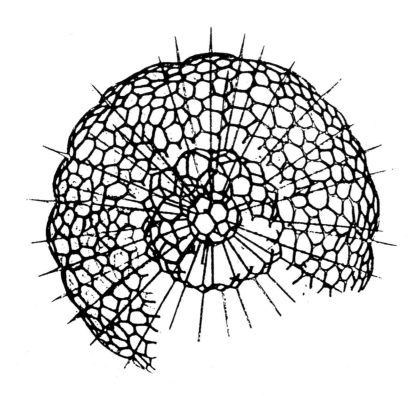

Figure 1.3 Skeleton of the
radiolarian *Actinomma
arcadophorum*.
(from E. Haeckel,
Challenger Monograph,
London, 1887)

has been acquired thereby. It is on such structures, and on their major elements, that attention will be focused.

Natural and man-made forms

Some of these man-made forms may be fairly closely paralleled by ones that occur in nature.[1] Indeed both the hut and the Sports Palace mirror, to some extent, the radial symmetry of such natural forms as the skeletons of microscopic water-borne radiolaria [1.3] and the surface pattern of a sunflower's florets. Yet all built forms also remain sharply differentiated from the world of natural forms even when they have been partly modelled on it. It is worth noting at the outset some of the principal differences.

Most obviously, perhaps, they are differences of scale and function. The closest natural parallels are almost invariably to be found on a scale several orders of magnitude less than that of the built form. The radiolaria, admittedly an extreme example, are so minute that the deposition of silica to form the skeleton is governed by the action of surface tensions and unaffected by gravity. Since it is no part of the skeleton's function to resist uni-directional gravity loading, its symmetry is fully three-dimensional, in conformity with the omni-directional action of the surface tensions. The symmetries of the hut and the Sports Palace, being related to the predominant gravity-loading which they must withstand, are only two-dimensional, existing only at right angles to the gravitational action. Where there is least apparent difference of scale – as when Smeaton modelled the external form of the third Eddystone Lighthouse on the well-rooted and tapering trunk of an oak tree [2] – there are usually other differences of function which are matched by very different internal structures. Here, the tree that served as a model had to do much more than simply resist gravity and stand up to the pressure of wind or waves. It was a living organism, with all the complexity of function that this entails; and that greater complexity of function was matched by a correspondingly greater complexity of internal structure which, as will be seen later, differentiates timber as a building material from other materials.

Even greater complexities of function and internal structure distinguish animal and human forms from man-made ones. Here we can draw meaningful comparisons only by isolating from the whole living organism a single bone, for instance, and setting it alongside a man-made latticed strut or trussed girder. This will still leave a difference of scale large enough, if it is not taken into account, to make the comparison seriously misleading.

Further differences between the two classes of form stem from the differences between the processes of construction and natural growth. These processes differ in themselves. They are also initiated and controlled in different ways.

Natural growth is a very complex process whereby a living organism, through such things as the imbibing of water and ingestion of food, slowly develops from the parent cells to its final size and form. This development may involve several or even repeated changes of form, as when the seed becomes the plant and the plant produces leaf and flower and fruit. Yet, even with these changes, the organism remains throughout both complete and individuated, usually with remarkable powers of regeneration if partly mutilated. The whole process is initiated and controlled by innate patterns of response to internal and external stimuli. The way in which it unfolds is influenced by environmental factors such as sunlight and mechanical stress, even by mental stress, as well as by the nourishment received. But it is not susceptible to direct control by the conscious will of the organism. None of us 'by taking thought can add one cubit unto his stature'.[3] Still less can other organisms modify their growth in this way. As D'Arcy Thompson has written, it is 'by one of the mysteries of biology' that 'where pressure falls there growth springs up in strength to meet it'.[4] Whatever this mystery may be, human choice plays no part in it.

Construction, on the other hand, proceeds by successive additions and approximations. At any intermediate stage in the process the form is an incomplete one or is, at least, essentially different from the final form. Often it needs temporary supports to hold it up. Each addition is composed of materials that are unchanged by the process of assembly or merely undergo simple chemical reactions (as in the setting of cements) or simple modifications of physical structure (as in welding).

Above all, the process is now initiated and controlled entirely from outside the structure itself. There is no innate programme, and if there is any corresponding mystery behind the initiation and control it is to be sought within our minds. For it is now by our deliberate volition that, where pressure will fall, construction is planned to spring up in strength to meet it. We, as designers, are now the choosers.

All these differences have had their effect on the pattern of development of man-made forms, though they have, of course, affected it in different ways. Broadly speaking, the scale and structural functions of these forms and the need to put them together piece by piece have limited fairly directly the ranges of possible choice in particular situations. The other functions to be met have, sometimes less directly, introduced further limitations. But, within these limitations, it is the choices made by designers, by their clients or patrons and often by others as well, that have led each time to the construction of one possible form rather than any other. The actual pattern of development therefore reflects all the variety of these choices as much as the wide diversity of scale and of structural and other functions.

Practical limitations on the choice of man-made forms

From the point of view of structural function and the need to put the form together, the designer's choices must embrace not only its geometrical configuration but also all the details of its construction and of the manner in which it is to be put together —though they will often, in practice, be made by more than one person for all but the simplest forms. Apart from the over-riding constraints imposed by the need to make the form stand, they will obviously be constrained also by the availability of materials and other resources and even by the time within which construction must be completed. These constraints operate in different ways, though, and are more restrictive or less so, according to the skill of the designer and the magnitude of the task. The greater this skill in relation to the demands made on it by the magnitude of the task and by the limited availability of materials and other resources, the greater the real freedom of choice. Moreover this freedom can usually be exercised in choosing the basic geometrical configuration, in detailing it, in deciding how to put it together, or in any combination of these.

The prototype of the hut in Figure 1.1 almost certainly typified the minimal choice that results from little or no relevant prior experience in the arts of construction and from a complete dependence on readily available naturally occurring materials that can be used almost as found and put together with unassisted human effort. Indeed it probably resembled the bird's nest or the spider's web in having been arrived at only through millions of almost blind trials and millions of errors. Its twentieth-century builders did at least know how to fashion another kind of shelter, as seen in the background, and may have chosen quite freely and deliberately to build in this way. But it is probable that, having made the initial choice, they were unable to depart significantly, with any assurance of success, from the known prototype.

A very different example of minimal choice is presented by a great bridge such as that over the Golden Gate [1.4]. Here a much higher order of skill was taxed to the utmost to devise a means of spanning a gap greater than any attempted before, and over deep water open to the sea at the entrance to a great natural harbour. With the materials available, only the suspension form could even carry its own weight. No other basic choice was possible. A wide area of freer subsequent choice did remain, but only in the detailing of this basic form and in the elaboration of the method of erection.

Both these examples are highly untypical of the vast majority of construction. Usually there is, and has been, much more freedom of choice at all stages of the design process. Even an unobstructed enclosure as large as Nervi's smaller Sports Palace need not, in 1957, have been a shallow ribbed dome carried on external

Figure 1.4 Golden Gate Bridge, San Francisco (Strauss). (Source: author)

raking supports. Several other basic choices of form would have met the essential requirements. Other ways of detailing the chosen form were also possible, as shown by the changes made in the larger Sports Palace only a year later. *A fortiori*, Torroja, like Nervi one of the most skilful designers of this century, certainly need not have built a fairly small coal bunker in the form of a reinforced-concrete dodecahedron [1.5]. More conventional choices of form would have met the more mundane requirements as efficiently and probably more economically, so that this choice of something completely new was almost a blatant assertion of freedom.

Figure 1.5 Coal Bunker,
Instituto Eduardo Torroja,
Madrid (Torroja).
(Source: author)

Figure 1.6 Chephren
Valley Temple, Giza.
(Source: author)

Other aspects of choice

Torroja wished, in fact, to do more than provide for the economical storage of coal. He wished also to demonstrate the ability of reinforced concrete, still architecturally an under-exploited material in 1951, to meet a simple everyday requirement with a new and striking elegance. This was not a structural objective in the first place. Nor, to return to a much earlier period, can it have been primarily for structural reasons that the builders of the Valley Temple of Chephren at Giza [1.6] chose a form that could more easily have been constructed in timber and then went to the immense trouble of realizing it in the most intractable material they had – a hard granite.

This secondary nature of structural requirements is typical of most construction undertaken primarily for purposes of shelter or, more generally, spatial enclosure. The precise structural requirements arise only out of decisions to meet the primary ones in a certain manner. Thus the constraints on choices of structural form so far considered operate, initially, only as limitations on the kinds of initial choice that are possible. Where the constraints are fairly loose, thanks to ample resources and skill in exploiting them in relation to the primary requirements, the initial choice of form may even be made with little explicit consideration of the structural implications. This has frequently happened. There have also been occasions when the structural implications have been similarly ignored more through ignorance of the constraints and an unwillingness to accept past achievements as marking limits of practicability. Complete or partial failure has then sometimes resulted. But

Figure 1.7 Dome of the Pantheon, Rome. Originally the coffers would have been decorated, probably with bronze rosettes or stars in the centre, the eye had a bronze cornice as seen in the cross section in Figure 7.7, and the vertical bars around it that are also seen in the cross section appear to have supported a large open bronze star or similar feature through which the light would have filtered. (Source: author)

some important new developments have been stimulated thereby.

This emphasizes again the wide variety of choice that has led to the actual pattern of development of structural forms on which we can look back today. This development is but one facet of the much broader history of building. We cannot build without thereby creating a structure, but that structure may be at the heart of the basic concept or only peripheral to it as something that cannot be ignored.

The masonry or concrete dome and vault, with or without ribs, and their reinforced-concrete counterparts are, for instance, important space-enclosing structural elements, and their development as structural forms will be

Figure 1.8 Mosaic
decoration of the central
dome of the Catholicon,
Daphni. (Source: author)

r 7. This development was,
being wholly a structural one.
with, prompted and stimulated
overshadowed by, many non-
es and overtones.

: dome-shaped huts like that in
no doubt, constructed in this
use it was easiest thereby to

fashion some kind of shelter from readily-available saplings, twigs, and leaves. Much the same
thing was probably true when similarly shaped
huts were first built of mud bricks or small
stones. But the subsequent rapid development of
the true dome in the first century AD had more
diverse roots. While it was greatly facilitated by
the discovery and perfection of a concrete that

Figure 1.10 Santuary dome, Shah Mosque, Isfahan. (Source: author)

could be cast into any shape to give a nearly monolithic mass, it was almost certainly encouraged to an even greater extent by what were, in the first place, completely non-structural objectives. These were a desire for large unobstructed monumental enclosures to serve the cult, ceremonial, and more worldly needs of a great empire and a parallel desire to exploit aesthetically a new structural possibility [1.7].

By the time that the concrete dome led, in turn, to true domes of fired brick and cut stone, the form had thus acquired, simply as a three-dimensional space-enclosing shape, many layers of symbolic overtones as ancestral home, mausoleum, temple, and baldachin and finally, through its appropriation to Christian uses, as a

frequent distinctive feature of martyrium and church in the Eastern Empire.[5] Indeed it is largely these overtones or associations that seem to have stimulated that further development and led designers to face the constructional and structural problems that arose through the need to use other materials. Certainly these associations, or more purely aesthetic considerations, seem to have been primarily responsible for the continued choice of the form once the main practical problems had been solved. This is reflected both in the manner of its use and in its detailed mutations or decoration.

In the later Byzantine church, for instance, it became the central focus and climax of the spatial composition – the vault of heaven from which Christ looks down as ruler of all [1.8].[6] In fact the typical cross-in-square form as a whole (with the dome over the centre of the cross) then became so hallowed that it was meticulously reproduced in small churches and chapels cut into solid rock in spite of the structural absurdity of this [1.9].[7]

A closely related symbolic ideal was pursued in the centralized church of the Italian Renaissance,[8] and something similar again in Islamic mausoleums and mosques. The inflation in the latter, from the fourteenth century onwards, of the brick outer shell of the dome into a bulbous form [1.10] derives, no doubt, from earlier timber outer shells or double domes. But, as another structural absurdity, it can have been undertaken only for the sake of its associations with the bulbous form of earlier ceremonial tents or a visual delight in the resulting silhouette.[9]

Ribs seem to have been introduced into Roman concrete domes and barrel and groin vaults largely for their usefulness during construction, but the diversity of emphasis in their subsequent use rivals that in the development of the dome.

They were wholly embedded in the concrete in the Roman vaults and hidden behind its rendering. Any possible value of a less utilitarian nature was thereby lost. In most subsequent applications they did either show on or stand out from the surface of the finished form. While still serving, usually, as an aid to construction, and to a greater or lesser extent as stiffeners and collectors of loads, they thus created a prominent linear pattern on the surface with obvious potentialities as decoration or as a basis for visually articulating the whole structure.

The decorative potentialities were exploited at an early date and in a very direct manner in Islamic domes and dome-like vaults, both Moorish [1.11] and Iranian. Indeed, since the domes and vaults were mostly quite small and should have presented no great problems if constructed without ribs by methods that had long been in common use, it seems likely that the ribs were introduced primarily for their value as decoration. The structurally gratuitous intricacy of their interlacing, and the fact that similar geometrical interlaces were of much earlier origin, unconnected with building, and soon became one of the favourite Islamic decorative motifs, both increase this likelihood. A dual ancestry of the pattern is probable – part functional in basketry and part fanciful in the superimposing of regular geometric figures and linear grids. Its earlier uses in an architectural context were chiefly on flat surfaces. The Romans used it for floor mosaics as in the House of Livia on the Palatine and on the soffits of flat or almost flat ceilings as at the Temple of Bacchus and in the Great Altar at Baalbek, and it was introduced into Islamic architecture at least as early as the beginning of the eighth century in marble window grills of the Great Mosque of Damascus. When applied in the form of ribs to a dome, its structural reference was almost inescapable, but Islamic builders seem mostly to have deliberately played this down.

Figure 1.11 Sanctuary vault, Great Mosque, Cordoba. (Source: author)

Certainly, like the Romans, they did not develop a whole structural system from the rib.

Gothic builders in the thirteenth century did just this and much more besides. Even if the introduction of the rib into Romanesque church vaulting in the late-eleventh and twelfth centuries owed a few hints to Islamic precedent or that of Armenia (which is uncertain), it was far from being a straightforward borrowing. A typical Romanesque vault was the groined vault formed by the interpenetration of two barrel vaults and not a dome. The groins were awkward to form at the best of times and tended to assume unsightly double curves when the bays over which the vaults were erected were not square. The ribs served primarily to simplify construction and to make it easier to do away with the unsightly double curves. But if this last was, in a rather negative sense, an aesthetic objective, it was initially a self-defeating one since the prominent diagonal ribs introduced a new visually disruptive element into the typical Romanesque nave. That was, in turn, eliminated by carrying all the ribs down to the floor as shafts attached to the walls and piers and then by progressively transforming these walls and piers into a glass-filled exposed skeleton of arches, columns, and flying buttresses.

The result now was an entire structural system in which structure, construction and visually expressive form were, for a hundred years or so, virtually indistinguishable. It was also much more than this, because the primary objective was neither to create a new structural form nor merely aesthetic. These aims, if they consciously existed, were subservient throughout to an unusually clear idea in the minds of influential men, such as Abbot Suger of St Denis, of what the finished building – the House of God – should be like.[10] All the skills of designers and builders and sculptors and stained-glass workers were directed to realizing this idea, and they produced one of the most powerful and moving symbols that architecture has to show [1.12].

In terms of objectives, the recent development of large thin reinforced-concrete shells and related forms resembles more closely the first- and early second-century development of the more massive plain concrete dome – though it has been more diversified, which makes any generalization about it more hazardous. Certainly many of the uses to which the newer forms have been put have much in common with the more worldly Roman uses for markets, baths, garden pavilions, and the like. There has also been the same excitement in the forms themselves and their potentialities simply as shapes, quite apart from their practical usefulness and economy. Structural necessities and problems of design and construction have intruded themselves more insistently because the newer forms have been so much more slender. The intrusion has, however, been balanced by a much deeper insight into these necessities and by a much greater technical mastery over materials and ways of using them. This insight has, on the one hand, even led, with little prompting from the parallel interests in similar forms shown by Gaudi and some constructivist artists, to the emergence of at least one virtually new shape: the hyperbolic paraboloid. On the other hand, it has still left room for a good deal of purely aesthetic choice, particularly where there has been a willingness to pay for rather more than the most economical conventional solution to the functional requirement, in order to obtain something new and distinctive.

Sometimes an aesthetically satisfying shape has been sought primarily through clarity of structure, as something that can be sufficiently expressive in itself and capable also of giving a recognizable identity to a building and a sense of place to its surroundings. Nervi's Sports Palace may be taken as an example of a successful outcome of this approach. It could be criticized by a structural purist for a certain arbitrariness in the pattern of the ribs, but it came as close as any structure of its time to achieving that unity of structure, construction, and visually-expressive form previously referred to as characteristic of the best Gothic vaults.

Figure 1.12 Amiens Cathedral. (Source: author)

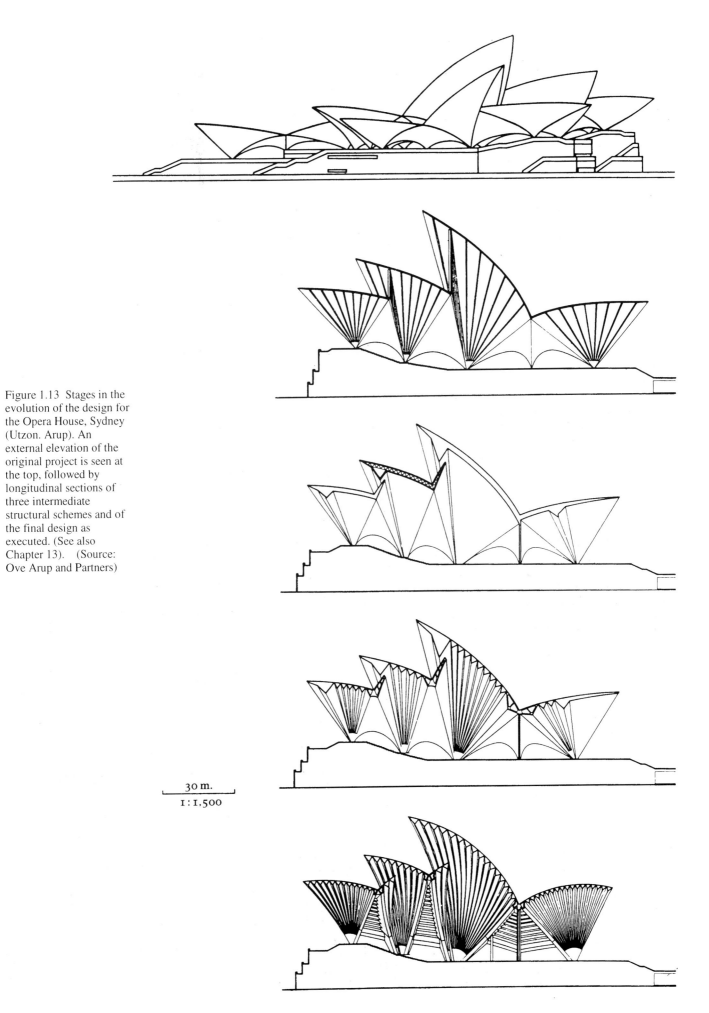

Figure 1.13 Stages in the evolution of the design for the Opera House, Sydney (Utzon. Arup). An external elevation of the original project is seen at the top, followed by longitudinal sections of three intermediate structural schemes and of the final design as executed. (See also Chapter 13). (Source: Ove Arup and Partners)

30 m.

1 : 1,500

At another extreme though, a concern for appearance has ranged from a Platonic search for geometrical purity and simplicity, as seen in Saarinen's Kresge Auditorium [7.25], to a rather naive, though sometimes inspired, allusion to some other form, as seen in his TWA Terminal at Kennedy Airport [13.8] and in Utzon's Sydney Opera House [1.13]. It has then sometimes so complicated the problems of structural design and construction that they could be solved, albeit still at considerable cost, only by departing radically from the simple shell-form. The wind-filled sails of the Opera House roof had to be constructed not as smooth continuous thin shells but as sets of ribs placed side by side to give a deeply corrugated cross-section capable of resisting the actions inevitably brought into play by the desired sharply pointed ridges.

Other structural forms have similarly developed in a rather wayward manner as a result of repeated shifts in emphasis in the non-structural objectives. The chief difference between the developments of the forms just considered and those of forms associated more with multi-cellular and multi-storey buildings is that, especially when the structural requirements have not been critical, the emphasis has been more likely to veer, with the latter, towards convenience of plan and general habitability than towards purely aesthetic preoccupations or symbolism. In recent years, for instance, a number of factors have together given rise to vastly increased demands for heating, lighting, air conditioning, and other services such as power and water. The need to accommodate these and make provision for possible further growth to meet unpredictable future demands has become a major constraint on the choice of structure. On the other hand, an increasing number of structures now serve purposes that can be quite simply defined in terms of the performance required, which, even if it is not primarily a structural one, is fairly directly reducible to one that is. Examples of these are water-retaining structures, cooling-towers, and most bridges. Appearance may also be important in their design [1.14], but in many cases the form has been chosen almost entirely for structural reasons.

If, in much of what follows, the structural aspects of choice are discussed in detail and other aspects are mentioned only in passing and much more briefly, this emphasis should not be misinterpreted. The prime purpose of the book is to consider the development of new forms as a continuing process from the structural point of view, and the emphasis will usually be necessary in order to do justice to this. It seems unnecessary, moreover, to dwell on aspects which have already been discussed at great length from many different points of view by architectural historians and critics. It is hoped that it will be sufficient to provide occasional signposts to and reminders of the vitally important wider contexts within which the structural development has taken place and been exploited.

The development of structural forms

Figure 1.14 Water Tower, Virginia. (Source: author)

In outlining the development, the emphasis will also, as stated at the outset, be mostly on structures and structural elements that have marked significant steps forward in widening the range of possible future choice. Much less will be said of those phases in the total pattern of development that were important chiefly for the new ways in which they exploited existing structural possibilities. To a large extent, the fifth century BC in Greece and much of the Renaissance and Baroque [1.15] in Western Europe were important in this way. Supremely important as they were architecturally, a good deal less attention will therefore be paid to their achievements than to those of, for instance, the first and early second century AD in and around Rome, the thirteenth century in France, and the nineteenth and twentieth centuries in Western Europe and the United States. Most space will be given to developments of the last 200 years, because of the rapid increase in the pace of development then and because of its obviously greater relevance to our situation today. But even here the emphasis will be on new structural developments whatever their origin rather than on their architectural exploitation however masterly and influential [1.16]. Purely civil engineering structures other than bridges will not be considered in any detail.

Development is, of course, a historical process. New developments build upon those that went before. But there are so many cross-currents, borrowings, influences, and interactions that there is no single linear progression. Possibilities opened up by some new development at a certain time and place may be seized only much later elsewhere. Where and when depends on many things such local resources, fluctuations in prosperity and the objectives and priorities of those in control. Several possibilities that have arisen at different times and in different places may also lead to a single new development that brings them together. This is particularly true of the way in which developments in elemental forms such as arches and domes, beams and slabs, columns and walls permit developments in complete structures in which they are used in combination

For all these reasons a simple history treating everything that happened at a particular time together would be far from straightforward and would run to excessive length. A more thematic approach will therefore be adopted.

In the main body of the book, the development of all the principal structural elements will be traced first and, only after this, the development of different broad classes of complete structure. This will sometimes involve looking more than once at a single structure, first at its elements and then at the whole. Also it is only possible in terms of what are really very recent abstract concepts of the ways in which those elements carry or weigh down upon or tend to push aside one another. But it has the further merits that it is usually easier to understand the structural behaviour of the part than of the whole; to understand then the ways in which it could, in practice, be used and combined with other elements; and to appreciate, thereby, one aspect of the freedoms of choice possessed by designers at different times. The principal elements will be considered in Part 2 (Chapters 6 to 10), and complete structures in Part 3 (Chapters 11 to 15). In these chapters, dates and periods will be referred to only where this is necessary to establish significant relative sequences. Where they are more precisely known, the dates of particular structures are given in their index entries.

As a basis for tracing these developments of elemental forms and complete structures, the next chapter of this introductory part will outline, without mathematics but with concrete illustrations of the behaviour of a few typical forms, the modern concepts of structural actions just referred to. Chapters 3 and 4 will then look at the capacities and limitations of the materials available at different times, and at the constraints on choices of form that have stemmed from the requirements of the construction process. Finally Chapter 5 will attempt to provide a fuller and more explicit framework for the discussions to follow by exploring further the relationships between structural form and form in its purely geometrical aspect. It will suggest some basic classifications of elemental and near-elemental forms.

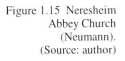

Figure 1.15 Neresheim Abbey Church (Neumann). (Source: author)

Figure 1.16. Falling Water, Bear Run, Pennsylvania (Wright). (Source: author)

Readers who already have a good understanding of modern structural theory and practice will, of course, already be familiar with most of the ground covered by Chapters 2 and 5 and part of that covered by Chapters 3 and 4, if not always with the particular approach that has been adopted here. They may therefore prefer to turn straight to the remaining chapters. Those with a purely historical interest may also find it more profitable, at least at a first reading, to omit Chapters 2, 4, and 5, and to refer to them only as need arises for the clarification of particular points. To make this easier, these chapters have been more liberally signposted with sub-headings than the rest of the book. As an alternative reference, the last chapter is followed by a glossary of all the technical structural terms used.

The concluding Chapter 16 that constitutes Part 4 will return to the processes of choice that have led to particular forms being built at particular times and places. In a wide-ranging review, it will look especially at the increased freedom of choice that has come about in recent years through increased understanding of structural actions and recent vast increases in computational power, and at some related changes in the processes of design.

Chapter 2: Structural actions

Clearly an essential structural requirement for all the forms with which we shall be concerned is that they shall stand and not collapse. Occasionally this will suffice. Usually, if the primary function is to be adequately served, it is also important that local failures and local or more general deformations shall be kept within acceptable limits even when they do not jeopardize the stability of the structure as a whole, and that the way in which it moves when subjected to moving or rapidly changing loads should be tolerable. It is not sufficient, for instance, that a tall building shall not collapse completely. Each floor, wall and stairway must, in normal circumstances, remain intact. If it cracks locally or deflects, it must not do so in such a way as to disrupt services, cause excessive damage to finishes, or interfere in any other way with normal use. Likewise, if the whole structure sways in the wind, it must not do so to an extent that causes discomfort or alarm. These requirements are met by ensuring that active loads are balanced throughout the structure by the resistances opposed to them, that the loads passed on by the structure as a whole to the foundations are similarly balanced, that there are adequate margins of strength and stiffness under normal circumstances in all structural elements and their interconnections, and that the energy imparted by alternating loads like the wind is safely dissipated.[1]

Loads and their effects

Active and reactive loads

Several types of load must be provided for. Some may be described as active and others as reactive. Until the earlier part of this century all loads were of one or other of these types. There is now a further class of loads that also react to the active ones but that are deliberately introduced for this purpose and might themselves be described as active in this respect.

Among the normal active loads, gravitational self-weight is always present [2.1]. Since it depends solely on the form of the structure and the materials of which it is made, it is unchanging and is normally referred to as a dead load. Imposed or live loads are those imposed on the structure by its users or its environment. They do change with time and they differ significantly from one another in their effect according to the rapidity and the regularity and frequency with which they change. They differ also in the extent of their dependence on the form and materials of the structure and some of them are further dependent on characteristics determined by the manner of its construction. They include the loads imposed by furniture in a room or by goods stored in a warehouse, by people or vehicles on a bridge, by the water pressing against the upstream face of a dam [2.2] or by the wind, and the loads resulting from rapid displacement of the whole structure as in an earthquake. Within a structure there are also the loads produced by changes in temperature or humidity and even by the setting and hardening of cement when the expansions and contractions to which these tend to give rise are restrained.

Balancing the net effects of most of these active loads at the points where the structure is supported are the normal reactive ones. They also frequently depend on the precise manner of construction, as well as, of course, on where we choose to consider the structure as terminating and its support as commencing. They may be the supporting loads where the foundations meet the natural earth. Or they may be the loads imposed by adjacent parts of the structure on any element which we choose, notionally, to isolate.

Loads that are deliberately introduced to counter directly some of the active loads or their effects include the internal air pressure that supports some fabric structures and the controlled fluctuating loads that, since the later 1970s, have been designed to reduce the movements associated with certain fluctuating dynamic loads.

Associated movements and deformations

All types of load tend to produce movement in the direction in which they act. The structure – or any part of it – and the earth on which it stands resist simply by being there as an obstacle in the path of the movement. No structural element is completely rigid though. Each gives way to some extent and it is through this limited giving-way or deforming that the resistance is developed. The term 'stiffness' denotes the resistance that is developed by a

Figure 2.1 Verrazano Narrows Bridge, New York Harbour, under construction (Ammann and Whitney). (Source: author)

given deformation and 'strength' the maximum resistance that can be developed. Both can have different values for the same element according to the types of resistance and deformation considered and the measures adopted.

Dynamic and static loads

Most loads on built structures change slowly enough to allow the resistances that are developed by deformation to keep pace with them. They are described as static loads.

Rapidly applied or rapidly changing loads have different effects because the deformations cannot so easily keep pace with them. Rapid changes are called for, involving significant accelerations. These bring into being, temporarily, another type of resistance – the inertial resistance to such acceleration that depends both on the acceleration and on the mass that is being accelerated. They may also result in the point of balance being overshot. The whole behaviour can be very complex, but its outcome depends

chiefly on the relationship between the rapidity of application or change in the load and on the natural period of vibration of the element or structure on which the load acts, this period being dependent, in turn, on both the mass and stiffness of the element or structure.

When the time taken for the load to reach its peak is commensurate with the natural period, and particularly when it is less than this period, the load is normally said to be dynamic. At worst, the effect of a load that is instantaneously applied and then maintained will be twice that of a load of the same magnitude applied slowly, and will be seen as a vibration of large amplitude about the final point of balance. Rhythmically repeated or periodic loads can have even more damaging effects if their repetitions chime in with the natural vibrations. A state of resonance is then encountered, with each successive application of the load reinforcing the effects of the previous applications.

Fortunately, static loads are the most important ones for most structures. Dynamic and

periodic loads of comparable magnitude are usually of sporadic and infrequent occurrence except for such loads as those due to traffic on short-span bridges. There are nevertheless many structures for which they cannot be ignored.

The commonest such dynamic load is that of the wind, which arises because the free passage of the air is obstructed. It therefore depends primarily, at a particular site, on the external form and orientation of the structure, and the ease with which some of the air can pass through it to help equalize the pressures on

33

different faces. Noticeable dynamic effects are usually confined to structures with long natural periods of vibration, measured in seconds rather than tenths of a second. Such structures include tall slender buildings, chimney stacks, masts and the long slender decks of suspension bridges. The effects can be greatly magnified by resonance if the cross-section is one that leads to a regular formation and shedding of vortices roughly in step with the natural vibrations [2.3].

Much less common are the dynamic loads due to earth tremors which, like wind loads but unlike most other loads, act horizontally as well as vertically [2.4]. They are essentially inertial loads acting on all parts of a structure above the ground as a result of very rapid displacements of its foundations. They are therefore dependent initially on the mass of the structure and on its distribution. Subsequently, because of the rapidity of the earth movements and their repetitive nature, they depend also in a very complex way on the periods of vibration of the whole structure and its elements, on whether these respond primarily by bending (like most modern structures) or by rocking (like many early masonry structures), and on the ability of the structure to dissipate the energy fed into it without undergoing serious local failures.

Internal actions

Tension, compression, bending, torsion, and shear

If, for simplicity, we concentrate now on static loading, there is one further distinction that is worth making at the outset between the possible types of load on individual structural elements. This is the distinction between tension, compression, bending, torsion, and shear. In illustrating this distinction [2.5], it is convenient to adopt the usual convention of representing each load by an arrow pointing in the direction in which it acts. Thus a simple force is represented by a straight arrow and a pair of equal and opposite parallel forces (tending to cause rotation and known as a couple) by a curved arrow.

Tension, seen in (a), is a pulling-apart. Compression, in (b), is a pushing-together. Bending, in its purest form as in (c), is the action that results from applying equal and opposite couples in its own plane to an element such as a prismatic bar. It is also, as in (d), the principal action resulting from applying transverse forces to such an element at some distance from its supports. Torsion is the twisting action that results, as in (e), from applying equal and opposite couples to the ends of a similar element in planes at right angles to its axis. Shear is an action tending to cause slipping of one part of an element on another as in (f). Each type of action may also occur at joints between elements, as may easily be visualized by imagining a cut and a joint near the middle of each of the elements illustrated.

These distinctions are important because the capacities of elements and joints to resist the loads upon them vary considerably with the type of action. Elements of relatively large cross-sectional area are required, for instance, to resist bending or torsion. Appreciable cross-sectional area is also required to resist compression because any departure from axiality in the load will result in bending that will increase the non-axiality and may lead to buckling. A non-axial tensile load, on the other hand, tends to realign the element with the load so that much smaller cross-sections, determined only by the tensile strength of the material, are adequate. Shear loading calls for adequate cross-sectional area on the planes on which the shear acts. Usually it is associated, in practice, with bending as in Figure 2.5(d) and must then be resisted by the transverse cross-sections of the element. Pure shear arises chiefly in panels or walls, where it is equivalent to a uniform diagonal compression plus an equal uniform compression at right angles to this as seen in (g). The cross-sections must therefore be sufficient to take the tension without cracking and the compression without buckling.

Compression can easily be transmitted at joints simply by providing enough bearing area

Figure 2.4 Earthquake damage, Gemona. (Source: author)

Figure 2.5 Types of
loading (see text).
(Source: author)

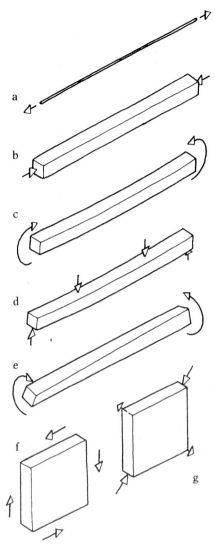

on two abutting surfaces. Tension is much more difficult to transmit. It calls for a much more positive type of joint such as that produced by welding, or a lapped, pinned or bolted one in which the tension is transmitted at some point or points by shear or friction. Transmission of shear calls, in turn, for an adequate surface contact plus some means of holding the surfaces firmly together. This may be either simple compression or some positive tensile connection. Transmission of bending or torsion calls, usually, for a joint able to resist compression, tension and shear. A welded joint serves this purpose as do some glued joints, but other suitable joints can be rather complex.

Actions of the human frame

In our own bodies the bones of the skeleton resist mainly compression and only a limited amount of bending; tensions are taken by the ligaments and muscles as represented in some of Leonardo's drawings by threads or copper wires [2.6]. The joints of the skeleton are all articulated, being effectively pinned or of the ball-and-socket variety to allow more or less free rotation in one plane or in any plane. They are thus capable of transmitting compression and

some tension and shear, but only limited amounts of bending and torsion when the limits of free rotation are reached. Bending actions have therefore to be resisted by the whole frame, with the bony skeleton, held together by the ligaments, taking the compressions and the muscles taking most of the tensions. Since, however, the muscular tensions are not merely passively developed as a wholly predetermined response to external actions but are under our own direct control, the structure is a highly versatile one capable of acting in many different ways and of exerting loads as well as simply resisting them and being one. Even to maintain a static balance we usually have a choice between adjusting our stance or adjusting the magnitudes of the internal tensions and compressions [2.7].

Actions of the built structure

The built structure lacks this choice. As we have already seen, all the relevant choices were already made before or during its construction by those responsible for designing and building it. Its elements were then given a fixed configuration, and the manner of its construction determined the initial state of its internal actions. All its subsequent behaviour, including the changes wrought by time, depends solely on external circumstance, even though it may be capable of responding to different circumstances in a variety of ways and each of these may be indeterminate in the sense that we, as outside observers, can never know enough about it to predict exactly what it will do. What is important, if it is to stand and serve its purpose, is that it shall have the capacity to resist in some acceptable manner all the loads that it is likely to be called upon to bear.

Structural requirements

The basic geometrical requirement

To have this capacity is in the first place a geometrical requirement. The individual elements must be so disposed and joined that together they constitute a stable system and not just a loose assemblage such as the human skeleton would be without the muscles attached. They must do so in the three-dimensional space inhabited by the structure and in the face of all likely loadings. It is much simpler though, by way of illustration, to consider only two dimensions and simple gravity loading.

Figure 2.8 shows a few of the simplest possible ways of supporting a weight (represented by the large circle) by one, two, three, or four structural elements either above a fixed base or below a fixed overhead support. In the sketches single or double full lines represent the individual elements; small open circles represent pinned joints allowing free rotation; and solid black represents rigid joints allowing no rotation. All joints are assumed to be able to resist either tension or compression.

Merely from a consideration of the possibilities of movement, it is clear that configurations (a), (a') and (d), (d') would be unstable in the sense that they would be incapable of maintaining the weight in its initial position. If we ignore for the moment the elements shown by chain-dotted lines in (f) and (f'), all the other configurations would be stable, but without any surplus source of stiffness. From the equivalence of (b) and (c), (b') and (c'), (e) and (f), (e') and (f'), it can be seen that the manner of jointing can be as important as the number and configuration of the elements.

*The complementary needs
for adequate strengths and stiffnesses*

If the geometrical requirement is satisfied, it is merely necessary, in addition, that the individual elements and joints should have the requisite strengths and stiffnesses. In simple structures such as those just considered, we can easily see which elements will tend to lengthen or shorten and which will tend to do one or the other and bend at the same time. From this we can see which will be subject to tension, which to compression, and which to a combination of one or the other and bending. All those subject only to tension have been represented by a single line and all others by a double line. In some more complex structures the nature of the actions may be less intuitively obvious. In such cases it must be determined by considering explicitly the balance of the forces at work – or, as it is usually called, their static equilibrium. Explicit consideration is always necessary if we wish to determine precisely what the requisite strengths and stiffnesses are.

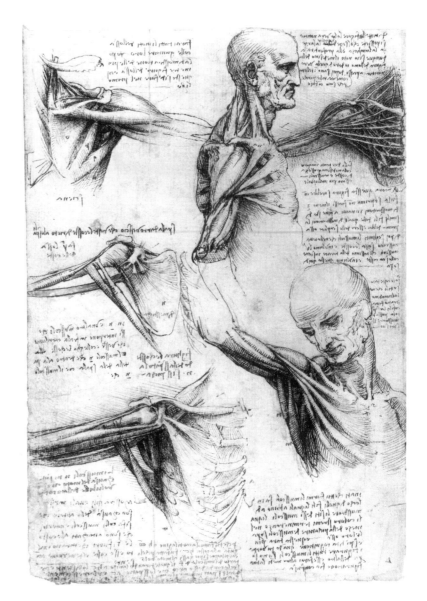

Figure 2.6 Drawings by
Leonardo da Vinci of the
human shoulder.
(Anatomical MS.A, fol. 4,
Royal Library, Windsor,
No. 19003 V)
(Reproduced by courtesy
of Her Majesty
The Queen)

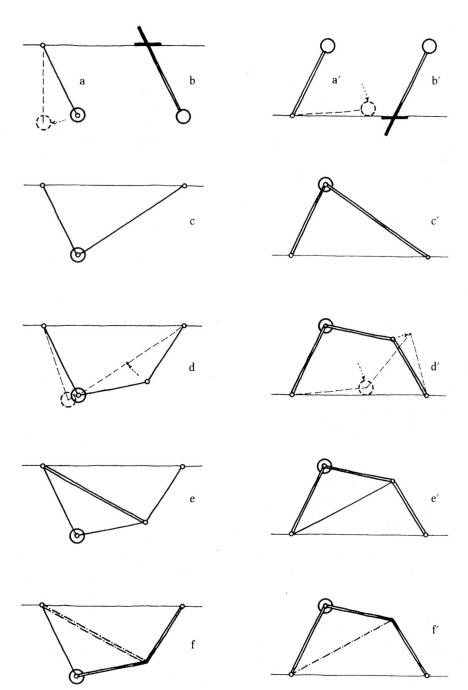

Figure 2.8 Geometric
aspects of static
equilibrium (see text).
(Source: author)

Figure 2.7 Detail of the
frieze from the
Mausoleum of
Halicarnassus, now in the
British Museum, London.
(Source: author)

Static equilibrium

Figure 2.9 shows the forces at work in a few of the simple structures of the previous figure. The forces represented by the heavy arrows in the sketches marked (1) are those acting on the weight and on the supports, and those similarly represented in the sketches marked (2) are those acting on the bars. Actions and reactions have been distinguished by the filled and open heads to the arrows. The arrows marked w denote the weight; those marked t and t' denote tensions in the bars and corresponding actions or reactions on the weight and at the supports; and those marked u and v denote additional forces brought into play in configuration (c).

All that is necessary for the maintenance of configuration (a) in a state of static equilibrium is that the forces t should be of the same

magnitude as w. Since they are already col-linear, action and reaction would then, in each case, be equal and opposite and there would be no resultant force to induce movement.

At (b) however, equilibrium is impossible without a prior rotation of the bar. Whatever the magnitude of the force t, it is not collinear with w. There must therefore be some unbalanced resultant action on the weight. If we assume that the bar and the support are strong enough to permit only a rotation about the support, this resultant must act at right angles to the bar. Knowing this, we could if we wished find its magnitude by the construction shown in lighter broken lines and known as the parallelogram of forces. In this construction the arrows denoting the forces w and t on the weight become two sides of the parallelogram and the resultant is similarly represented by the arrowed diagonal,

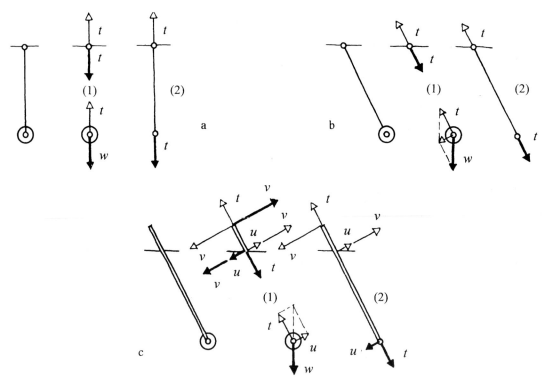

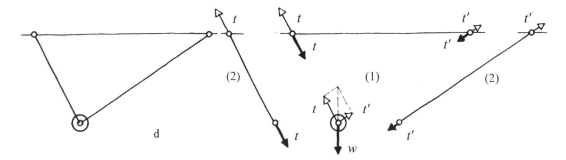

Figure 2.9 Forces in static equilibrium (see text). (Source: author)

with the relative lengths of the lines denoting the relative magnitudes of the forces.

At (c) and (d) equilibrium is re-established by the additional forces *u* and *t'* balancing the unbalanced resultant force on the weight at (b). The additional force *t'* is provided and carried to the support in the same simple manner as *t*, but *u* can be provided only by a bending action in the supporting bar which calls for a more complex set of reactions at the support. These must include a reaction *u* parallel to the force on the weight and, in addition − since this pair of forces alone acting on the bar would cause it to rotate − another pair of reactions *v*, *v* with an equal tendency to rotate the bar in the opposite sense. Under the action of these two pairs of forces (forgetting for the moment the tension *t*) the bar is like a balance arm. The condition of balance is therefore that *u* and *v* shall be inversely proportional to the effective arms of the balance which are known as their moment arms. Each force multiplied by its moment arm is known as its moment, and this condition can be stated alternatively by saying that there shall be no resultant moment.

These simple structures together exemplify

all the primary conditions of static equilibrium: actions and reactions at any point must be equal and opposite (or, more generally, all forces meeting at a point must have a zero resultant); and sets of parallel forces acting on an element must, as well as having a zero linear resultant in every direction, have no resultant moment. The same conditions apply even in the most complex situations and they apply not only to the total forces acting on an element but also, on a smaller scale, to the forces acting anywhere on a small part of it. Like the geometrical requirements considered earlier, they must be satisfied in all three dimensions and not just in two.

The equilibrium of some structural elements acting in simple tension or compression

To illustrate the application to real structures, it is nevertheless easier to concentrate again on ones that can, for most purposes, be considered as two-dimensional.

The simplest types of behaviour to understand are those which, like those in Figure 2.9(a) and (d), can be visualized as direct tensions or compressions with or without shear but with no bending or torsion. Two structures acting in this way are the flexible hanging chain or catenary and the arch composed of rigid voussoirs, the one acting in tension and the other in compression without bending. These will be considered first, followed by the beam acting primarily in bending and finally the domical shell acting simultaneously in compression and tension but again without bending.

Figure 2.10 Detail of an arch, Old Gorhambury, near St Albans. (Source: author)

The catenary and the arch

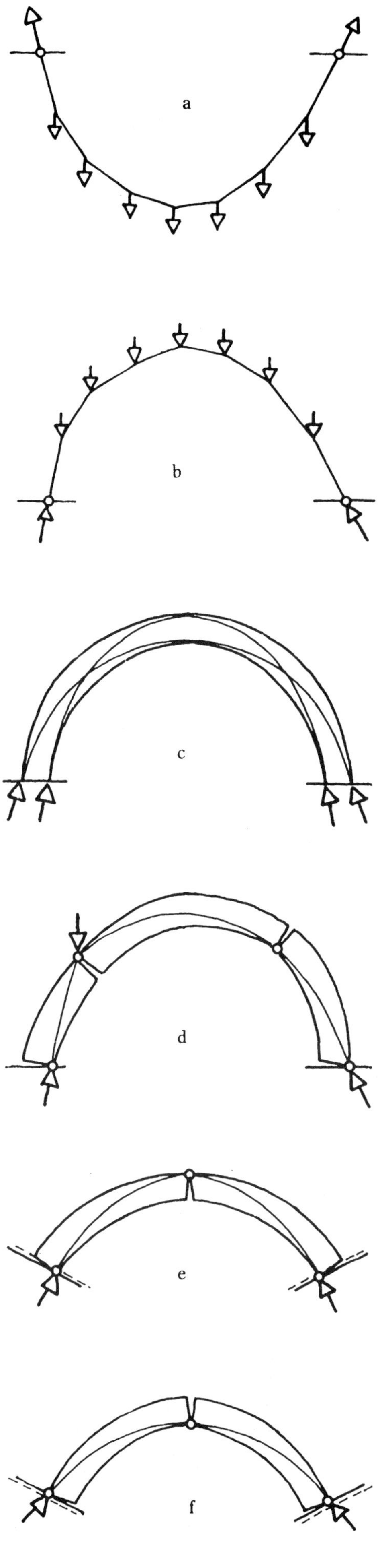

Figure 2.11 Static equilibrium of the catenary and arch (see text). Continuous lines drawn within the arches (c) to (f) are possible thrust-lines. Dashed lines in (e) and (f) denote initial positions of the arch supports. (For clarity, all the weights and other loads are shown at (a) and (b) as concentrated at a few points. In practice the self-weight, at least, would be fairly uniformly distributed so that the lines of action would be made up of smooth curves broken only where there was a concentrated load. Figures (c) to (f) take this into account.) (Source: author)

From the point of view of the geometric conditions, the catenary and arch are analogous to the simple forms seen in Figure 2.8 (d) and (d'). These have both been shown to be unstable. But, whereas the analogue of the arch would fall to the ground if loaded as shown, that of the catenary would merely change its shape as shown by the dotted lines. In other words, the flexible catenary is self-equilibrating [2.1], whereas the voussoir arch is not. At least it would not be if the joints between the voussoirs were frictionless pins as in Figure 2.8(d'). In practice this defect has to be overcome by giving an appreciable depth to the voussoirs and their abutting faces. The arch is then not only capable of carrying a variety of loads and undergoing small movements of its supports but is able, unlike the catenary, to do so without changing its shape. It already contains within itself, in effect, many different arches, although, as their range is still limited, collapse can still occur by rotations about the extremities of some of the joints, coupled perhaps with relative slips of some of the voussoirs. Such rotations are clearly visible at the crown and left-hand haunch in Figure 2.10 and there is a small slip to the right of the crown.

The associated internal forces will be distributed over the finite areas of the cross-sections of either the catenary or the arch but will be statically equivalent to single concentrated tensions or compressions. These can be represented by their lines of action just as we have used arrows to represent other forces. At any point they will act in one direction or the other according to whether we consider the action on that part of the structure lying to one side of the point or on that lying to the other side. In the catenary, with its minimal cross section, the line of action is very easily visualized because it necessarily follows closely the structural form itself [2.11(a)]. In the arch it is the inverse of the line for the corresponding catenary [2.11(b)] and must, for stability, fit somewhere within the depth of the voussoirs. The different thrust lines in Figure 2.11(c) could arise through variations in the live loads or solely in the inevitable slight yielding of the supports. Instability through hinging rotations alone as in Figure 2.8(d') can occur only if the thrust line comes close to the intrados (inside) or extrados (outside) at four or more points [2.11(d)]. Sliding of one voussoir on another may, however, hasten it by reducing the effective depth. This can happen where the thrust line is oblique to a joint as seen to the right of the crown in Figure 2.10 or when a voussoir splits locally under the concentrated load at a 'hinge'.

Two further points may be noted. The first is that the possibility of equilibrium in the systems shown in Figure 2.11 (a) and (b) is not dependent on any particular horizontal separations of the support reactions. Equilibrium could still be achieved if these moved somewhat apart or towards one another. For the catenary, this would happen only if the supports themselves moved. For the arch, though, it need entail no more than different distributions of the reactions over the widths of the supports as in Figure 2.11(c), and both thrust-lines shown there could be in equilibrium with the same set of applied loads. This is one example of what is called statical indeterminacy, referred to below.

The second point is that the conditions of equilibrium call, in each case, for inclined reactions at the supports. The reactions marked are those on the catenary and the arch. On the supports they will be inclined inward for the catenary and outward for the arch, and, in order to provide them, the supports must give way to some extent to develop their resistances. In neither case, for the reason just mentioned, will this necessarily jeopardize the equilibrium; but the consequences are worth looking at. The supports of the catenary will move towards one another. As can easily be verified by holding one end of a heavy chain and moving it slowly towards a fixed support at about the same level at the other end, starting with it as taut as possible, this will reduce the inward pulls and will thereby contribute directly to re-establishing equilibrium. The supports of the arch, on the other hand, will move apart. If they were truly pinned, this would have the opposite effect of increasing the outward thrusts and calling for

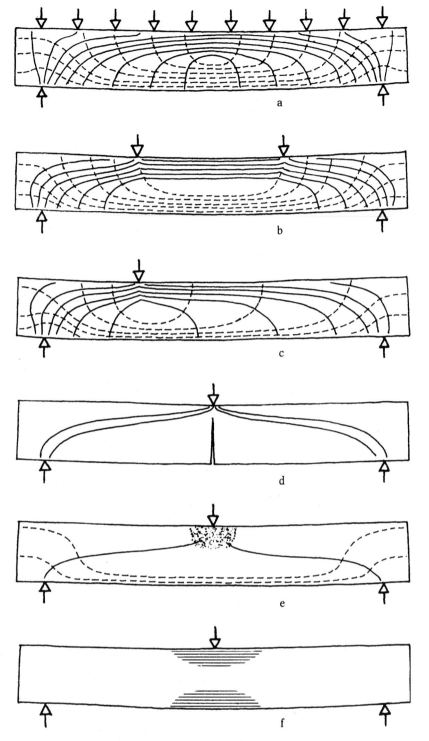

still more outward movement and a further increase in thrust. The compensating factor is the depth of the arch ring and the consequent statical indeterminacy just noted. If there are no relative slips of the voussoirs, the separation of the supports is accommodated by hinging rotations roughly as in Figure 2.11(e) and the thrust line moves as far as possible towards the intrados, thereby reducing rather than increasing the thrust. Any movement of the supports towards one another, due to other actions, would, conversely, be resisted by rotations in the opposite directions as in Figure 2.11(f) and an increase in thrust.

Figure 2.12 Internal static equilibrium of a beam with principal compressions (arch actions) shown by full lines and principal tensions (catenary actions) shown by broken lines. (Source: author)

Equilibrium calling for both tension and compression

For purposes of a general understanding rather than routine calculation of strength or stiffness, more complex actions can best be visualized in terms of the lines of action of the direct tensions and compressions brought into play. This approach is completely general and will be adopted again in considering a wider range of forms in Chapter 5. It usually involves looking at the distribution of the actions over the cross-sections and not merely at their resultants as with the catenary and arch. The local intensities of distributed forces are known as stresses. Tensile and compressive stresses always act at right angles to the cross-section considered, and shear stresses act tangentially to it. If we consider a cross-section in an arbitrary direction, there will normally be either tension or compression plus shear acting on it. Shear is, however, equivalent to a combination of tension and compression acting in other directions as shown in Figure 2.5 (f) and (g). By a suitable choice of the directions of the cross-sections at every point throughout a structure or structural element – varying the choice as necessary from point to point – we can therefore eliminate the shears and arrive at patterns of tensile and compressive stresses only. These stresses are called principal stresses. They always act at right angles to one another. Lines indicating their directions through the structure or element will be referred to as isostatics. They are akin to the catenary curve or the straight line of a tie when they relate to tensile stresses and to the thrust line of the arch or that within a strut when they relate to compressive ones. The difference that must be borne in mind is that they represent only the local directions of the continuous distributions of stress and not their resultant actions.

The simply supported beam

The simply supported beam or lintel (free to rotate about the supports at its ends and to expand or contract longitudinally) may be taken as a simple example of an element acting partly in bending and partly in shear. Figure 2.12 (a) to (c) shows that, in carrying any particular set of loads, it is equivalent to a family of arches associated with an orthogonal family of catenaries. These 'arches' and 'catenaries' readjust themselves within the beam to each change in the loads. In parts of the beam where the effective loading reduces to two equal and opposite end couples (as in the central part of (b)), the action is one of pure bending and the isostatics run parallel to one another. Elsewhere there is also a shearing action and this shows itself through the intersection of the isostatics. The downward-bowing deformation (shown grossly exaggerated) results from the small deformations of the material of the beam that are necessary everywhere to develop the stresses.

Provided that the material can develop these stresses without cracking or crushing or excessive yielding, collapse of the beam cannot occur. Unlike the arch it will not push aside its supports because the thrusts of all the internal 'arches' are exactly balanced by the inward pulls of the 'catenaries'. A failure of the material in tension would, however, disrupt the 'catenaries' and leave only a flat thrusting arch [2.12(d)]. Collapse would then ensue unless some impediment to the spreading of the base of the arch prevented it [2.13]. A failure in shear could also permit collapse by allowing one part of the beam to slip or drop away from another.

Because some materials – notably concrete – are much weaker in tension than in compression, and therefore weaker also in shear, they are now usually reinforced against both these types of failure. Over-reinforcement

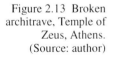

Figure 2.13 Broken architrave, Temple of Zeus, Athens. (Source: author)

makes a crushing failure [2.12(e)] more likely. The joggling of the voussoirs of a flat arch was a much earlier attempt (as mentioned in Chapter 6) to prevent a shearing type of failure by preventing the slipping of some voussoirs relative to the others.

Collapse through excessive yielding of the material arises only with materials like mild steel which neither crack nor crush when they reach the limits of their useful strengths but merely deform more and more with little change in the resistance they offer. This results in a loss of stiffness of those parts of the beam where the stresses are highest, such as the shaded areas in Figure 2.12(f); and hence in large rotations over short distances almost as if hinges had been inserted there.

The domical shell and the dome

The thin domical shell is an element in which, ideally, there is no bending (and therefore no variation in stress through the thickness at any point), but one in which the principal tensions and compressions are distributed continuously over a three-dimensional surface.

Under symmetrical loading such as self-weight, the stress pattern for such a shell of hemispherical form is as shown in Figure 2.14(a). One set of principal stresses acts radially from the crown to the base. These stresses are compressive throughout and may be likened directly to the thrusts in a large number of arches into which the shell might be split by cuts on vertical radial planes. Such arches, with a common keystone but otherwise acting independently, would, however, be unstable unless they were much thicker than the shell need be. They would develop hinging rotations much as in Figure 2.11(e) in the upper part and an extra hinge at the outside at each springing. The upper parts would then fall towards the centre while the lower parts would burst outwards. The other set of principal stresses in the shell acts circumferentially to prevent these movements. These stresses are also compressive near the crown but become tensile nearer the base where there is the tendency to burst outwards.

Other distributions of load and other surface geometries will call for different patterns of the primary 'arch' compressions and for different patterns of stresses acting at right angles to hold the 'arches' in equilibrium. With one proviso though, there will always be such a pattern capable of supporting the loads even in a paper-thin shell. If the shell has this minimal thickness there will, however, be one pattern only for a particular shape and distribution of load. In contrast to that of the arch in Figure 2.11(c), this action is therefore said to be statically determinate. The proviso is that, with a smooth surface, there must be no highly concentrated loads acting at a point or along a line. This proviso is necessary because there cannot be

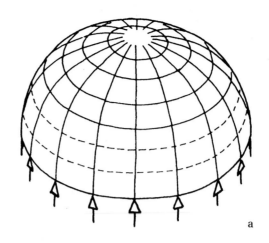

a

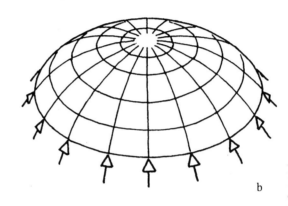

b

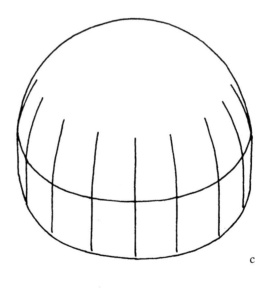

c

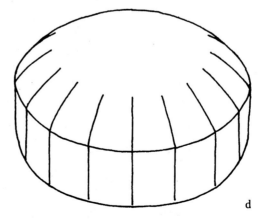

d

Figure 2.14 (a) and (b) Internal static equilibrium of domical shells with principal compressions shown by full lines and principal tensions by broken lines; (c) and (d) Typical crack patterns. (Source: author)

equilibrium between any such load acting at an angle to the surface and stresses acting only in the plane of the surface. If such loads were present they could be carried only by a bending action or by a thickness sufficient (like the depth of the voussoir arch considered previously) to contain a surface with the necessary sharp change in slope. Conversely, a thin shell coming to a point or ridge (like the shells initially proposed for Sydney Opera House) should have a concentrated load at this point or ridge.

As with the arch, though, it is necessary to look further at the dependence of this pattern of stresses on the reactions provided by the supports and on any associated movements of the supports. Because the pattern now involves horizontal stresses around the base as well as the 'arch' compressions, it calls not only for appropriate reactions to these 'arch' compress-ions but also for a contraction or expansion of the entire support to match exactly that required to develop the horizontal stresses. If both these requirements are not met, the action must change. The shell must be thickened to permit this and the action will become statically indeterminate.

To achieve the pattern of stresses shown in Figure 2.14(a) in a hemispherical shell would require only vertical reactions at the base, but it would be necessary to reinforce the lower part of the shell to carry the horizontal tensile stresses, and in addition to prestress (or tighten) this reinforcement so that it took the stresses without any expansion of the base. To achieve the corresponding pattern shown in Figure 2.14(b) for a much flatter shell would require not only inward reactions around the base but also a contraction of the whole support to match the contraction associated with the horizontal compression in the shell. This might call for a tension ring here, prestressed even more to give the necessary contraction.

If, in the hemispherical form, the rein-forcement were omitted, the lower part would crack radially as in Figure 2.14(c). The inde-pendent arches thus formed would then exert outward thrusts on the supports and any giving way to them would lead to further cracking. Such cracking may be seen in nearly all early domes where it has not been obscured by patching or rendering [2.15]. If, in the flatter form, there were no tension ring or equivalent provision, the spreading of the support as it took the horizontal components of the 'arch' thrusts would lead to a similar radial cracking as in Figure 2.14(d). If this cracking is not to pre-cipitate the collapse of either form the partial arches to which it gives rise must have the greater thickness typical of the masonry dome. Where such domes have remained standing in spite of extensive cracking, it is because they are not thin shells but have thicknesses approaching

Figure 2.15 Cracking around the base of a concrete semidome, Baths of Trajan, Rome. (Source: author)

the depths of the voussoirs of masonry arches of similar span.

Buckling

Before turning from these examples of individual structural elements to consider how they interact in complete structures it is desirable to look further at the possibility of buckling that arises when they are subject to compression. In elements acting wholly in tension (like the straight tie or catenary) a small disturbance leads, as we have seen, only to a small displacement to another equally stable form. In elements where the action is partly compressive the corresponding displacement may, however, introduce a bending moment that adds to the displacement. The commonest instance is the buckling of a slender strut where the bending moment can arise solely from a slight initial eccentricity of the compression. But a similar out-of-plane buckling can occur in other slender elements like beams and shells that act partly in compression. Figure 2.16 shows an instance of this in which the beam has buckled laterally under the compression developed in the upper part of the central region as seen in Figure 2.11(b). To guard against buckling, there is a need for adequate bending stiffness or other stiffening restraint.

Structural interdependence

This need for restraint is but one instance of the high degree of interdependence among the elements that make up most whole structures. In noting a similar interdependence among the elements of our own bodies and those of animals D'Arcy Thompson wrote that

> Muscle and bone . . . are inseparably associated and connected; they are moulded one with another; they come into being together, and act and react together. We may study them apart, but it is as a concession to our weakness and to the narrow outlook of our minds.[2]

Arches and domes, beams and slabs, and their supporting columns and walls do not usually come into being together. To that extent it is more justifiable to study them apart. It is also illuminating to do so because it is easier then to understand their essential characteristics and behaviour in a manner which is not inseparably tied to particular contexts of use. But we cannot properly understand the whole structure without considering how these elements act and react together.

Statically determinate behaviour

In purely geometric (or more precisely kinematic) terms, their interdependence means that the elements must be assembled in such a way that each has, potentially, the support that it needs to maintain it in static equilibrium [2.8]. When the number, configuration and interconnection of the elements will permit only one pattern of equilibrium to be established, all the reactions between them will be determined solely by the geometry and the loads. The behaviour is then said to be statically determinate.

Statical determinacy of the overall behaviour is most easily recognized in structures composed, as in Figure 2.8 (c), (c') and (e), (e'), solely of bars able to carry only direct tension or compression and simply pinned together. It calls then, in a two-dimensional structure, simply for a fresh pair of bars to establish each new node as the apex of a rigid triangle, or, in a three-dimensional structure, for three fresh bars to similarly establish each new node as the apex of a rigid tetrahedron.

Since the number of elements is only just sufficient for stability, the failure of any element through cracking, crushing or yielding of its material under excessive stress will lead to collapse of the whole structure. This is the penalty of statical determinacy. On the other hand, no element has a preordained space into which it must fit. Thus there is no such thing as initial 'lack of fit' and no stresses will be induced by imposed deformations such as thermal expansions or contractions or relative settlements of the foundations. The structure is free to 'breathe'

Figure 2.16 Lateral buckling of a slender reinforced concrete beam. (Source: Building Research Establishment)

Figures 2.17, 2.18 Cracks (accentuated for clarity of reproduction) showing the directions of principal compressive actions, S. Spirito, Florence. (Source: author)

Statically indeterminate behaviour

Statical indeterminacy, with more than one potential pattern of equilibrium even for a single loading pattern, can arise in various ways if we take a statically determinate structure as the starting-point. It can result from the introduction of an extra bar like that shown chain-dotted in Figure 2.8 (f) and (f'), since this bar may or may not participate in supporting the loads. It can result from modifying a joint so that it can transmit a bending action as well as a direct tension or compression (the rigid joint in (f) and (f') might, alternatively, be regarded as the redundant element introduced into the statically-determinate structures of (e) and (e') in the same figure), or so that the line of action of the reaction can vary as in Figure 2.11(c). It can result from modifications analogous to these such as adding an intermediate third support to the simply-supported beams of Figure 2.12 (a) to (c). Or it can result from multiplying the number of bars indefinitely so that they become continuous with one another and lose their separate identities – though the indeterminacy then will relate to the internal action of the resulting continuous form and not necessarily to its overall action.

When the behaviour is statically indeterminate the establishing of one pattern of equilibrium rather than another will depend on the relative stiffnesses of alternative paths for

carrying the applied loads to the foundations. Wherever there are alternative paths, the loads will clearly tend to follow the stiffer ones rather than those that give way more readily.

Such a structure will not necessarily collapse when a single element fails. The result will usually be no more than a redistribution of the internal actions, though the degree of indeterminacy will be reduced and the possibility of collapse thereby brought nearer. Some previously available modes of action will be lost. The loss may not, however, be permanent. The open joints of the arch in Figure 2.11(d) might close again if the load were removed in time to forestall collapse, and this would re-establish the indeterminacy of the original uncracked form.

But nor is the statically indeterminate structure free to 'breathe' in the same way as statically determinate one. All imposed deformations, including those due to thermal expansions or contractions or relative settlements of the foundations, will modify the pattern of equilibrium in the same way as deformations resulting directly from the loading. This may be troublesome. Differential settlements and thermal expansions and contractions, for instance, often lead to cracking. But it may also be turned to advantage, as when under- or oversized elements are deliberately introduced to bring about a beneficial change in the pattern of equilibrium. This is one basis of prestressing.

In practice, virtually all standing structures are statically highly indeterminate and would be found, if it were possible to inspect them closely enough, to have cracked or yielded in numerous places. This is merely part of a continuing process of adjustment to continually varying imposed loads and deformations. It is not usually harmful in itself, though the long-term consequences of cracking may be if the cracks let in water or, through being blocked by debris or deliberately filled, they become progressively wider.

Cracks and deformations as clues to structural actions

It is worth noting finally that, since cracks and deformations are the results of internal structural actions, they can also serve as visible clues to these actions.

Deformations tend to be the less revealing of the two because they are usually small and can be positively identified only if the original undeformed geometry is known. There is no such difficulty with cracks. In structures made of materials that are much weaker in tension than in compression – particularly in vaulted structures of masonry or concrete – they can be very revealing. It is necessary only to remember that they occur more or less at right angles to the principal tensile stresses in the uncracked form. And since these act at right angles to the principal compressions, the cracks themselves indicate directly the directions of these principal compressions. Where, for instance, the principal compressions are those radiating from the crown of a dome under the action of its own dead weight, the cracks follow their radial directions. The cracking seen in Figure 2.15 is a simple example of this. The more extensive cracking seen in Figures 2.17 and 2.18 is a more complex and revealing one. Here the cracks provide evidence not only of the radial compressions in the central dome and its immediate supports [2.17] but also of the thrusts exerted by it on the vaulting systems of the aisles [2.18].

This ready legibility of the structural behaviour of masonry means that masonry structures provide the best introduction to a wider understanding for those coming to the subject afresh, which is one justification for the attention given to them in later chapters.

Chapter 3 : Structural materials

The basic requirements for structural materials, as for the elements made from them, are adeuate strength and stiffness. It is also highly desirable that deformations under normal loading should disappear with the load. Without this characteristic – known as elasticity – repeatedly applied loads would lead to cumulative increases in deformation and to collapse simply as a result of this.

As properties of a material rather than of an element, strength is measured in terms of the maximum load per unit area (known as stress) that can be resisted, and stiffness in terms of the stress that is developed by a given deformation per unit of length (known as strain) in the direction of the stress. For most materials strengths vary according to the type of loading. The values most commonly measured are for

Figure 3.1 Local splitting and spalling of a corner pier in a facade tower of the Cathedral of Angra do Heroismo, Azores, as a result of high eccentric compression during an earthquake.
(Source: author)

loading in one direction only, either tension or compression, in a broad simulation of loads that occur in practice. Brittle materials like stone, brick and concrete are strongest in compression and are chiefly so used, so it is usually their compressive strength that is measured. Other materials like steel are equally strong in tension, and they are usually tested in tension. It should, however, be noted that when brittle materials fail under an applied compression they actually do so by splitting under the smaller tension that always arises at right angles to the compression, so that the measured strength is really an indirect measure of tensile strength [3.1]. For both types of material, deformations increase more rapidly as the limit of useful strength is approached. The behaviour is therefore best represented by curves showing how the stress is related to the strain at all points in the range.

Figure 3.2 reproduces a few such curves for materials in common use today. The precise significance of some of these curves will be discussed further below. It should be noted, though, that most of them flatten before the limit of strength is reached, and that this flattening usually denotes a loss of elasticity. In general, therefore, only the strength developed up to this point is available to carry normal loads. Where, as for mild steel, it is clearly defined and there is a long flat response, it is known as the yield strength. The maximum value of the strength is known as the ultimate strength. The absence of a flattened portion to the curve means that the material is brittle and will crack or crush suddenly without prior warning. In compression this tends to happen at a stress about ten times the tensile strength, though not necessarily for cast iron whose brittleness varies with its precise constituents and the way in which it is made.

Important associated properties are density, response to changes in temperature and humidity, and ageing characteristics including durability. All structural materials expand slightly when heated, and some, notably concrete and timber, expand also when they become damp. These expansions are reversible but they give rise to stresses when restrained and may then, as noted in the previous chapter, lead to cracking which is irreversible. With steel, very high and very low temperatures can, in addition, lead to major changes in the internal structure. Low temperatures make the material brittle, and high

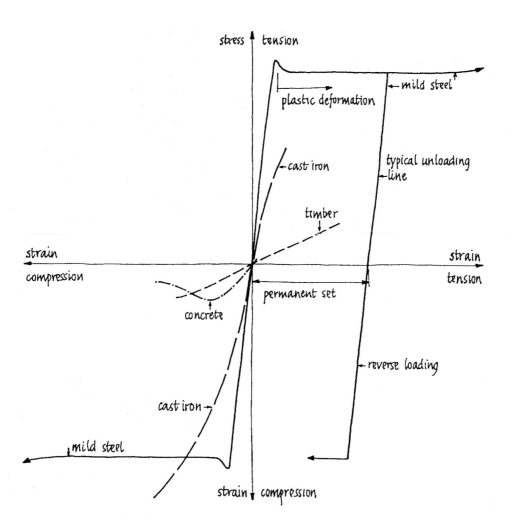

Figure 3.2 Typical stress-strain relationships for common structural materials. (Source: author)

Figure 3.3 Approximate ranges of strengths, density, and stiffness for materials available up to 1850, up to 1930, and up to 1990. (Source: author)

temperatures make it excessively ductile. By ageing is meant an irreversible change in the material itself. This may occur independently of the environment as, to a large extent, does the setting and hardening of modern cements. It may be favourably influenced by the environment as is the hardening of lime mortars. Or it may be unfavourably influenced as when timber decays due to biological attack or masonry suffers surface disintegration due to weathering.

For most materials density is important chiefly in relation to strength, but its importance varies with the type of structure and the nature and magnitude of the live loading. It directly determines the self-weight or dead load. Frequent significant variations in loading in relation to the maximum are usually undesirable since they may lead to cumulative damage known as fatigue. Similarly, large wind-loads in relation to vertical loads are undesirable since they will accentuate any existing tendency to buckle under the vertical loading even if they do not overturn the structure as a whole. From both these points of view, therefore, the greater the dead load and hence the greater the density, the better. On the other hand, if the dead load constitutes a large part of the total load, it will also make the major contribution to the total stresses. This is undesirable where these stresses approach the limits of strength of the materials

as in some large modern structures. From this point of view a low density is desirable, or rather a low ratio of density to strength provided that this does not entail too low a stiffness. It may also be desirable under earthquake conditions since a high density will result in high inertial loads.

Other characteristics are important because of their effects on the whole process of construction – on the ease with which it can be carried out and on its costliness in terms of natural resources and human effort. Some materials can be used as they occur in nature, but differ widely in their availability and the ease with which they can be obtained and cut or shaped. Others are artificial and differ even more widely in the ease with which they can be made and shaped or moulded. Most place some restrictions on the ways in which elements made from them can be joined together and hence, fairly directly, on the classes of structural form for which they can most appropriately be used.

Properties such as strength, stiffness and density have always operated in the same way in determining the suitability of particular materials for particular structural purposes, whether or not their relevance was fully understood. The availability of materials with particular sets of properties has, however, varied widely from place to place and from period to period, as have the essential skills in using them efficiently – in overcoming, for instance, the difficulties of making certain types of joint. Choices of structural form have, therefore, been constrained in a very varying manner by the need to give substance to the basic concept in some particular material or combination of materials.

Figure 3.3 gives a broad picture of one major aspect of this constraint by showing ranges of densities, stiffnesses and usable strengths available at different times as measured in modern standard tests. The usable strengths for ductile materials are plotted as the yield strengths. Even though it is not possible on the same scale to show the much higher strengths now attained by steel strands and prestressing bars and by composite strands of carbon and other fibres, the vast increase in the range of strengths that resulted from the earlier development of other man-made materials like steel and concrete is clearly apparent. The changes that took place within each of the periods identified are, however, ignored. Likewise the fact that, for materials like stone, brick and timber, the available strengths could vary considerably from place to place. To fill in the picture, it will be convenient to look in turn at each of the most commonly used materials.[1]

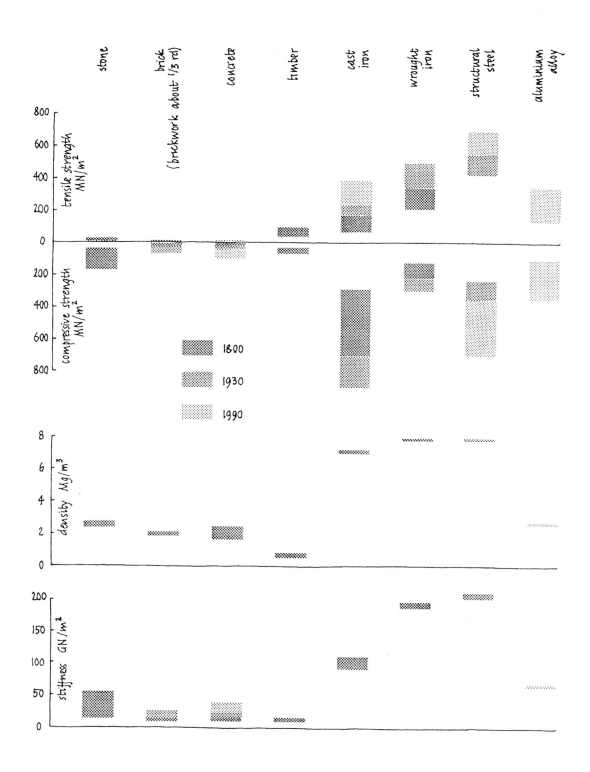

Stone

The earliest structural materials were probably stone, mud, reed and the lighter kinds of timber. Usually only one or two would be available at any locality. Stone was found most frequently in association with mud or timber, though in Upper Egypt it is found alongside both mud and reed. As found, it is much the most durable of the four and therefore tended to be preferred, where it was available, for those structures considered to be of greatest importance. It nevertheless varies greatly in hardness and internal structure, and hence in the ease with which it can be extracted from the solid mass and cut to shape when it is not possible simply to select usable blocks from the loose material produced by natural weathering or from boulders in a river bed. Geologically, these differences correspond to the different ways in which the rock has been formed.

Primary, or igneous, rocks have been formed by solidification of molten magma either well below the surface or closer to it as a result of volcanic action. With the exception of light volcanic deposits such as tufa, these rocks are mostly crystalline and very hard. Lacking any natural beds or cleavage planes, they cannot easily be split apart. They are therefore difficult both to extract and work. This has led in the past to a restricted use of them for structural purposes and to an emphasis, when they were used, on maximum simplicity of form and on any expedient which would simplify or reduce the labour that had to be expended. Granite is the chief rock of this kind to have been used structurally, especially by the ancient Egyptians. The Valley Temple of Chephren at Giza [3.4] which has already been mentioned in Chapter 1 [1.6] is a good example of this use. The choice was almost certainly made for the sake of durability. Apart from the basic simplicity of the finished form, the detailing of the masonry of the wall is most revealing. The bed joints are horizontal but not always continuous; the rising joints are sometimes inclined; and both they and the top surfaces of individual blocks sometimes show re-entrant angles. To the right of the doorway, every block also shows a re-entrant angle in plan. All these apparent anomalies result from an attempt to minimize the effort of dressing the somewhat irregularly sized blocks received from the quarry by dressing the mating faces largely *in situ* as they were fitted together and leaving the dressing back of the exposed faces until the wall was complete.[2]

Secondary, or sedimentary, rocks form the other main group. They are largely derived from the deposition, either on land or under water, and the subsequent reconsolidation of disintegrated or decomposed primary rocks. They may be further divided into sandstones or limestones. Sandstones are usually the harder of the two since their main constituent is simply the weathered fragments of primary rocks. Both, however,

can usually be cut or sawn without too much difficulty. The fact that they have been deposited in successive layers means, moreover, that they can be split fairly easily in the direction of the bed. They therefore lend themselves much more readily than the primary rocks to use in the form of pre-cut squared or otherwise precisely shaped blocks. According to the particular deposit from which the stone is taken, these blocks may be very large in size or may have to be quite small. Some deposits will yield blocks that can be used as lintels to span considerable openings or as large monolithic shafts or columns, but others will yield only much smaller blocks that call for a different approach to construction.

Metamorphic rocks form a third group derived from the primary or secondary rocks by heat, pressure or chemical action. Of these, marble, derived from limestone, is the only one

Figure 3.4 Chephren Valley Temple, Giza showing blocks of granite dressed largely *in situ* as they were fitted together. (Source: author)

Figure 3.6 Foundations of the sanctuary of the Temple, Kom Ombo. (Source: author)

of much structural importance. Its relevant characteristics are similar to those of a good limestone.

The compressive strength of all types of stone varies considerably with the precise source of the stone, but is generally highest for granite and least for the poorer sandstones and limestones. Tensile and shear strengths are low, the latter particularly along the cleavage planes of the sedimentary rocks. Use in the form of lintels and other elements subject to tension or shear may therefore be restricted by strength limitations as well as by possible difficulties in obtaining large blocks. Use in columns, walls or arches is unlikely to be so restricted provided that the compression is fairly uniform. An ideal thick prismatic column subject to perfectly uniform axial compression would not, for instance, reach its limit of strength until it attained a height of several thousand metres. Non-uniform stress can easily lead to local splitting where the stress is highest, and thereby to much earlier failure.

It follows that the strength and stability of

Figure 3.5 Stonehenge, near Salisbury. (Source: author)

structures of any size built of stone masonry, particularly those with wide spans and those with the weights concentrated on relatively small piers or columns, can best be ensured by a choice of form that limits all main actions to compression and by a technique of fitting the blocks together that distributes the stresses as uniformly as possible. Since there will only be compression to transmit, the chief requirement is for a good uniform bearing between blocks. This can be obtained either by careful dressing of the mating surfaces or by interposing a layer of another material. Provided that this layer is relatively thin, its strength, as measured under the usual test conditions, is not of great importance: it is so confined that, though it may crack, it is unlikely to be squeezed out or lose its bearing capacity. In much early masonry the greatest reliance seems to have been placed on accurate dressing, very thin layers of mortar

51

sometimes being used more as a lubricant to assist in manoeuvring the blocks into place than to improve the bearing. Most later masonry, however, made use of gypsum or lime or pozzolanic mortars. Sheets of lead were even used at intervals in the construction of the main piers of the sixth-century church of Hagia Sophia in Istanbul.

Ideally, the joints should all be at right angles to the compression, though there will always be enough friction to resist some obliquity of load. To give further stability where the direction of loading might be expected to vary (for instance in the event of earthquakes) or where it was bound to be oblique, various devices have been adopted. Dowels were used to locate the lintels and prevent lateral movement on the columns in the Chephren Valley Temple and they continued in use to locate the drums of later multi-drum columns. Projecting tenons served the same purpose at Stonehenge [3.5]. Between the lintels themselves, dovetail cramps were used or the blocks were directly dovetailed to one another.

Such dovetailing or cramping was also common in the rising joints of foundations and walls [3.6]. Frequently, however, its purpose may have been merely the temporary one of holding the blocks together during construction. Where friction was not considered adequate to prevent subsequent lateral displacement, oblique rising joints or more complex forms of bonding like that of the polygonal masonry of Figure 6.6 were one safeguard. Another, seen in the best classical Greek masonry, was to use the squared blocks of finely dressed ashlar but to tie them together in both walls and entablatures by means of iron cramps of double-T form. These were cast into their chases in molten lead as a protection against corrosion. Similar cramps continued to be used to a more limited extent and often with less protection in later masonry.[3]

Even in the best masonry of these kinds subjected only to pure compression in a wall or pier, the strength would, however, be well below that of an ideal monolith on account of the weakening effect of the inescapable vertical or inclined joints in each course of a wall or pier. To reduce this weakening, care was taken to break joints in successive courses. Some of the tension that arose at each joint at right angles to the primary compression was then transferred by friction to blocks above and below the joint.

In rubble masonry, and in the rubble that often filled the space between ashlar facings, there is no such bridging and therefore very little strength unless the whole mass is bound together by a tenacious mortar to create a form of concrete.

Brick

In the alluvial valleys of the Nile, the Tigris and the Euphrates, mud, straw and reed have been continuously used for nine or ten millennia to

construct simple dwellings and similar structures, and they are still so used in other parts of the world. The mud, rammed hard to consolidate it, dried by the sun and where necessary given a protective covering against rain, is quite able to resist the modest compressive loading. When used in the mass, however, it is slow drying and, even with the addition of straw, tends to be cracked by the inevitable drying shrinkage. It is therefore more convenient and more satisfactory to form it first into small lumps and to allow these to dry at least partially before using them [3.7]. Sun-dried mud bricks of high quality and regular size were in use from at least the third millennium B.C. Set in a mud mortar, often with reed mats incorporated at regular intervals in the horizontal joints, they permitted the construction of massive walls, pyramids, and ziggurats of considerable height, and of vaults and arches several metres in span. Their chief drawback was their poor durability when subject to alternate wetting and drying. This defect was fairly soon overcome by firing at a temperature of around 1000°C to give the burnt brick we now know. It tended to be used originally just as a facing or in other situations where the sun-dried brick was inadequate, but it became a major structural material in Roman times.

Burnt bricks differ from natural stones chiefly in three respects. Their individual compressive strengths are very dependent on the quality of the clay used and the conditions of firing, but never equal those of the better stones. They are essentially small units, considerably smaller than the sizes of stone normally used. And, although the process of manufacture readily yields a fairly uniformly shaped product of convenient rectilinear form, this has usually lacked the trueness of surface and accuracy of shape and size that would be needed to dispense with relatively wide mortar joints. Such joints make it difficult to attain a compressive strength of the brickwork as a whole more than about one-third of that of the individual bricks. This strength may therefore be a greater limitation on design than it is with the best stone masonry. However, the small size and wide joints also make it easier to construct curved forms like arches and vaults [3.8]. The curvature is easily taken up by slight tapering of the joints. For such forms the compressive strength is not critical and the lower density, as compared with stone, is an advantage. For these reasons, brick arches and vaults have often been built over stone piers, brick being used again for infilling walls between the piers [3.9]. The tensile strength of the brickwork is usually negligible, though there are exceptions where bricks with a large surface area in relation to thickness have been set in a good pozzolanic mortar.

Closely related to bricks are a number of other fired-clay products such as the roofing tiles. These seem, in some places, to have been made before the burnt brick itself, and probably

led to the adoption of a large flat form for the Roman brick. Similar flat tiles have been used in more recent times for a special form of vault construction that will be described in Chapter 7. Various forms of hollow pot and perforated brick have also been used, chiefly to reduce self-weight or to provide better thermal insulation than is obtained with solid masonry.

Concrete

Today concrete is a clearly defined material formed by chemical action on a mixture of water, cement, sand, and a coarser aggregate such as river gravel, the mixture being controlled within close limits to give the desired strength. It is, in this sense, a fairly new material. If, however, it is regarded more generally as any artificial conglomerate formed by binding together a mass of hard coarse material by some kind of cementitious material, it is much less clearly defined and much earlier in origin. It shades off on the one hand (in late Roman times) into a type of brickwork containing as much mortar as roughly coursed broken brick, and on the other (especially in the Romanesque and Gothic periods and later) into the more loosely bound and often much weaker and more deformable rubble fillings of ashlar-faced walls and piers.

As a structural material in its own right, it almost certainly developed from such rubble fillings when a good pozzolanic mortar was first regularly used by the Romans in place of a much less satisfactory straight lime mortar.[4] A lime mortar gains strength only by carbonation of the lime as a result of contact with the atmosphere. It therefore does so very slowly in the centre of any large mass, and large proportions of the lime may remain unchanged even for thousands of years. A pozzolanic mortar contains a further active ingredient – a compound of silica or alumina – which reacts directly with the lime independently of the atmosphere. It hardens much more rapidly, attains higher strengths, and is much more durable in the presence of water. The further ingredient may be a natural earth or volcanic ash or an artificial material such as crushed underfired brick, tile, or potsherds. The Romans made extensive use of the volcanic clay or tufa from near Pozzuoli in the bay of Naples, and it is from this that the name derives. Having realized the qualities of the mortar, they progressively reduced the thickness of the outer facing of the wall until it became no more than a thin skin, half-a-brick or so in thickness, with just occasional full bricks projecting into the main mass to give a good bond [3.10]. Or, chiefly in foundations and vaults, they dispensed entirely with a permanent facing and used only temporary timber formwork to contain the concrete mass until it hardened sufficiently to stand on its own [3.11].

The chief modern innovations were the nineteenth-century development of Portland

Figure 3.10 Detail of construction, entrance to the island villa, Hadrian's Villa, Tivoli. (Source: author)

Figure 3.11 Concrete annular vault, Temple of Fortune, Palestrina. (Source: author)

cement as an alternative to lime-pozzolana mixes,[5] and the pre-mixing of all the ingredients of the concrete in contrast to the earlier practice of placing alternate layers of coarse aggregate and mortar and then mixing these more or less adequately *in situ*. Coupled with these innovations has been the close control already mentioned.

The useful strength, as with brick and stone masonry, is largely in compression. It is now possible to achieve strengths approaching those of the best natural stones and therefore considerably in excess of those of most constructed masonry. High strength depends, however, on good compaction as well as on a good mix because, as usual in compression, failure takes place by splitting. The density varies with both the degree of compaction and the density of the aggregate. Usually it is similar to that of brickwork, but considerably lower densities can be and have been achieved through the use of natural or artificial lightweight aggregates.

Since the material is initially plastic, jointing presents few problems when the join is subject only (or very largely) to compression and the concrete is made or cast *in situ*. However, modern applications mostly call also for considerable tensile strength in the finished element and may call also for precasting. These necessitate the incorporation of tensile reinforcement and give rise to what is really a new material – reinforced concrete.

Timber

All the main materials still to be considered, including reinforced concrete, make good the principal shortcoming of those considered hitherto in having tensile strengths to match their compressive ones. Until the latter part of the eighteenth century the only such material

Figure 3.12 Log construction, Kostroma. (Source: author)

that was available in quantity and in suitable sizes for the construction of roofs, floors, and similar structural elements was timber. It was, therefore, very widely used in conjunction with stone, brick and, to a lesser extent, concrete, as well as being used on its own.

Much as with other natural materials, its use has, of course, reflected the availability of suitable species. Indeed in the past it probably reflected local availability to a greater extent than did the use of some other materials because of the difficulty of transporting long lengths over considerable distances. In a given locality, the ease with which it could be exploited wherever it grew also led, again and again, to the excessive depletion of the better supplies. This sometimes necessitated significant changes in the structural forms adopted. The European hardwoods, for instance, have virtually disappeared as structural materials, and it had

already become difficult, in the later Middle Ages, to find the tall, straight and mature trees that had been selected earlier. Today these timbers (such as oak and chestnut) have been replaced, for nearly all structural purposes, by softwoods of the pine family.

In one respect, timber is unique among structural materials. It is the only one which is the direct product of natural growth, and which is subject as it grows to structural actions comparable with those to which it is afterwards subjected in its use in construction. There are two important differences, though, between the tree and the timber it yields. First, the tree is alive; the timber taken from it and re-used is dead. This leads to significant changes in the internal structure which should be taken into account if the timber or the structure made from it is not to be seriously weakened. In particular it leads over time to a considerable loss of moisture which results in pronounced shrinkage, especially across the grain. This can tighten joints but can also lead to splitting and serious distortion. The timber also becomes more vulnerable to fire and to various forms of attack and decay. Second, as part of the tree, it was adapted in a highly efficient way to the particular structural demands then made on it. Essentially these were to resist axial compression plus a varying amount of bending, and they were met by a closely packed system of fibres running in the direction of the resulting stresses. When the timber is cut, these fibres can still resist either axial compression or the axial tension which arose previously from the bending action, but they are much weaker in resisting stresses such as compression across the grain and have hardly any strength to resist tension across the grain or shearing along the grain. Timber is therefore a material whose strength depends very much on the way in which it is used and on the grain. It can, for instance, vary greatly from one piece to another of the same species according to the straightness of the grain and the presence or absence of knots.

Structurally its greatest drawback until recently was the difficulty of making joints capable of transmitting and developing the tensile strength of the fibres. The transmission of compression is much easier.

We thus find that where timber of large enough girth was far more readily available than stone it was sometimes used in almost the same way as stone. Logs cut to length and notched at each end were simply placed on top of one another to create walls. In place of the bonding typical of masonry, the logs were interlocked at the notches with those of adjacent walls to give stability [3.12].

A few joints used in the past to transmit direct tension or the tensions associated with bending are shown diagrammatically in Figure 3.13 (a) to (c) and (e) to (l). In all of them the tension was developed by shear along the grain, and failure tended to occur as indicated by the

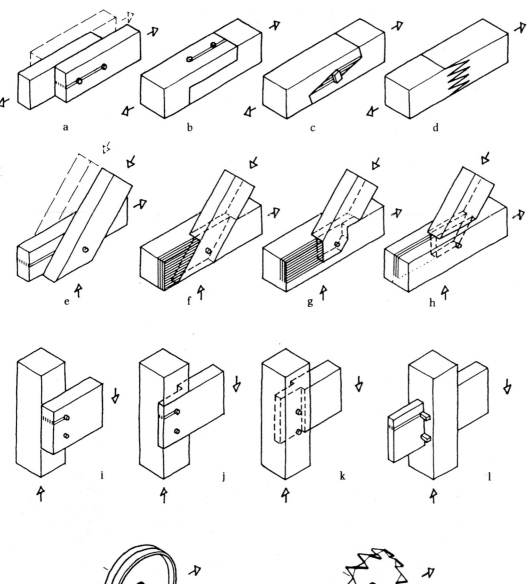

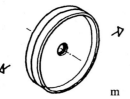

Figure 3.13 Typical joints in timber (see text). (Source: author)

shading if the fixings are simple bolts, wedges or treenails (slightly tapered round wooden pegs). To avoid this, it was usually necessary to make the member under tension project some distance beyond the point of connection, which led to waste of material and a cumbersome joint. If, in addition, it was desired to keep the members in the same plane (as in (b), (c), (f) to (h), and (j) to (l)) considerable cutting was required, which weakened both members at a critical point. Considerable ingenuity was devoted to devising intricate ways of fitting members together to resist pulling apart from almost any direction, the notching in (g) being a very simple example of this.[6] But the result was almost invariably a further weakening of one or both members.

Today it is more usual to lap the members (as in (a), (e) and (i)) instead of halving (as in (b), (f) and (j)) or morticing and tenoning (as in (h), (k) and (l)). This leads to some undesirable eccentricities of loading unless one member is doubled (as suggested by the lighter broken lines in (a) and (e)); but it avoids the weakening introduced by cutting. To obtain a more effective transfer of shear at the joint, it is also usual to interpose special metal connecting devices between the members. These either fit into pre-cut grooves (m) or are provided with teeth which bite into both members (n). By distributing the stress more evenly, they largely eliminate the risk of failure that existed with the earlier counterparts.

A completely different approach has become possible through the introduction of new synthetic glues and new cutting tools. Figure 3.13(d) illustrates a highly efficient and economical tension joint in which closely meshing teeth are glued together. The other main application of the gluing technique is to the lapped joint for the transmission of shear and thereby also of tension. Starting with only relatively thin sections of timber in readily-available short lengths, it has always been possible in principle to build up any desired section by multiple lamination with staggered joints along the length. This was done in one way or another over a long period, but previously with mixed success because of the difficulty of ensuring durability and good transfer of load. With the glues now available the full potential strength of the complete cross-section can be developed and, provided that all the joints are well lapped, there is no limit to the finished length of the member. Nor must it be straight. If all thin sections are bent before gluing, it is equally easy to build up members with a permanent curvature or twist. The result is virtually another new semi-artificial material. This is capable both of providing members of a size and quality once, but no longer, available directly from the best forest trees, and of doing so with a less restricted choice of form. In the example illustrated in Figure 3.14, up to 2500 individual thin laminae made up the total-cross section of 50×35 cm. and made it possible to achieve a radius of curvature as low as 2.5 m. at the same time as a twist of about $14°$ per m.[7]

Iron and steel

Metals other than iron and steel have been, and are, used structurally. Bronze was used for the trusses of the portico of the Roman Pantheon and aluminium has been used to a limited extent in recent years where its light weight is an advantage. Cast iron, wrought iron and steel have, nevertheless, been much more important hitherto and steel continues to be so.

With very rare exceptions, iron occurs in nature only in chemically combined forms, usually as one of its oxides. To extract it, it is necessary to heat the ore in the presence of what is known as a reducing agent to remove the oxygen. The reducing agent is carbon, originally charcoal and now coke, which also serves as the fuel to produce the necessary temperature. If the temperature is just sufficient to extract the iron but not sufficient to melt it, a soft and fairly pure form is obtained. This is the blacksmith's iron of Antiquity and the Middle Ages. If the temperature is higher, as has been usual since the fifteenth century, the molten iron takes up further carbon from the fuel, but can be run off much more conveniently into moulds to produce bars, which are known as pigs. These can be simply re-melted and the material recast in fresh

Figure 3.14 Glued laminated timber arched ribs, Hölderin Haus, Maulbronn. (Source: Professor Klaus Linkwitz)

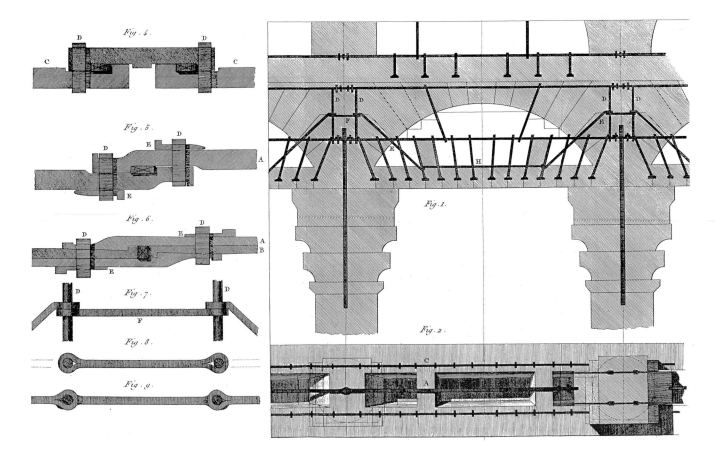

moulds having the forms of the desired elements or components. It is then known as cast iron. Alternatively, most of the carbon taken up from the fuel can be removed by passing the pigs through a further process. Whereas cast iron made in this manner is hard and brittle with a tensile strength only about one fifth of its compressive strength because of the carbon trapped in it as flakes of graphite, the purified material is softer again with similar strengths in tension and compression and, like the earlier product, can be easily forged. It is known as wrought iron. Steel is also produced by passing the pigs through a purification process, but it retains slightly more carbon than wrought iron and is less soft. It can, nevertheless, be rolled at a red heat into various shapes and can, although to a more limited extent, be formed when cold. Its strength and other properties can be considerably modified by the addition of small quantities of other metals. These additions produce what are known as alloy steels, or by names descriptive of their special properties such as high-tensile, high-yield, stainless, and weathering steel. Recent developments have been ductile cast iron and cast steel. Small additions of manganese and cerium in the first make the graphite inclusions spheroidal in shape so that they do not cause embrittlement and reduced tensile strength. In the second, the desired properties and freedom from undesirable casting stresses are achieved partly by suitable heat treatment

Until cast iron became cheap enough to be used structurally in the eighteenth century, only the much softer and easily forged material was so used and always in a secondary role.[8] It was used chiefly for cramps and tie bars in masonry construction, and for bolts and straps for jointing timber. Some eighteenth-century applications are illustrated in Figure 3.15. The details at the upper left show tie-bar joints comparable with the timber joint of Figure 3.13(c), the lowest of the three being for the tie bar that encircles the intermediate dome of the French Pantheon. Tie bars of considerable size were forged when their tensile strength could not be provided in any other way, but few indisputable instances of large forgings for other structural purposes are known.

In the late eighteenth century and the first half of the nineteenth, the use of cast iron expanded rapidly. Being cast, it could easily be made to assume a great variety of forms, with the sole restriction that large variations in the thickness were undesirable since they could lead to marked differences in the rate of cooling and, thereby, to a greater risk of hidden defects in the casting. Because this risk always existed to some extent, and because of the weakening effect of the high carbon content, the material was more suitable for columns and arches subject largely to compression than for beams, and was best avoided for members that were subject to pure tension.

To a large extent it was used as a fairly direct higher-strength incombustible substitute for timber, and the methods of jointing

Figure 3.15 The use of iron tie rods and cramps ir the Pantheon, Paris, from J. Rondelet, *Traité théorique et pratique de l'art de bâtir*, vol. 4, part 2, Paris, 1814. (Source: author)

frequently followed, with little change, those used for timber. Figure 3.16 shows the use of dovetails, mortices and wedge-shaped keys in the first major cast-iron structure in 1779. Even in 1851, the standard joint between the columns and the trussed beams at the Crystal Palace [4.11], though perfectly adapted to its structural function and to the requirements of full prefabrication and rapid assembly on site, still remained, in principle, the same as the timber joint of Figure 3.13(l). The wrought-iron or hardwood keys marked T performed the same function as the wedges inserted to tighten the latter joint. To transmit loads between columns or other members in pure compression, it was necessary to provide an even bearing-surface and this was done either by machining the castings to fit accurately together or, as in masonry construction, by interposing a layer of a jointing compound.

First wrought iron and then steel progressively displaced cast iron from about the middle of the nineteenth century because of

Figure 3.16 Detail of jointing, Iron Bridge, Coalbrookdale. (Source: author)

their improved performance in tension and freedom from hidden defects, and because they lent themselves to more efficient processes of production of the sort of members required in large quantity to serve as beams, columns, struts and ties. These members could be rolled in a more or less continuous process and then cut to length as required instead of having to be individually cast. Steel ousted wrought iron by the end of the century partly because different methods of purification enabled it to be produced much more economically in the large billets necessary for efficient operation of the rolling-mills. The change from individual casting to continuous rolling meant, of course, the loss of much of the earlier freedom to choose almost any form. Members had to be of constant cross-section and limitations had to be imposed on the available choice of cross-section. Other sections could, however, be built up, usually from standard flats and angles [3.17].

The marked difference in performance between the early form of cast iron (now known as grey cast iron) and mild steel (the usual structural grade) is clearly shown by the relevant curves in Figure 3.2. If cast iron is overloaded, it breaks suddenly. If mild steel is loaded beyond its elastic capacity or yield strength, it merely continues to deform and thereby usually sheds some of the load to another part of the structure. On removal of the load, as represented by the descending line at the right, it does not completely recover its initial state but it reverts to elastic behaviour with what is known as a permanent set. The permanent set is equal to the so-called plastic deformation undergone after reaching the yield strength.

The significance of this difference in performance is twofold. In the first place, a member made of the old grey cast-iron fails suddenly and without warning. A steel one fails gradually and with ample warning. (It should be noted, though, that its ductility will be reduced by very low temperatures and that sudden failure can also occur through the progressive growth of some surface defect when similar loads are applied and removed many thousands or millions of times.) Second, the ductility of steel makes it possible to construct structures which are statically highly indeterminate without this leading to extensive cracking as the structure settles down and expands and contracts with changes in temperature. This is not possible with cast iron, but is possible with steel because, instead of cracking, the steel merely yields locally and then resumes its elastic behaviour.

Until this advantage of ductility (which applies also to wrought iron) was fully appreciated, joints similar to those used with cast iron continued to be employed. Except where pinned joints were essential for structural reasons or to facilitate erection, the earlier types of joint soon gave way, though, to riveted joints [3.17]. For most purposes these have now, in turn, been largely superseded by welding or by a combination of shop welding and high-tensile bolts for completion of the joint on-site.

Riveting and bolting both call for the use of ancillary members such as brackets, cleats, gussets and cover plates through which the loads are transferred from one main member to the other. Rivets are inserted red hot so that, in cooling, they draw the members tightly together. Ideally, the load is then transmitted by friction at the meeting faces: if this fails it is carried by shear within the rivets. Bolts function in a similar manner except that, as they are now used, it is possible to control the tension to ensure that the load is properly transmitted by friction without slip.

Welding is a fusion process and is capable, therefore, of directly connecting two members and of directly transmitting tension, compression or shear as if the two members were one, more in the manner of the glued timber joint of Figure 3.13(d) than in that of any of the other joints in that figure. It calls for ancillary members, if at all, only to provide additional local stiffness or to permit the joint to be completed by bolting where this is more convenient.

Because brackets and cleats are flexible and rivets may permit some relative slip, riveted joints cannot, and fully bolted joints may not, give a completely rigid connection at which no relative movement is possible. The fully-welded joint and some partly-bolted joints do, however, give such a connection and provide full structural continuity in tension, shear or bending as well as in compression. Welding also permits a very unobtrusive and elegant type of joint without the great sacrifice in performance that, until very recently, was always the price of a similar elegance in timber construction. Its chief drawback is that the high temperatures reached can lead to undesirable distortions if precautions are not taken to avoid these. It calls also for careful control to ensure that fusion is complete.

The new ductile cast iron and cast steel have offered a further possibility of considerable value in tensile structures and space frames. Like earlier cast joints, joints of these materials can be produced in quantity off site and they can often be more precisely shaped than wholly welded joints to transfer loads in the most direct, elegant and efficient manner. The connection between casting and structural member can be completed by either welding or bolting. Cast steel has also brought back the older freedom of choice of form for complete members without the earlier penalty of embrittlement. Extensive use for both joints and whole members was pioneered in the Centre Pompidou, Paris [3.18].

The ease with which wrought iron and steel can be formed is another consequence of their ductility. Relatively heavy structural members were and are usually formed by repeated rolling at a sufficiently high temperature to make the material even more ductile. The ductility at

Figure 3.17 Detail of construction, Eiffel Tower, Paris (Eiffel). (Source: author)

normal temperatures is sufficient, though, to permit not only such simple processes as the punching of holes in these heavy members for rivets or bolts but also the cold drawing of rolled bars of a centimetre or so in diameter into wires of much reduced diameter. Since cold drawing permits no recrystallization of the material, it causes the existing crystals to be stretched out in the direction of the load they will have to carry later, rather like the fibres in a straight piece of timber. It can thereby increase several times the original tensile strength in this direction. High-strength wrought-iron wires were first produced early in the nineteenth century as a alternative to bars or chains where there was a purely tensile load to be carried. High-strength steel wires have been produced from the latter part of the same century and now attain strengths of some three times that of the original strength of the steel.

Unfortunately, high ductility also has its drawbacks, notably a poor performance in fire. Wrought iron and steel do not burn, as timber does, but they lose much of their stiffness when exposed for some time to temperatures approaching those used for hot forming. They then distort much more if the normal loads continue to act, and the structure may collapse, partly as a direct result of these increased distortions and partly because of unfavourable consequential changes in the modes of action of the loads. Both materials also rust if exposed unprotected to normal atmospheric conditions. (Cast iron does so much less and wrought iron less than mild steel.) Paint or some other thin surface treatment will give adequate protection against rusting if it is well maintained, so that structural form is influenced only to the extent that there must be access for painting and complete sealing-off of inaccessible surfaces. Some alloy steels even require no protective coating since one is produced by natural weathering. Something more is needed, though, to give protection against collapse in fire where

Figure 3.18 Detail
of cast steel joints,
Centre Pompidou, Paris.
(Source: author)

the risk of this would be unacceptable. In the past it was usually a substantial casing of concrete, which, regarded solely as an insulator, added much more to the self-weight of the structure than it did to its strength. Such casings are still used, but greater advantage is taken now of their potential contributions to strength and stiffness. The result may be a composite form of construction in which the resistance to the loads is fully shared between the steel and concrete.

Alternatively much lighter casings or intumescent paint are used or protection is given in some other way such as filling tubular members with water.

Reinforced concrete

Reinforced concrete might, from one point of view, be regarded as a development of the form of composite construction just mentioned. It differs from this encased steelwork chiefly in that its steel content is distributed through the concrete as bars of, at most, a few centimetres in diameter. These are assembled into cages or meshes before the concrete is poured around them. Yet, even when so assembled, they are very flexible and hardly capable of carrying their own weight over any distance without excessive deformation.

In its nineteenth-century beginnings it is better regarded as a development of earlier practices of making good the tensile deficiencies of stone masonry by means of embedded iron cramps and tie bars. In the Neo-Classical architecture of eighteenth-century France, such reinforcements were used so extensively in the flat-arch forms that imitated monolithic Classical lintels that the proportion of iron to stone approached that in some reinforced concrete beams. The first example will be referred to in Chapter 8. In the later example from the portico of the French Pantheon seen in Figure 3.15 the manner in which the iron is used to resist both tension and shear while the stone resists only compression might almost be seen as a clear anticipation of reinforced concrete.

The change from reinforced masonry to reinforced concrete came about around 1850, largely as a result of the invention of Portland cement. This gave a much better concrete than most of those known previously, and one which gave complete protection against corrosion to the embedded iron – the greatest drawback of the earlier reinforced masonry having been the tendency of the iron to rust and, in so doing, to split the masonry. Of the patents taken out at this time, that of W. B. Wilkinson [8.11(b)] came closest to specifying the material now known.[9] Development continued, particularly in France, England, Germany and the United States, up to the beginning of the present century.[10] In the course of it, numerous types of reinforcement were tried, but the convenience and other advantages of plain round bars or similar bars with an indented surface eventually led to their universal adoption. It was greatly assisted by two further characteristics of the concrete. One is the slight shrinkage which it undergoes on setting and hardening. This causes it to grip tightly the bars embedded in it. The other is that it expands and contracts with changes in temperature to almost the same extent as steel. The shrinkage is, however, liable to cause cracking across the width of a member, and care has to be taken to distribute the rein-

Figure 3.19 Piers of the Tay Road Bridge. Dundee, under construction (Fairhurst). (Source: author)

forcement in such a way that no cracks are wide enough to expose it. Provided that this is done and that the reinforcement is everywhere adequately covered by the concrete, it can be placed wherever it can most advantageously resist the tensile stresses while the concrete resists all or much of the compression. By varying the amount of reinforcement and the strength of the concrete, it is possible also to match closely the required strength and stiffness of a member without being restricted to a narrow choice of cross-sections. Even with a fixed cross-section, it is possible to achieve a wide range of strengths and stiffnesses.

Since neither the concrete nor the reinforcement has any useful strength initially, substantial formwork is required both to impart the desired shape and to give support without excessive deflection until the concrete has hardened enough to bear its own weight. The need to provide this formwork is, in practice, the main limitation on the forms that can be economically adopted. It must usually be built up from timber or steel. Two approaches are possible. All the formwork may be built up on site and the concrete cast directly in it as in earlier unreinforced concrete construction. Or individual elements or their components may be cast elsewhere and then assembled after the concrete has hardened.

If the concrete is all cast on-site, the only joints will be between individual reinforcing-bars and between successive 'pours' or 'lifts' of the concrete. The bars are either welded together or lapped a sufficient distance to develop their individual strengths by the natural bond created by the shrinkage of the concrete around them. Joints between successive pours of the concrete will not be monolithic but they can develop the full compressive strength if enough care is taken. Full continuity is achieved by carrying sufficient reinforcement across the join. Figure 3.19 shows vertical reinforcement projecting from one pour of the pier on the left to give continuity with the next. On the right-hand pier, the steel formwork is in place for the next pour, and it will be seen that the joins have been accentuated by slight grooves in the otherwise flat surfaces of the piers.

If components or complete elements are precast, a modified procedure has to be adopted. One possibility is to treat them almost as if they were made of timber or cast iron and to adopt similar jointing details. More commonly, except in prestressed construction, they are made with reinforcement or special steel inserts projecting from the ends. These projections are then looped round one another or welded or bolted together when the units have been assembled, and the joint is finally packed with a small quantity of mortar or site concrete.

Occasionally, however, there is a need for joints that allow relatively free rotation at a support. Unless a pinned bearing is introduced, these joints have to be made by means of a very considerable reduction in the cross-section of the concrete and by the concentration of the whole of the steel reinforcement into this small section. The stiffness in bending is thus greatly reduced, but the compressive load can still be taken by the reinforcement plus the remaining closely-confined concrete.

Prestressed concrete

Ordinary reinforced concrete is less than an ideal structural material chiefly in that the concrete cracks whenever it is subjected to appreciable tension. Reinforcement prevents this cracking from leading to failure of the composite member, but the need to limit the width of the cracks in order to protect the steel means that the stress in the reinforcement must be limited. High strength steel cannot, therefore, be used to full advantage as reinforcement. Similarly, the strength of the concrete itself is obviously being wasted to the extent that it is being subjected to tension rather than compression.

Prestressing overcomes these drawbacks by applying a compressive 'prestress' to the concrete at some stage of construction, sufficient to ensure that it always remains in compression under subsequent loading. This prestress may be applied either by tensioning the reinforcement against the concrete or by exerting external forces on the concrete, by jacking against the abutments of an arch for instance. Both operations are tantamount (as already noted in Chapter 2) to introducing deliberately under- or oversized elements into a statically indeterminate structure: the reinforcement is stretched

Figure 3.20 Detail of reinforcement and ducts for prestressing cables in a precast unit for a pre-stressed-concrete bridge. (Source: author)

to fit a concrete encasement for which it would otherwise be too short, or the arch is squeezed into a gap that is made too small for it. In the latter case the technique is not restricted to prestressed concrete, so it will be more appropriate to consider it in the next chapter. In the former case the reinforcement may be stretched either before the concrete is cast around it (pre-tensioned) or after the concrete has been cast and has hardened to some extent (post-tensioned). When pre-tensioned, it is initially anchored to the formwork and then cut loose after the concrete has set and gripped it firmly. When post-tensioned, it is usually threaded through ducts previously cast into the concrete to receive it and then, after tensioning, anchored against specially strengthened anchorage blocks at the ends of these ducts. Examples of such ducts may be seen in Figures 3.20 and 8.16 where they may be distinguished from the normal reinforcement by their larger diameter and curved profiles. The special helical reinforcement of the anchorage blocks may also be seen in Figure 3.20.

There is always some loss of the initial prestress. If the reinforcement is pre-tensioned, the loss arises in three ways: the concrete shrinks on setting, it shortens elastically as it takes up the compression when the temporary anchorages of the reinforcement are cut, and it slowly shortens further (or 'creeps') under this compression. If the reinforcement is post-tensioned, no loss occurs through elastic shortening and the loss through shrinkage may be largely eliminated but the loss through creep remains. Early attempts at prestressing in the late nineteenth and early twentieth centuries failed partly because the inevitability of these losses was not appreciated and partly because, even if they had been appreciated, the concrete and reinforcement of the time were not really suited to the demands made on them. The fact that the reinforcement is both prestressed and stressed further by subsequent loading means that it should have a higher strength before yielding than normal mild-steel bar. High strength and high quality are also required of the concrete to resist the additional compression and to minimize the shrinkage and creep. Within the last seventy years, the technique has developed rapidly as more suitable materials (such as the high-strength cold-drawn wires referred to above) have become available.[11]

Partly because it calls for concrete of high quality, prestressing has frequently been associated with precasting which usually permits a closer control over the mixing and placing and compacting of the concrete. Precasting is almost essential when the reinforcement is pre-tensioned. It is not essential when the reinforcement is post-tensioned, but another reason for its use then is that post-tensioning itself provides one of the best methods of joining precast units. It is necessary merely to butt them closely together (with a mortar or similar packing if necessary) and to arrange the ducts so that the prestressing cables or bars pass over the joints and, when stressed, tie the units firmly together with their abutting concrete faces wholly in compression. Any shear stresses on the joints can then be resisted by friction.

New composite materials

Alongside the materials discussed so far and a few others to which passing reference has been made are several newcomers which are less directly derived from basic resources. All are valuable chiefly for their high tensile strengths which are attained by spinning or drawing the primary constituents into thin fibres that are free of the defects that reduce the strengths of larger cross-sections. Since these thin fibres cannot be used by themselves they are bundled together and protected by a sheath or they are encased in a plastic matrix. The chief primary constituents are glass, carbon and aramid, a synthetic polymer. Fibre composites of all three have ultimate tensile strengths comparable with or even exceeding that of high-tensile steel wire. They are also considerably lighter and the composite cables or strands made from them will not rust and are probably more durable. But they are less stiff and less ductile. Indeed the stiffness of steel is close to the maximum that is physically attainable, so that it is unlikely to be widely superseded where stiffness is as important as strength and where a considerably reduced density is insufficient compensation for reduced stiffness. Elsewhere they have opened up new possibilities such as the development of new composite materials analogous to reinforced concrete but reinforced with much finer strands of glass or carbon and the development of new fabrics with greatly improved strengths, durability and translucency. By the mid-1990s, only the latter possibility had been exploited to a significant extent using woven mats of glass fibre coated with Teflon or silicon. PVC-coated polyester has been more widely used as a cheaper alternative for smaller, less permanent, structures.[12]

Chapter 4: Construction and form

In Chapter 1, the process of construction was contrasted with that of natural growth. Whereas growth is the continuous development of an organism that is complete and individuated from the start, construction is a discontinuous process of putting together. In it, at any intermediate stage, the form is either incomplete or essentially different from the final one.

The only man-made structural forms on the scale with which we are concerned here that are not so constructed are those cut from solid rock. One example has already been illustrated in Figure 1.9. Figure 4.1 shows a large temple-like facade cut from a natural rock face. Such forms

Figure 4.2 Rainbow Arch natural bridge, Utah. (Source: author)

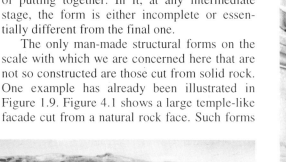

are closely akin to some of the results of natural weathering and erosion [4.2] in being similarly revealed, like a figure emerging from a block of marble under the sculptor's chisel, as something that was always there, entire, within the mass.

The naturally weathered or eroded form is, of course, the product of innumerable chance events in the initial formation of the rock and its adjacent strata and in the subsequent action on them of changing internal pressures, of wind and water, and of heat and cold. The unusually perfect form and impressive size of the Rainbow Arch [4.2] must, for instance, owe something to the chance occurrence at a particular point of a favourable internal structure of the rock mass – to favourable alignments of the planes of weakness along which exposed faces would most readily spall off – and to the equally chance occurrence at the same point of a meander in one of the small tributaries to the Colorado River. Its man-made counterpart, cut or excavated in solid rock, must respect the internal structure of the rock mass in order to minimize both the effort of cutting and the risk of unintended splitting and spalling. To this extent, chance events long ago act as a constraint on the choice of the form to be adopted in a particular situation. They are no different, though, from the constraints that are imposed by the working characteristics of any natural material on its shaping for use in another place for structural purposes. In the past, moreover, they have had the effect chiefly of limiting the practice of cutting or excavating the structure directly from the rock to places where it is both easily cut and evenly textured. The form could then be chosen with complete freedom provided that it would be stable and strong enough when fully revealed. *A fortiori* it would then be stable and strong enough at any intermediate stage of cutting unless the unwanted rock was cut away in an almost perversely inappropriate sequence. In fact, the forms chosen have usually been very closely modelled on existing built forms.

The need to assemble a built form piece by piece (or one pour of wet concrete after another) means, on the contrary, that an assurance that it will be stable and strong enough when finished is not a sufficient guarantee of adequate stability and strength at intermediate stages of construction. To avoid the risk of a collapse during construction it may be necessary to accept some limitations on the choice of the final form. Clearly these limitations will be related to the sequence of construction envisaged. This sequence will also affect the manner in which the loads are finally distributed through the structure, so that it may be necessary to take it into account also when considering the behaviour of the finished structure.

Other limitations on the choice of form that are directly related to the process of construction stem from the need to shape or fabricate the component pieces or elements whether on site or elsewhere, to set them in place if they are not fabricated in their final position, to join them together, and to make such measurements as are necessary to ensure that the structure does have the desired form. The methods adopted will, in all cases, influence the effort that has to be expended and hence the cost of the work. Some of them, notably the methods of joining the pieces together, will also have a considerable further influence on the final stability and strength.

When, therefore, a structure is to be constructed rather than cut directly from solid rock, the designer's freedom to choose its form is subject to a very complex and interrelated set of constraints stemming in different ways from the possible processes of construction – these constraints being additional to those already considered that stem from the structural requirements to be met by the finished form and from the characteristics of the available materials. To the extent that designers find themselves having to work in particular technological situations that are not of their own making – situations in which particular techniques of construction are known and practised and particular human skills and mechanical aids are available – these constraints have operated in the same continually changing manner as those

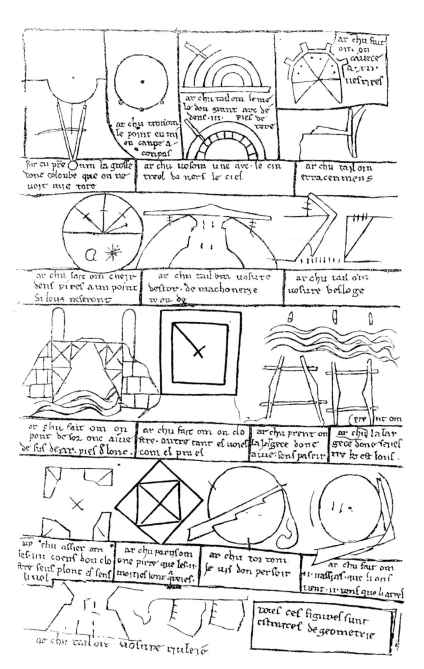

imposed by the available materials. Designers have, nevertheless, usually been better placed to develop the techniques that they have found and to invent new ones than to improve upon the available materials. The development of these techniques is thus more closely linked to the development of new structural forms than is the development of new or improved materials, so that it will be appropriate to defer some of the discussion to later chapters. Here we shall confine ourselves to a more general discussion of the whole construction process viewed as an essential part of the designer's total concept of a new form. It will be seen as something that must be taken into account in the choice of the form itself, but something to which he or she may, in some respects, be able to make a positive contribution. It will be convenient to look first at those constraints which stem directly from the operations of the construction process rather than from their structural implications.

Figure 4.3 Medieval builder's geometry. (From MS. Français 19093, Bibliothèque Nationale, Paris, with added emphases)

Operational aspects

Setting out

It would be an exaggeration to say that no structure can be 'designed' through the very process of building it. But it is generally true that design must precede construction even if it sometimes keeps only a little way ahead as the work advances. The exceptions are not of sufficient importance to detain us here. The form must first be conceived as a unique spatial configuration. The concept must then be conveyed to the builders by means of drawings, models or some other kind of unambiguous specification. Finally the form must be faithfully reproduced on the correct scale.

At first sight this may not appear difficult, but even the precise conveying of intentions from designer to builder can be far from easy if the form cannot be defined simply by straight lines and circular arcs. Greater difficulties can arise when the builder has to reproduce the form on a large scale in three dimensions in the empty air.

For a considerable time, accurate measuring-tapes for linear measurements and theodolites for angular measurements in any direction have been a great help. Now lasers are even more helpful and convenient to use. But even they do not eliminate the greater difficulties of setting out forms whose geometry is not very simple. Only with the help of computers has it become possible to define with sufficient precision the spatial coordinates of ribs like those illustrated Figure 3.14 or those of large cable nets and to determine desirable cutting patterns for tensile membranes.

In the past there were no such aids. Without them and with little more than a few measuring-rods of a standard length (not subdivided as today), a large compass like a sergeant major's pace-stick, and some pegs and lengths of cord, it was almost essential to keep to forms that could be set out on the ground by the simplest geometrical constructions [4.3] and whose elevations could be controlled in equally simple ways.[1]

The way in which a particular structure must be set out is, however, closely related to the way in which it is to be built. This, in turn, will obviously be related to the choice of materials and the skills available for handling and shaping or working them, though it will be not wholly determined by these. Only when every major element of the structure is constructed *in situ* without the use of centring or formwork and without prefabrication of its components in a way that predetermines its finished form will the full difficulties of setting-out by direct measurement at the building site have to be faced. Any use of centring or formwork or any prefabrication of major elements or components will clearly shift some of the emphasis to the centring or formwork or to the prefabrication process. This will relax some constraints that stem directly from the difficulties of setting-out, but will substitute others in their place.

The use of in situ *centring or formwork*

To the extent that *in situ* centring or formwork is used, the job of defining the desired forms by direct measurement in space is, for instance, replaced by that of constructing the centring or formwork to the appropriate shape and dimensions. This can usually be done, in part at least, on the ground. From that point of view it is less important to keep to the simplest possible geometry.

Since, however, the timber or steel which is normally used for the purpose comes in straight lengths and flat boards or plates, it is still desirable, in order to make the most economical use of it, to keep to configurations with simple plane surfaces, or at least to ones whose surfaces

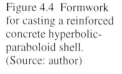

Figure 4.4 Formwork for casting a reinforced concrete hyperbolic-paraboloid shell. (Source: author)

can be swept out by straight lines. Part of the attraction of the hyperbolic paraboloid from the 1940s onwards was that, despite the fact that it is a doubly-curved surface convex in one direction and concave in another, it could be defined by sets of straight lines [4.4].

Any centring or formwork, even when economically designed, tends, moreover, to be expensive in both materials and labour in relation to the direct costs of whatever is built on it: it contributes nothing, of course, to the finished structure. Wherever possible therefore, it is also desirable to design the structure so that it can be re-used a number of times with the minimum trouble. In designing modern reinforced-concrete buildings, for instance, this frequently means standardizing the forms of beams, slabs, columns, or walls so that these repeat from floor to floor and possibly from bay to bay, and ensuring that no details of the design interfere with the removal of the formwork after the concrete has gained enough strength to be self-supporting.

Tower-like structures with reinforced concrete load-bearing walls – including the central lift-shafts of tall buildings – lend themselves particularly well to the repeated re-use of vertical formwork. Provided that the designer ensures that there are no horizontal excrescences from the wall faces, as well as keeping the plan unchanged throughout the height [4.5], the formwork can be simply slid up from one level to the next. A long barrel vault can be similarly constructed in sections on much narrower centring or formwork which is periodically slid sideways, provided that the profile remains the same throughout. The main arches of the Roman Pont du Gard [4.6] were obviously constructed in this manner. Each arch consists of several independent parallel arch-rings. Assuming that these were constructed in turn, only one would have needed the centring at any time.

The use of prefabricated elements or components

To the extent that whole structural elements such as beams, columns or trusses are pre-fabricated, the job of defining by direct measurement in space the configuration of these elements is replaced by whatever is entailed in their fabrication and erection. Since the whole essence of prefabrication is that it should be done wherever is most convenient, it usually permits the freest possible choice of form for the individual element within the limitations imposed by the characteristics of the materials used, the ways in which they can be shaped or worked and the need to consider the economy of the construction process as a whole. Considerations of economy do not so much restrict the basic freedom of choice as make it desirable to limit the number of different designs that will be called for.

Figure 4.5 Sliding formwork in use for casting the walls of Highpoint One, London (Lubetkin, Arup). (Source: Avery)

When only smaller components of the whole elements such as the individual struts and ties of a truss or the individual voussoirs of an arch are prefabricated rather than complete elements there will, however, be a similar freedom of choice without a need for independent setting out only if certain conditions are met. It is essential for the components to be fabricated to a high degree of accuracy and for them to fit together in such a manner that there can be no variation in the geometry of the element when it is assembled. Given these, erection should amount to no more than the equivalent of fitting together the pieces of a child's construction toy. Otherwise it will be necessary to check by measurement the correct location of each component as it is set in place or to use centring and to accept the limitations on the choice of form of the element that each entails.

It is difficult, therefore, to generalize very widely about the desirable constraints on choice of structural form when the process of construction includes a measure of prefabrication.[2] This difficulty is accentuated by varying requirements for structural continuity in different types of structure, by the varying ease with which the required continuity can be achieved, and by the varying importance of factors that have no direct structural relevance but that may still call for attention in the total design concept. Such factors may include a high premium on speed of

Figure 4.6 Detail looking
up to the soffit of one of
the second-tier arches,
Pont du Gard, Provence.
(Source: author)

construction even at the expense of increased
cost, a social pressure to remove as much work
as possible from the building site to the
preferred environment of a factory or a pressure
to do likewise in order to obtain better control
over fabrication. It will also obviously be more
desirable to minimize the variety of prefabricated components on a small project than on a
larger one since the same degree of repetition
could be achieved with a larger variety on the
larger project.

Even the basic choice between constructing
a structure entirely on-site and either wholly or
partly prefabricating it must take into account
the materials to be used and the degree of
technical competence that has been reached in
building with them. With cast metals, prefabrication has always been essential. It is also essential with large cable nets and fabric membranes
in order to achieve the desired final form and
distribution of stresses. With natural materials
like timber and stone that have to be cut to
shape and with other metals its convenience has
usually outweighed any disadvantages. With
reinforced concrete, however, its advantages
and disadvantages have always been more
evenly balanced, and with unreinforced concrete
or brickwork it is virtually useless because of
the difficulty of handling finished units of a
worthwhile size without damaging them.

The first material to be prefabricated for
structural use was probably timber, because of

Figure 4.7 Pont du Gard,
Provence.
(Source: author)

the ease with which it can be cut and the convenience of doing this on the ground. Indeed all timber must be at least partly cut to its finished form before being set in place, even if it is cut piece by piece immediately before this is done.

In the past, when only solid timber was used, the prefabrication of all the components of a large truss or frame was advantageous chiefly in simplifying all subsequent operations including the transport of the timber if all cutting was completed before it was taken to the building site. The individual posts and beams or struts and ties would be cut on the ground and fitted to one another with joints designed for easy dismantling and easy subsequent reassembly in some convenient sequence. This might, for instance, mean allowing some free play of a tenon in its mortice. Eventually the free play would be eliminated by inserting bolts, wedges, or treenails, and only when this had been done would the precise configuration be established: it would not, in other words, have been uniquely predetermined. But, since the structural forms adopted were not critically dependent for their overall strength or stability on a precise unique configuration, the slight doubt about what the actual configuration would turn out to be was not of great importance. Prefabrication of the components still made it unnecessary to take the sort of precise measurements above ground that might otherwise have restricted the choice of configuration. A very free choice was possible, coupled with an equally free choice of form for the individual components. This allowed full advantage to be taken of the available shapes and sizes of timber. Since each component was individually cut and fitted to its neighbours, it was not even of great importance to minimize the variety of form.

Today, with solid timber much less readily available in large sizes, the equivalent forms are likely to be large built-up members such as beams, arches, trusses and portal frames, prefabricated as complete components or complete prefabricated trusses or frames. They are made in jigs from machine-cut timber and the need for the jigs means that variety in the choice of form may be costly. Nevertheless, the relatively high cost of the timber itself still makes it desirable to make the greatest use of the available sizes – now usually the standard lengths, widths and thicknesses supplied by the mills – even if this calls for some variety.

The practice of pre-cutting blocks of stone to fit closely together was almost certainly a later development. The potential advantages would, however, have been apparent from the experience acquired in building simple structures with naturally weathered blocks or boulders of convenient size [6.1 and 10.2]. There is clear evidence that, in some of the earliest surviving structures built of finely-dressed stone, the dressing was done partly *in situ*. As already noted, this is true of the masonry of the walls of the Chephren Valley Temple at Giza [3.4]. It is probable, on the other hand, that the columns and lintels of the same structure [1.6] were fully dressed before final erection. It is even more probable that a large proportion of the monolithic columns and lintels in structures of later periods were similarly prefabricated, often at a distant quarry.

Whenever complete elements were so prefabricated, there would be a very free choice of their form, limited only by some standardization of those overall dimensions that were important in fitting the elements together. When, as with an arch, prefabrication as a single

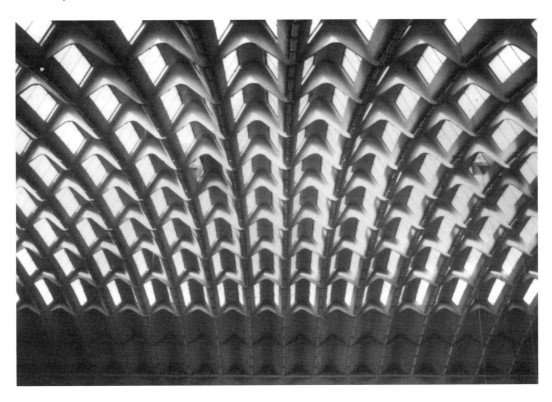

Figure 4.8 Vault of an exhibition hall, Turin (Nervi). (Source: author)

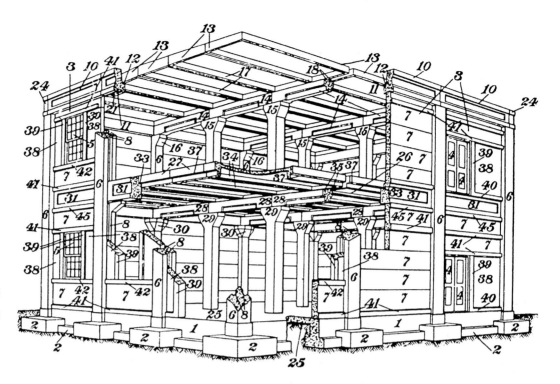

Figure 4.9 Patent drawing of the Conzelman system of precast-concrete framed construction, 1912. (From the paper by Peterson cited in note 3)

unit was not possible, other factors would have had to be taken into account, leading to a more restricted choice of form as typified by the strictly semicircular profiles of the arches of the Pont du Gard [4.7]. There is no doubt there that the voussoirs were fully dressed on their mating faces and on the soffit before erection. To ensure that they would fit closely without the need for mortar, they were laid out on the ground as a final preliminary check and individually lettered and numbered to indicate their final positions. Some of these markings (FRD IV etc.) can still be seen in Figure 4.6. This procedure would have obviated the need to ensure that every voussoir was cut to an absolutely identical pattern, but it would still have been far more convenient to be able to mark them out identically for cutting than, for instance, to mark them out according to a varying curvature of the soffit. A further reason for adopting the simple semicircular profile would be the convenience of setting-out and making the centring needed for erection.

An interesting parallel to the arches of the Pont du Gard is provided by the main vault of Nervi's Exhibition Hall in Turin illustrated in Figure 4.8. Here the material is reinforced concrete, partly precast and partly cast *in situ*. Precast concrete has as long a history as reinforced concrete itself, having always had a number of advantages, of which the greater ease with which quality could be ensured and safely put to the test was particularly important in the earlier days.[3] An early twentieth-century patent for a fully-precast framed structure is shown in Figure 4.9. This closely followed the precedent of earlier prefabricated construction in heavy solid timber. Nervi was one of the first to break away from this precedent, recognizing that if the structure as a whole made sufficient repetitive use of identical components or elements these could, individually, be given quite complex shapes. Provided that enough castings are made from each mould to cover its cost, curved forms and even trusses can be precast quite economically. Whether it is desirable to take advantage of this freedom depends, of course, on the purpose that the structure is to serve: there are many cases where a simple rectilinear form would be preferable.

To roof the Exhibition Hall, a corrugated barrel vault was chosen, with each corrugation built up largely from precast voussoirs. These were set slightly apart from one another, and joined by a cement-mortar packing. So joined, they served also as formwork for the casting of *in-situ* ribs at the tops and bottoms of the corrugations. Re-usable centring was again used to support the voussoirs initially. It was now mounted on rails so that it could be periodically shifted laterally to construct a fresh section of vault, though it was now wide enough to support simultaneously several parallel rings of voussoirs.

There were two main differences as compared with the Pont du Gard from the point of view of non-structural constraints on the choice of form. The first was that the profile of the vault was controlled solely by the centring. Since the voussoirs did not butt tightly against one another, their adjacent faces did not have to be exactly matched: the normal accuracy of casting was adequate. The second was that, though all voussoirs at a particular height in the vault were of the same form, this form could be very freely chosen and could quite reasonably be varied slightly from one voussoir to the next in each ring. This permitted a progressive reduction in the depth of the corrugations towards the springings of the vault.

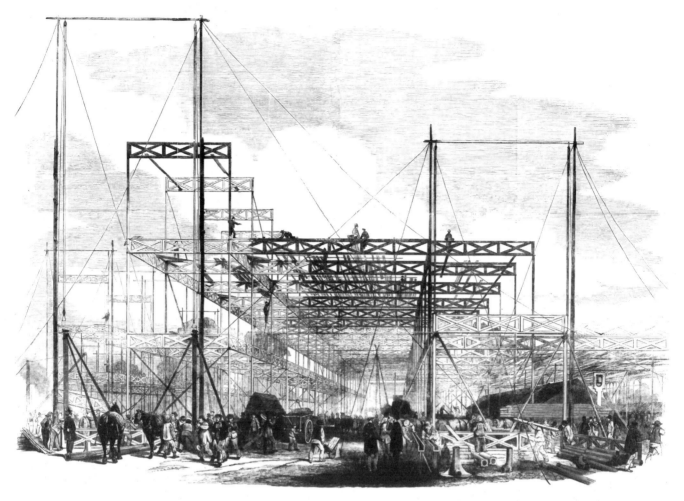

Consciously or unconsciously, this freer approach to prefabrication in reinforced concrete looked back more to earlier practice with cast iron.[4] There was then no alternative to pre-asting. The chief constraint on the choice of form for element or component stemmed from the behaviour of the metal on cooling which made sharp corners and large variations in thickness undesirable. The patterns from which the moulds were made (the pattern being a replica of the final form and the mould a negative impression of it in tightly compacted sand) could be fairly easily fashioned in timber to any desired form. The only further constraint was, therefore, that the maximum number of castings should be made from each pattern. This meant simply that structures should be so designed that they could be put together from standard components or from the minimum number of specially shaped or sized ones.

An outstanding example of what could be achieved by the full acceptance of this discipline, and still an object lesson for the rationalization of the construction process today, was the first Crystal Palace, constructed in Hyde Park in less than six months [4.10 and 15.11]. The whole structure was planned on a modular grid. This meant that there were only two different storey heights and three different beam spans. Apart from the side cladding, flooring and roofing, there were only seven types of

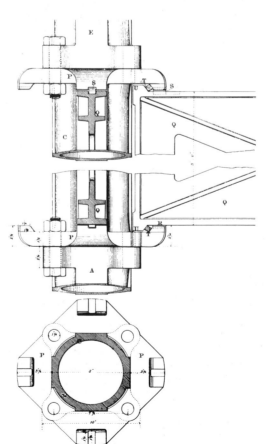

Figure 4.10 The Crystal Palace, London under construction. (Paxton, Fox and Henderson). (From *Illustrated London News*, 1850)

Figure 4.11 Standard joint between columns and beams, Crystal Palace, London (Paxton, Fox and Henderson). (From C. Downes and C. Cooper, *The building erected in Hyde Park for the Great Exhibition of the works of industry of all nations*, London, 1852)

component: column lengths, column base plates, column connecting pieces, trussed beams, trussed arches (for the transept only), and screwed rods and connecting bosses to provide diagonal bracing. Most of these were single castings and the remainder were fully prefabricated from wrought iron or timber. Variations in loading on beams of the same span and on the columns were provided for by variations in material or thickness without any change in the overall dimensions or end detailing. The connecting pieces could thus be fully standardized. Finally, everything was detailed to facilitate the simplest possible erection procedures. The standard joints between the columns and the trussed beams carrying the floors and roof [4.11] allowed, for instance, just enough freedom in seating the beams to compensate for slight deviations from the nominal dimensions. To tighten the joints when the beams were in place, it was merely necessary to drive in the wrought iron or hardwood keys marked T.

Intermediate possibilities

As a final example of the wide, and ever-widening, range of possible approaches, we might look briefly at the somewhat ill-defined borderline between prefabricated and *in-situ* construction. In the 'lift-slab' technique for multi-storey construction in reinforced concrete,[5] floor slabs are cast one above another directly on the foundation slab and separated only by thin membranes. They are then jacked up the columns to their final positions [4.12]. This almost eliminates the need for formwork, but calls for unbroken runs of columns throughout the height of the building and for simple beamless slabs of uniform thickness throughout. In a second set of techniques, which might be grouped together under the general name 'extrusion',[6] the whole structure is fabricated at ground level (starting with the top storey) for a multi-storey building, or near one abutment (starting with the opposite extremity) for the superstructure of a bridge. It is then progressively jacked upward or across the void as fabrication proceeds. The constraints on the choice of form that these techniques impose are more complex, but they will usually be satisfied if the main vertical supports of the building are continuous throughout its height and if the cross-section of the bridge superstructure is constant for its whole length.

Other operational aspects

These, then, are the major constraints stemming directly from the processes of construction. There are others. We have, for instance, largely ignored the need for adequate access and working platforms for all *in-situ* operations including assembly, the need for handling, lifting or jacking equipment [4.13], and the further need for well considered sizing of components to make efficient use of the equipment that is available. For some types of structure, particularly large bridges, it is becoming increasingly common to construct expensive new equipment (such as that required for some of the extrusion techniques just mentioned) in order to permit the adoption of a particular construction process. It is then necessary to balance carefully the cost of this equipment against the savings to be expected from using it. This will be easier when the design is prepared by the contractor than when it is prepared independently. More emphasis should also be placed on the importance of designing the joints between large prefabricated components so that they can be completed easily and safely under actual site conditions [4.14], and on some of the differences in approach that become necessary with the very largest modern structures. It is necessary now, though, to look at the closely associated structural implications of the construction process.

Structural aspects

When we turn to these structural implications, some are obvious; others perhaps less so. It is

Figure 4.12 Lift-slab construction, Coventry. (Source: Hevilift Ltd)

obviously essential to ensure that the incomplete structure will be stable at all intermediate stages of construction, and equally essential to ensure that every element will be strong enough to bear all loads that will act upon it during construction. These are absolute requirements which must always be satisfied. It may be less obvious (though it follows from what was said in Chapter 2 about deformations) that the internal distribution of loads and stresses in a structure on completion of construction, its stiffnesses, its precise geometry, and perhaps also its strength and stability, will depend on the manner and sequence in which it is put together unless it is truly statically determinate. This dependence is an inescapable consequence of the deformations that are always associated with the taking up of load, of their time dependence with materials like concrete, and of the stresses induced by any lack of fit. It clearly cannot be considered as a requirement; but it must be taken into account. It may be passively accepted, in which case it is merely necessary to ensure that the final condition of the structure is acceptable. Alternatively – as when a structure is prestressed – it can be deliberately exploited in the total design concept as a means whereby the final condition can be favourably modified.

Ensuring adequate strengths during construction

Ensuring that every element of the structure is strong enough to bear all loads associated with

Figure 4.13 Last stage of construction, Rhine Bridge, Emmerich (German Federal Ministry of Transport). (Source: author)

the construction process will call for special attention chiefly with prefabricated units which must be hoisted, jacked, transported or otherwise moved to their final positions. In the lift-slab technique, for instance, the inertial bending stresses due to the jacking-up of the slabs could easily exceed the stresses expected after construction is completed. Either these inertial stresses must be minimized by a very close control on the jacking, or additional reinforcement must be provided to resist them. Similarly, when precast load-bearing wall panels are lifted into position, tensile stresses will arise and tensile reinforcement must be provided which might be superfluous under the final compressive loading. When construction is carried out entirely on site, the need will be more for temporary supports such as centring and formwork and for the planning of the construction process itself (including the striking of temporary supports, the stacking of materials, the siting of cranes and hoists, and, of course, any prestressing operations) so that the safe loads on the as-yet incomplete structure are not exceeded.

Figure 4.14 Detail of closure of the deck truss, Rhine Bridge, Emmerich. (Source: author)

Ensuring the stability of the incomplete structure

Ensuring that the incomplete structure is always stable is relatively easy when the supporting elements are vertical piers, columns or walls and the spanning elements are prefabricated beams, lintels or slabs. It is then largely a matter of proceeding with the erection in the right sequence: the choice of structural form can be made with only its ultimate stability in mind. There are cases, however, where the right sequence is inconvenient or even precluded by other requirements. The lift-slab technique will serve again as an example. To enable the slabs to be freely jacked up the columns, the latter must stand completely free of the vertical infills

construction is likely to be the most critical requirement governing the choice of the form. All multi-span arch bridges constructed before Perronet's bridge over the Seine at Neuilly show one very obvious consequence of this. The piers are much wider than would be necessary simply to carry the weights of the arches since they were also required, during construction, to resist the thrust of a single-arch, unbalanced, as yet, by that of the arch still to be built on the other side [4.7].

The use of centring for arches and vaults

The use of centring is probably the simplest expedient for constructing an individual arch or vault of any size, but it does create its own

Figure 4.15 Deformations of an arch on removal of the centring as shown in Perronet, J. R., 'Mémoire sur le cintrement et le décintrement des ponts, et sur les differens mouvemens que prennent les voûtes pendent leur construction', *Mémoires de l'Académie Royale des Sciences*, 1773. (Source: author)

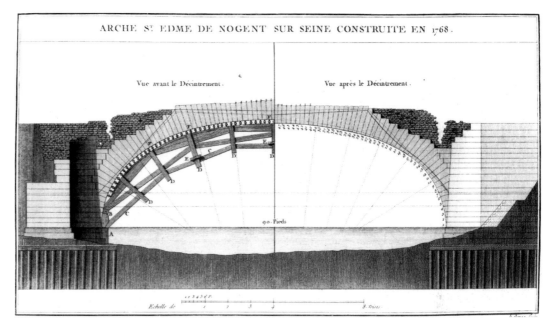

which will, in the completed structure, provide the necessary lateral bracing against a possible sideways collapse. Here stability must be ensured, at certain stages of construction, by a carefully planned sequence of jacking coupled with the use of temporary diagonal bracing. Such temporary bracing may be similarly required in some fully framed multi-storey structures when, for other reasons, erection of the frame proceeds well in advance of the addition of the vertical infills or cladding on which the lateral stability will ultimately depend.

When horizontally-thrusting spanning elements are introduced, and especially when their supports are inclined against their final thrusts, stability during construction is less easily ensured and the necessity of avoiding a collapse then may call for some modification of the structural form merely to provide an increased margin of stability. Indeed, unless there is a very extensive provision of temporary supports coupled with a choice of erection sequence guided by considerations of stability rather than convenience and economy, the stability during

problems. The most fundamental is that it introduces an inevitable indeterminacy into the structural action as long as it remains in place and assists in supporting the dead weight of the arch- or vault-to-be as well as carrying itself. This indeterminacy will be the more marked when large quantities of mortar are used, especially when the entire arch or vault is of concrete cast *in situ* with initially no stiffness or strength and a tendency to shrink as it gains both.

Ideally, from this point of view, no wet materials should be used. The entire form should be built up, like the arches of the Pont du Gard, from closely fitting precut or precast voussoir-like units so that the centring need serve merely as a temporary prop while these were set in place. The one remaining difficulty then would be to construct the centring so that it had the desired profile *after* being progressively deformed by the weights of the voussoirs. This difficulty could be avoided by using a cord to sweep out the desired curve for the final correct positioning of all the voussoirs, and to construct the centring to a slightly lower profile. Each voussoir would then have to be supported on it by wedges which would be adjusted as necessary when the last one was in place. The centring could then be freed by knocking out the wedges.[7]

Where there is a substantial use of mortar or *in-situ* concrete, the deformations of the arch or vault as it hardens and shrinks and takes up the load from the centring will be considerable. Moreover, unless the casting of a concrete arch or vault is completed before any substantial hardening can occur, the deformation of the centring itself on completion of casting will depend on the sequence of casting since the sections cast first will be less ready to follow subsequent deformations. These factors greatly add to the difficulties both of safely easing and removing the centring and of obtaining the desired final profiles [4.15][8]. In the past the recommended practice was to construct such a vault 'without Intermission' and to ease the centring gradually.[9] In addition, both arches and vaults were given ample thickness to render any deviation from the intended profile relatively unimportant. With modern thin shells, the casting sequence may have to be considered as an integral part of the design and the deformation of the centring may have to be closely controlled and taken into account during its construction in order to avoid errors in the final profiles that could lead to buckling.

A last drawback is the costliness of centring. One way of reducing this – by repeated re-use – has already been noted. Another is to restrict its use to that part of the arch or vault for which it is essential. The Pont du Gard again furnishes an example, since the projecting blocks on which it was carried can still be seen a little above the springing of each main arch [4.7]. Yet another is to design the arch or vault itself so that its full thickness or cross-section is built up only by a number of successive additions, each of which contributes to supporting those added later. One variant of this procedure has long been adopted in Iran, where timber for centring is scarce [14.3]. Another was adopted by Torroja for the construction of a large reinforced-concrete arch near Zamora. Here, in an adaptation of the earlier Melan system, the steel centring became, as construction proceeded, part of the permanent reinforcement.[10] This called for the construction process to be considered as an integral part of the design.

Construction without centring

The cost of centring and the indeterminacy of structural action to which it leads can, of course, be avoided by doing without it. In unreinforced masonry, this becomes possible only with forms that can be built throughout from a firm base provided by what has been built hitherto. Each successive brick or block must then be set in place in such a manner that, until construction is further advanced, it will be held there by

Figure 4.16 Vault of a storeroom, Ramesseum, Thebes. (Source: author)

Figure 4.17 Detail looking up into one section of the vault in Figure 4.16. (Source: author)

Figure 4.18 Brick vaults in a room opening off the west gallery, Hagia Sophia, Istanbul. (Source: author)

friction, by a mechanical keying action, by the bonding action of a rapidly setting mortar, or by the arching action of a few such bricks or blocks set in place together. In the last case, some temporary support may be necessary until all the blocks constituting the arch have been set. Friction can be relied upon only when the bedding angle is fairly small. Centerless construction in stone has, therefore, been limited to spires and similar sharply pointed vaults. Bond is more effective when there is a large surface area in relation to the weight of the unit as in a flat brick or tile.

In some of the earliest brick vaults, such as those seen in Figures 4.16 and 4.17, bond was assisted by scoring the surfaces deeply with the fingers when the bricks were made to give also a mechanical key. These vaults were constructed as continuous barrels by first building an end wall (since collapsed in the example illustrated) and then inclining each successive ring of bricks back towards it. Each brick was pitched on edge to give the maximum surface in contact with the previous ring. A similar technique has been widely employed for the centerless construction of vaults of more domical form over square and other mostly non-circular bases [4.18].

The easiest form of vault to construct without centring is the circular dome since, as Alberti noted, 'it is composed not only of Arches, but also, in a Manner, of Cornices'.[11] Alberti's 'Cornices' are the horizontal rings of Figure 2.14. In a completed dome these are subject to tension near the base. But, in any dome under construction with an open central eye for the time being, the rings near this open eye will be compressed and will take over the role of the missing upper parts of the vertical 'Arches' – even when they are at levels where tension will subsequently develop as construction advances above them. Temporary support is thus required, if at all, only for an incomplete horizontal ring or for a completed one whose mortar has not yet hardened sufficiently to take the compression without undue deformation. This is discussed further in Chapter 7.

In steel, reinforced concrete and prestressed concrete there is a further possibility of doing without centring by building out even free-standing arches, long-span beams and trusses as pairs of cantilevers, one on each side of the span. When completed, they meet at the centre.

In early applications of this principle to steel arch bridges [4.19 top], temporary wire cables were used as the upper tension members of the cantilevers. They exerted thrusts on the incomplete arches similar to those exerted, on insertion of the closing members at the crown, by the opposite parts of the arch. After closure they were rapidly slackened and then removed.[12] Closure of the arches of the Eads Bridge presented some difficulty because the feet were fixed so that there was no easy means of adjusting the gap at the crown. The pinned feet

of the Maria Pia Bridge, and of the similar but slightly later Garabit Bridge [14.10], avoided this difficulty. They permitted a very simple and precise adjustment by means of the temporary cables, and set a precedent which was followed in the design of most subsequent large arch bridges constructed in this way.

As an alternative, the temporary wire cables can be replaced by permanent members. This gives a pure cantilever unless a horizontal thrust is subsequently introduced at the crown. Earlier applications of this procedure were to large steel trusses in which more substantial permanent ties took the place of the temporary cables. More recent applications have been to long spans in prestressed concrete when it is not feasible or economical to use centring, and to cable-stayed girder bridges. In the former, box-sections are used, reducing in depth with increase in distance from the piers. They are cast *in situ* in formwork suspended from the cantilevers themselves with the prestressing cables running through the top slabs of the boxes and diagonally down into the vertical webs [4.19 bottom and 4.20] and serving also the role of the temporary cables used at the Eads Bridge.[13] In the latter, the permanent stays rise well above the girders, allowing these to be progressively extended outwards as fresh stays are added. This procedure is, however, complicated by the need for progressive adjustments of the tensions in the stays and is less straightforward than the use of temporary supports. This usually makes it more costly for moderate spans. The greater flexibility of the cantilevered arms of the incomplete spans can also present problems with longer spans, especially under high wind loading. Flexibility can be reduced by first completing the construction of any approach spans and can be further reduced by the use of tuned mass dampers to exert counter-forces to the wind excitation as was done in the erection of the main span of the Normandie Bridge [14.34].[14] This is one example of the deliberately introduced loads referred to at the beginning of Chapter 2.

The effects of deformations occurring during construction

An unusually clear illustration of the deform-ations that accompany the construction process is provided by the changing curve of the deck truss of a large suspension bridges when this truss is built out, piece by piece, from each tower [4.21]. If, on completion, it is to have the desired profile and to receive the desired uniform support throughout its length from the main cables, it must, while still incomplete, hang differently because it will then be exerting a very different load on the cables. Because this will stretch them less and cause a reduced sag at the centre, the truss must initially rise quite steeply at its advancing free ends. If it is relatively deep, its bottom joints must therefore

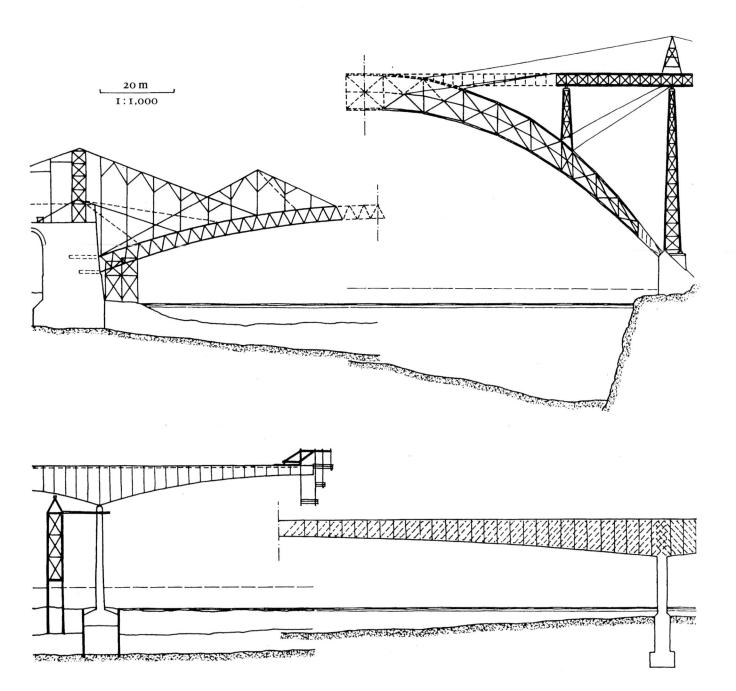

20 m

1 : 1,000

be left free until a late stage to allow the curve to change freely as the increasing load on the cables stretches them further. If, without any other changes being introduced in the construction process, these joints were bolted as the truss was extended, its stiffness would increasingly restrict changes in its own curve and in that of the cables and would lead both to different final profiles and to a different and non-uniform distribution of the support given by the cables. This, in turn, would alter the whole pattern of internal loads and stresses, giving higher maximum stresses for precisely the same dead load.

In most structures the deformations are less apparent but no less real, and they usually mean that the manner in which the completed structure finally supports itself is similarly dependent on the precise manner and sequence of construction. Consider the much commoner case of a

floor or bridge deck consisting of a reinforced concrete slab carried on steel beams in such a way that, in the completed structure, the slab and the beams bend as a single unit. If the concrete is cast *in situ*, its weight and that of the formwork can be supported, until it has hardened, either by the beams or independently of them. If it is supported by the beams, these will finally be stressed by the entire weight of the concrete. If it is supported independently, the concrete will, on removal of the formwork, carry part of its own weight and the stresses in the beams will be that much less.

More generally, when any structure or structural element is progressively stiffened as construction proceeds, the changes in stress and deformation resulting from the application or the removal of a given load will be progressively reduced. Thus even a temporary load removed before construction is completed may

Figure 4.19 Erection of the arches of the Eads Bridge, St Louis (top left) and the Maria Pia Bridge, Oporto (top right); and arrangement of prestressing cables (shown by broken lines) in the box cantilevers of the Medway Bridge, Kent (bottom left) and the Bendorf Rhine Bridge, (bottom right). (Source: author)

78

Figure 4.20 Erection of cantilever arms of the Medway Bridge, Kent. Since, in the completed structure, the pier was not required to be very stiff, additional supports were provided during construction for the inshore arms which were always further advanced than the river arms and therefore somewhat heavier (Freeman Fox). (Source: author)

Figure 4.21 Erection of the deck truss of the Forth Road Bridge, near Edinburgh (Freeman Fox). (Source: Sir Hubert Shirley-Smith)

leave its mark on the final structural actions.[15] In 'dry' construction (with materials like timber and steel) this 'memory' will be related only to the sequence in which the structure is built up, including the sequence in which members are connected to one another. In 'wet' construction (using mortar or concrete) it will be related also to the rate at which the 'wet' constituents harden and develop their strength and stiffness. It will be absent only when the completed structure is statically determinate in relation to the stresses considered. Its importance varies, however, with the importance of the final stresses and deformations either *per se* or as determinants of strength, stiffness or stability. This importance is greatest when the deformations might prejudice stability (as for thin shells as already noted in the discussion of the use of centring); when the stresses must be kept within prescribed limits and it is desirable to approach as close as possible to these limits (as for long-span suspension bridges); and when materials of different strengths are used in combination and it is desired to make the fullest use of the capacities of each (as, most typically, in reinforced concrete where the concrete alone also has very different strengths in tension and compression). It is relatively slight for many steel structures such as the typical frame of a modern multistorey building, because of the ability of the steel to yield harmlessly to relieve any locally excessively high stresses that result.

Prestressing

In constructing the stiffening truss of a large suspension bridge, the stresses are kept within the prescribed limits by following an appropriate erection sequence and by closely calculating the length of the main cables, the spacing along them of the vertical hangers, and the lengths of these hangers to give the desired final profiles when these elements are fully stretched by the weight of the completed deck.

The alternative way of obtaining a preferred initial distribution of internal stresses and acceptable initial deformations on completion of construction is to prestress. In terms of the construction process, this may be defined as to modify, by deliberately applying external forces during construction, the internal stresses that would otherwise develop. It is characteristic of prestressing that an external force (i.e. one not contributed by the weight of part of the structure itself) is applied at one stage of construction and then released at a subsequent stage to leave behind the 'prestress' by which it is 'remembered'.

In the discussion of prestressed concrete in Chapter 3, reference was made to two ways of doing this. In the first (applicable, as described, only to prestressed concrete but, with some modification, also to other materials), the prestressing force is applied to the reinforcement while this is still quite free, at one or both ends, to be stretched relative to the concrete. It is then bonded or anchored at both ends and, after this, the external prestressing force is released. In the second (applicable to any structural form that is not statically determinate), the force is applied to substantially-completed elements of the structure. Some restraint to relief of the resulting displacement or deformation is then introduced and, after this, the external force is likewise released. Either way, subject to the losses that result from the long-term behaviour of the materials under stress, the prestress remains because, in effect, incompatibly sized members have been stretched or squeezed to fit.

In a purely pragmatic manner, the advantages of prestressing were recognized 2000 years or more ago in such diverse contexts as the rigging of ships, the erection of tents, and the shrinking of iron tyres on wooden wheels to compress tightly together the pieces from which they were assembled. (With the tyres, cooling after an initial heating took the place of the application of an external force.) The driving in of wedge-shaped keys to tighten the joints in timber frames and to pre-tension iron tie rods may have been a medieval innovation.

For several reasons, however, the introduction of prestressing as a major constructional technique had to wait until the nineteenth century, and its full exploitation until even more recently. One reason was purely practical. It was the impossibility of applying and controlling external forces of a useful magnitude with the screw jacks and other means available before the development of the modern hydraulic jack. A second was the inability of designers, until the nineteenth century, to calculate the forces that it was desirable to apply, or the deformations that it was desirable to induce through their application, coupled with the lack of adequate means to measure the forces. A third, which delayed full exploitation for almost another hundred years, was that the greatest advantages were to be reaped only, as noted in Chapter 3, with modern steels and concretes.

The first of the two techniques of prestressing referred to above was, after a false start, successfully applied for a short time in the 1840s to the construction of trussed beams of cast and wrought iron.

The second technique was introduced immediately afterwards by Robert Stephenson in the construction of the Britannia Bridge to carry the Holyhead Railway over the Menai Straits [14.14].[16] The four spans were twin hollow box-shaped girders, one for each track.

In the completed bridge, all four spans were connected continuously together. This meant that the maximum bending-moments due to a train passing over, or to the wind, were considerably less than they would have been if the spans had been supported independently of one another. It was impossible, however, to fabricate the centre spans on staging *in situ*. They had to be prefabricated at a nearby site,

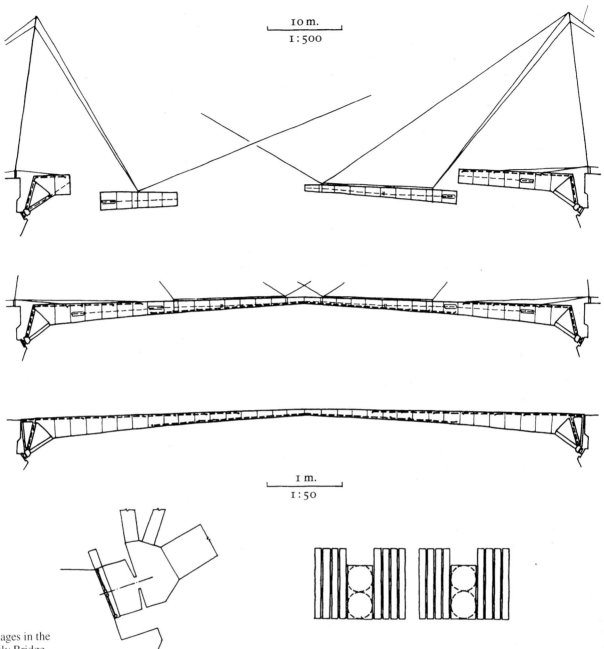

Figure 4.22 Stages in the
erection of Esbly Bridge
(Freyssinet). Temporary
prestressing-cables are
shown by light broken
lines, and permanent ones
by heavier broken lines.
Details below show one of
the main hinged supports
both in cross-section and
looking towards the
abutment.
(Source: author)

manoeuvred between the piers on pontoons and
jacked up into position, one span at a time. If
each of them had been set immediately on its
final bearings and if the end spans (which were
fabricated on staging *in situ*) had been fabricated on their final bearings, the advantages of
the continuity achieved by subsequently
connecting them together would have been lost
as far as the bending moments due to their own
considerable dead weights were concerned.
These moments and the associated stresses
would have been unaffected by the interconnection: they would have been a maximum
at the centre of the spans and zero (or almost
zero, depending on when the staging was removed from under the shorter outer spans) at the
ends.

To equalize as nearly as possible the
moments at the centres and at the ends for the

dead load as well as the live load, it was
therefore arranged that the connections between
the spans should be made with built-in
deformations. Or, to look at it in another way,
the connections were made in such a manner as
to largely eliminate the relative rotations of the
ends of the adjacent spans that would otherwise
have been built-in as a result of each span
having been independently self-supporting at the
time. The procedure adopted was to set one
centre span immediately on its final bearings,
but initially to jack up higher the farther end of
each of the other spans and to lower it to its
final position only after the connection to the
adjacent span had been completed.

Today, the widespread adoption of both
techniques owes much to the pioneering work of
Freyssinet[17]. His prestressed-concrete Esbly
Bridge [4.22], and a number of other bridges

over the Marne built to the same design about a hundred years after the Britannia Bridge, are outstanding examples of this.[18] The cross-section of each bridge consists of a number of identical I-section ribs side by side, connected together both by a continuous *in-situ* concrete topping to form the road surface and, at intervals, by transverse diaphragms. The depth decreases towards the centre to give a shallow arched profile, and the span is terminated at each end by a triangulated frame incorporating a reinforced-concrete hinge (the equivalent of a pinned joint) at the foot.

The degree of repetition involved made it economical to precast the ribs in short sections each about 2 m long. At the time of precasting, each section or 'voussoir' was prestressed vertically by jacking the flange parts of the mould apart against the web reinforcement. Next, groups of voussoirs were joined together to make larger units. The joints were packed with mortar and the voussoirs were held together by temporary prestressed cables running along the outsides of the webs. So assembled, these larger units were erected by the cantilevering method already described. As each span was completed, permanent cables were set in place in prepared ducts in the bottom flanges near the centre of the span and along the top flanges towards the supports, and these cables were tensioned as the tension in the temporary ones was released. Transverse pre-stress was similarly applied to link the adjacent ribs together. The span was then effectively a monolithic whole.

Had nothing further been done, however, the slight shortening along its length, restrained as it was by the hinges at the feet of the triangulated units, would have resulted in an increased longitudinal bending moment tending to depress the crown and calling for a greater depth there to resist it. To avoid this, the hinges themselves were displaced inwards by the second technique. Very flat hydraulic jacks, developed by Freyssinet himself, had, for this purpose, been placed between the hinges and the abutments. Flanking the jacks were hard concrete wedges which could be driven in from above [4.22 bottom]. It was, therefore, merely necessary to apply an appropriate pressure to the jacks to give the desired inward movement and then to drive in the wedges before releasing it. Since the jacks were made a permanent feature of the design, the possibility was also created of re-adjusting the state of internal stress at any future date in the event of a marked loss of prestress due to creep of the concrete or to movement of the foundations.

Chapter 5: Structure and form

It will be apparent already that there is no unique relationship between structure and form if, by form, we mean simply the geometrical configuration of the structure. Different structural actions in response to the same loading, different strengths, and different stiffnesses may wide-spanning structures, the scale can be increased beyond the point at which the strength or stiffness becomes critical only by changing the material or materials or by increasing the cross-sectional dimensions more than the others so that geometrical similarity is lost.

Figure 5.1 Sketches of bones (Galileo). (From the work by Galileo cited in note 1)

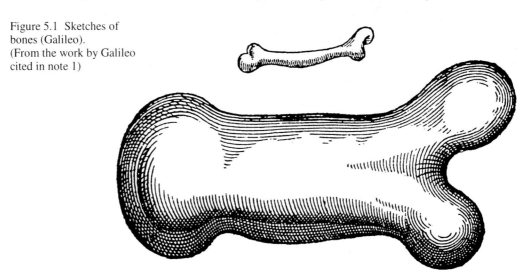

all be associated with a single visible external geometry. They will vary according to the materials used, the internal detailing if the cross-sections are not homogeneous, the manner and sequence of construction, and the scale. Moreover, in looking for the real structural form, it may first be necessary to peel off or look behind the skin that is normally seen if this is merely loosely wrapped over it to create a different visual impression, to hide services, or to provide insulation or other environmental control.

The different capacities of different materials and the relevance of the construction process have already been discussed. It is because of their importance that the internal detailing may also be important. The scale determines not so much the nature of the actions as the relationship between the loads to be carried and the available strengths and stiffnesses for a particular choice of materials and construction process. Self-weight, for instance, varies with the volume of material, whereas strength and stiffness vary only with the cross-sectional area if all dimensions are increased proportionately. When, therefore, self-weight is a significant part of the total load to be carried, as it is with many

That this empirical fact was recognized long ago is evident from the reference in Galileo's *Dialogues concerning two new sciences* to the practice of the shipbuilders of the Venetian Arsenal of employing

> stocks, scaffolding and bracing of larger dimensions for launching a big vessel than they do for a small one . . . in order to avoid the danger of the ship parting under its own heavy weight, a danger to which small boats are not subject.[1]

Later in the same work, Galileo makes his character Salviati illustrate the more general principle by sketching a human or animal bone and, below it, the change in proportions that would be necessary to give an increase in strength proportionate to the increase in weight to a bone three times as long [5.1]. One consequence is that, for a given choice of materials, there is a limit of size beyond which a structure of a particular form will be unable even to carry its own weight.

The geometrical configuration is, therefore, only one aspect of structural form; and it is be

Figure 5.2 Freeway interchange under construction, Los Angeles. (Source: author)

Figure 5.3 Inverted photograph of the funicular model for the main vault system of the Colonia Güell Chapel, near Barcelona (Gaudi). (Source Mas)

coming less and less the only determinant of the structural actions as the range of choice of materials and constructional techniques becomes wider.

It may still, however, be regarded as the primary determinant. It determines whether equilibrium is possible at all and, if it is possble, it determines which types of structural action are possible and rules out others. The choices of materials, internal details, and construction techniques develop some possibilities (or the only one if the supports and joints between elements are detailed so as to make the form statically determinate) and reject the remainder. In general, the larger the cross-sectional dimensions in relation to the others, the greater the variety of structural action that will be possible within a given form.

The basic curved shallow box form resting on isolated central supports selected for the structure in Figure 5.2, for instance, left open a considerable number of options for the final choice of structural form, though it did effectively limit these to some combination of bending and torsion and, for maximum economy, to one that exploited to the full the possibilities of structural continuity from one span to the next.

At the other extreme, a minimal cross-section in relation to its length for a linear element will virtually limit its useful action to that of a tie or suspension cable [2.1], while action as a strut, column or arch calls, as was noted in Chapter 2, for a somewhat greater cross-section in order to allow different internal thrust lines or develop enough stiffness in bending to prevent buckling. Thus, in designing some of his vaulted structures by the empirical method of hanging weights corresponding to the weights of the masonry from networks of

flexible cords and then notionally inverting these networks,[2] the Catalan architect Gaudi was forced to increase all the cross-sectional dimensions in recognition of the change from tension to compression. Had either his Güell Chapel or his Church of the Holy Family [5.3 and 5.4] been completed as intended (in masonry), they would, nevertheless, have very closely reproduced, with compression in place of tension, the actions of the funicular models. Together with recent cable and fabric structures more closely analogous to his models, they would then have been among the most 'legible' of recent structures. In steel or in reinforced or prestressed concrete, rather different possibilities, involving some flexural continuity between elements, might have been exploited.

To conclude this introductory part of the book, it will, therefore, be helpful to review the structural potentials of a wider range of basic geometrical configurations than was considered in Chapter 2, emphasizing the elemental forms and some trussed and framed analogues but looking also at the implications of combining them in different ways.

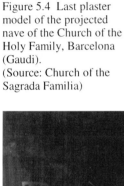

Figure 5.4 Last plaster model of the projected nave of the Church of the Holy Family, Barcelona (Gaudi).
(Source: Church of the Sagrada Familia)

Simple elemental forms

It is convenient to consider first those forms that are, in a geometrical sense, indisputably elemental. The simplest of them – the solid block with length, breadth and width all similar to one another – can act structurally in almost any manner consistent with its loading and support and will not be considered further here. All other forms are characterized by a single major axis or a single median surface, and have one or two relatively small cross-sectional dimensions. Figure 5.5 shows typical examples, numbered in brackets to indicate the subsequent chapters in which the actual developments of the forms will be traced.[3]

The basis of classification is again primarily geometrical, although, since the structural potential is being considered, it is also necessary to take into account the support provided and the manner of loading. The nature of the support is indicated by arrows denoting, as before, the directions of the support reactions. The loading is assumed, in all cases, to act vertically downwards and to be uniformly distributed over the surface. It could, therefore, be envisaged as gravitational self-weight. (The abilities of the forms to carry other patterns of load will be considered in the discussion that follows.)

There are two main geometrical distinctions apart from the obvious one between the essentially linear forms and the rest. The first relates to curvature. Of the curved forms in Figure 5.5, (f) to (j) are singly curved; (k) to (m) are doubly curved with both curvatures of the same sense (known as synclastic); and (n) to (p) are doubly curved with the two curvatures of opposite senses (known as anticlastic). The second, which is less clear-cut, relates to cross-sectional dimensions. The forms to the left of the vertical line are of potential structural value with almost negligible thickness; those above the horizontal line must have an appreciable depth if they have no curvature; and those to the right of the vertical line, though not incapable of acting in the manner assumed with an almost negligible thickness, do call, in practice, for a significant thickness to guard against buckling. It will also be seen, from the point of view of the uses that might be made of them, that some forms are capable only of transmitting loads, either directly (a) and (e) or indirectly (f), (j), and (q), but that the remainder are capable also of enclosing space.

In terms of structural actions, the main distinction is between those forms whose primary action is wholly tensile, those whose primary action is wholly compressive, and those whose action may be or must always be a combination of tension and compression. As in Figures 2.11, 2.12 and 2.14, the primary internal actions for the loading assumed have been indicated by sketching the directions of the principal stresses, using full lines to denote compressions and broken lines to denote tensions.

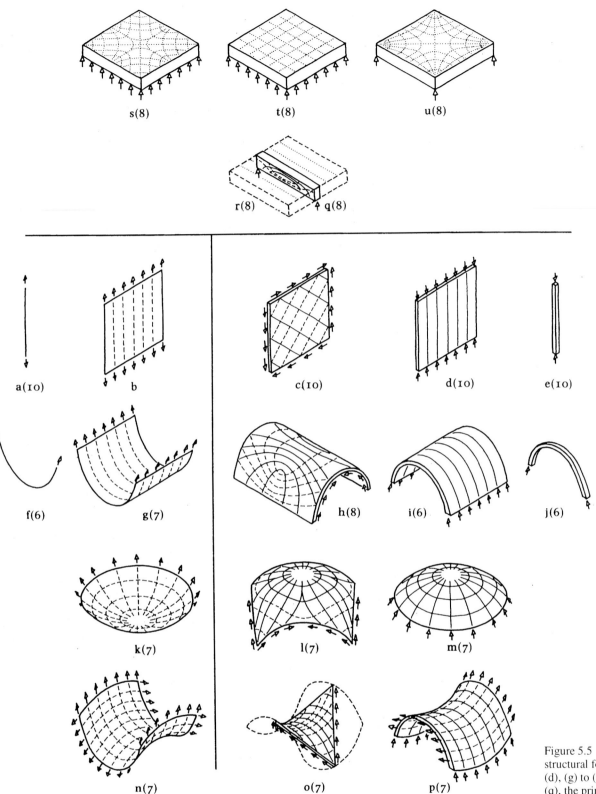

s(8) t(8) u(8)

r(8) q(8)

a(10) b c(10) d(10) e(10)

f(6) g(7) h(8) i(6) j(6)

k(7) l(7) m(7)

n(7) o(7) p(7)

Figure 5.5 Elemental structural forms. In (b) to (d), (g) to (i), and (k) to (q), the principal compressions for simple gravity or equivalent loading and for support reactions as marked are shown by full lines, and the principal tensions by broken lines. In (r) to (u), the directions of principal bending moments are similarly marked by dotted lines on the top surfaces. (Source: author)

A further distinction relates to the way in which the tension and compression, where the primary action calls for both, are distributed over the depth or thickness of the cross-section. For all the forms below the horizontal line, the primary action calls only for a uniform tension or compression throughout the thickness at any point in any given surface direction. For the forms above the horizontal line, however, the primary action, being a flexural one, calls for both tension and compression within the depth. For the linear form (q), it is the pattern of principal stresses within the depth that has been sketched as in Figure 2.12. For the surface forms (s) to (u), a complete picture of the pattern would have to show the directions of the stresses three-dimensionally. Since this is not possible, the directions of the principal bending-moments are plotted as dotted lines on the top surface. Each of these dotted lines may be read as standing for a pattern of principal stresses within the depth, similar to that sketched in (q).

Ties, struts, and columns

Referring to the forms now by the names that they are usually given as structural elements (i.e. names that denote the manner of loading and support as well as the geometrical configuration), the simplest are the tie (a) and the strut or column (e). Since the tie automatically aligns itself in the most effective way to carry the load, even when this includes a sideways disturbing force, it can be stressed to the full tensile capacity of the cross-section whatever its shape. This does not happen with the strut or column. If their loading is eccentric, or if there is a sideways disturbing force, the resultant secondary bending action leads to a less favourable re-alignment with a worse eccentricity of the primary compression. To avoid buckling, the cross-section must be increased to resist the bending rather than being stressed to its full potential capacity in pure compression. Since material farthest from the longitudinal axis will be the most effective in resisting secondary bending, the ideal cross-section is that of a thin circular tube if there is the same likelihood of bending in any direction.

Plane membranes and walls

The plane membrane (b) and wall (d) when loaded as shown in simple tension or compression are merely lateral extensions of the tie and column, and have similar characteristics, with one important difference. This is that the wall is a little less prone to buckle out of its own plane and will not buckle at all within that plane. As a bearer of vertical loads, it is an efficient structural element only when made of a material like masonry or concrete. There is then little to be gained by departing from a solid cross-section of constant thickness, though for a given thickness it may occasionally be worth curving it in plan to give more stability and a greater capacity to resist side loads [5.6]. However, compression flanges of box- or I-section beams are closely analogous to the wall in their behaviour, and for a large steel compression flange a ribbed or cellular cross-section is more efficient than a solid one [5.7].

The shear wall (c) differs from the membrane and vertically loaded wall merely in being subject to both tension and compression within its own plane. It must simultaneously resist both, and must, therefore, have a thickness comparable with that of a wall. The analogous vertical webs of large steel beams are best provided with rib-like stiffeners to resist any secondary transverse bending due to the compression [5.7].

Catenaries and arches

The catenary (f) and arch (j) were discussed in detail in Chapter 2. It was noted there that if each was given the minimum cross-section necessary to resist the tension or compression, a change in distribution of the loading would merely cause the catenary to change its profile slightly but would cause the arch to collapse. If it is important that a catenary should retain its chosen profile, this is best ensured by adding other bracing or stiffening elements as in a cable network. To ensure the stability of the arch calls, on the other hand, for an increase in cross-section similar to that called for in the strut or column, though the increase that would otherwise be required can be reduced by adding other bracing or stiffening elements as for the catenary. The increase needed will depend also on whether any resistance to tension is provided. If it is, it becomes unnecessary for the thrust line to remain always within the cross-section as it must do for the masonry arch. But the further it departs from this the more will beam action take the place of true arch action.

Singly-curved membranes and vaults

The similarly supported laterally extended forms (g) and (i) have similar characteristics, with the differences that, like the wall, the barrel vault continuously supported along its base (i) will not buckle laterally and its resistance to buckling under the primary compression can be increased by undulating the surface. This undulation will not affect the essentially uni-directional pattern of principal stresses on the surface, but will make it possible to adopt a reduced thickness as compared with that required for the free-standing arch. The singly curved membrane (g), if constructed as a true continuous membrane, would be too flexible for most structural purposes, so that it is preferable to construct it as a series of parallel catenaries spanned by horizontal slabs whose weight will limit and damp out any 'flutter' due, for instance, to changing wind pressure on the surface [7.30].

The overall depth of both these laterally extended forms makes them potentially capable also of spanning in the direction of the lateral extension if they are supported at the ends instead of along the longitudinal edges. Their action will then have some of the character of that of the beam. At (h), the arched barrel-vault is shown supported in this way. It is also possible to use it inverted (as shown for the membrane at (g)) provided that it is given a similar thickness. Either way up it will be subject to a combination of compressive and tensile stresses and the thickness is again necessary to resist buckling under the compression. The weight of the vault and any other load on it will be transmitted to the ends as a continuous shear like that in the shear wall, and ribs or end walls must be provided there to take it. They will also help to prevent buckling. (If, instead of spanning as shown, the vault is cantilevered at one or both ends, there is, of course, no shear at these free ends and no need there for ribs or walls: these are required only along lines of support.) Another difference, as compared with the longitudinally supported forms (g) and (i), is that the stability of the form is less dependent, for a given distribution of load, on the choice of a particular sectional profile. Even the V-shaped profile of what is commonly referred to as a folded plate will do [5.8], provided that the thickness is sufficient to withstand a limited amount of transverse bending.

Doubly curved membranes, vaults, and shells

Doubly curved surfaces are less easily deformed than singly curved ones. They cannot, like the latter, be simply unrolled, but must instead undergo considerable extensions or contractions or combinations of the two in different directions. The forms (k) to (p) are therefore inherently stiffer than the corresponding singly-curved ones. Also, as explained in the discussion of the dome in Chapter 2, the simultaneous existence, in all cases, of two potential load-carrying mechanisms – the two orthogonal sets of principal stresses sketched for each form – means that the close conformity to a particular profile determined by the loading that is required for a simple catenary or a slender arch acting wholly in compression is again unnecessary, although large concentrated loads are incompatible with continuous smooth curvatures. Subject to this proviso, but with only very small thicknesses relative to the maximum radius of curvature, equilibrium can be achieved between a particular loading and the internal actions for widely differing surface geometries. Similarly, a form having a particular surface geometry will be able to carry loads distributed in many different ways. All that is necessary to ensure stability is adequate strength everywhere to resist the different stress distributions, plus enough thickness to remove the risk of buckling under any compressive stresses.

The forms with the greatest inherent stiffness are, in general, the synclastic ones (k) to (m). In the domical vault or shell continuously supported around its base (m), the load may be carried entirely by compression [2.14(b)] or by a combination of compression and tension [2.14(a)]. A similarly-shaped membrane (k) must, however, act wholly in tension, because, if it is a true membrane, its negligible thickness will cause it to buckle under compression. It is difficult to ensure this purely tensile action if it hangs from the supports as shown and the only loading is that of self-weight plus varying and uncontrolled air pressures on both sides. Therefore, unless it is used for the storage of liquids or weighed down by a heavy cladding, it is best used inverted and given a tensile prestress throughout

ISOMETRICAL PROJECTION
PORTION OF ONE OF THE TUBES
OF THE
BRITANNIA BRIDGE.

Figure 5.6 Curved
boundary wall, University
of Virginia (Jefferson).
(Source: author)

Figure 5.7 Isometric
sectional view of a tube of
the Britannia Bridge,
Menai Straits, from E.
Clark and R. Stephenson,
*The Britannia and
Conway Tubular Bridges*,
London, 1850. In this first
example of such a form,
the material used was
wrought iron and not steel.
(Source: author)

Figure 5.8 Terminal
Building, National
Airport, Minneapolis.
(Source: author)

by maintaining an internal air pressure greater than any likely external pressure [5.9].

If the support of the domical shell is concentrated at a few points around its base as in (l), its action will still resemble that of the continuously supported shell near the crown but must be more like that of the end-supported barrel (h) where it spans, lower down, from one support to the next. A rib or a stiffening of the edge of the shell itself is then necessary to take the shear between each pair of supports.

Anticlastic forms, unless prestressed, will tend to act in compression in the directions of maximum arch-like curvature and in tension in directions of maximum catenary-like curvature ((o) and (p)). Ideally, if there is enough thickness to resist buckling under compression and if the surface geometry is so chosen as to call for principal tensions and compressions of equal and uniform magnitudes throughout, the saddle form (p) may be cut along lines midway between the directions of these principal tensions and compressions and will then transmit only continuous and uniform shears to the edge members (o). Again the membrane form must act wholly in tension (n), and this calls for tensile prestressing in the direction of the arch-like curvature. Such prestressing may be applied through continuous rigid edge members giving, if the curvatures are sufficient and all compression is eliminated, the stiffest form of membrane. Alternatively it can be applied by means of guy ropes as in a tent [5.10].

Beams

The beam of constant rectangular cross-section (q) is the simplest example of a form that, by virtue of a greater depth than that normally characteristic of a shell or membrane, can act within this depth as both arch and catenary [2.12]. It does, however, make very inefficient use of some of the material even when the bending moment is constant because, unless the

form is also acting partly as a strut, there will be no stress at all at mid depth and stresses of very varied magnitudes elsewhere. This inefficiency may be even worse if, as in Figure 2.12 (a), (c) and the outer parts of (b), the bending moment is not constant.

Greater efficiency can be obtained by concentrating the material where the stresses would, in this simplest form, be highest. We have already seen that the ideal form to resist secondary bending of a strut or column is a thin circular tube if there is the same likelihood of bending in any direction. A box-like form is similarly preferable to resist the secondary torsion in a beam due to loading that is transverse to its axis but displaced to one side. To resist a primary bending-action in one plane only – such as that produced by transverse loads that always act in a vertical plane through the axis of the beam – the most efficient cross-section is one that concentrates the resistance to the largest tensions and compressions in horizontal flanges at the top and bottom, and the resistance to the inclined and intersecting tensions and compressions at intermediate depths in vertical webs between the flanges, to give either a box-form [5.7] or an I-form.

Slabs

The slab spanning in one direction only (r) is merely the simple beam extended laterally. Its efficiency in resisting transverse bending may be similarly improved by adopting a cellular or ribbed cross-section, but it is worth noting that, as a solid element, it has a greater capacity to act also as a horizontal shear-wall (c). This capacity is valuable in some types of multi-bay structures to transmit wind loads from one line of supports to another.

To support it so that it spans in one direction only does, however, waste its ability to span similarly at right angles to this direction or, indeed, in any direction between. If this ability is exploited by supporting it so as to span simultaneously in two directions, more efficient use can be made of the material, and greater strengths and stiffnesses can be obtained for a given thickness. The internal actions in transmitting loads to the supports may then be roughly visualized as those of sets of very flat doubly curved shells and inverted doubly curved membranes within the thickness of the slab – the outward horizontal thrusts of the former being exactly balanced around the edges by the inward horizontal pulls of the latter if the supports provide only vertical reactions. These 'shells' and 'membranes' will, like those already considered, be capable of bearing loads distributed over the surface in many different ways. But, unlike the thin single shell or membrane, they will be able also to assume a variety of different surface geometries within the thickness of the slab in order to carry both continuously-distributed and locally concentrated loads. In other

Figure 5.9 Air-supported fabric covering for a swimming bath, Kesteven. (Source: author)

Figure 5.10 The prototype tent structure housing the Institute for Lightweight Structures, University of Stuttgart (Otto). (Source: author)

words, the thickness of the slab spanning in two directions confers on it a potential three-dimensional freedom of action to support different patterns of load that is analogous to the two-dimensional freedom of action conferred on the beam by its depth that was illustrated in Figure 2.12 (a) to (c).

The actual three-dimensional pattern of the principal tensions and compressions in any particular case is much more difficult to visualize fully. As already noted, it has therefore been indicated at (s), (t) and (u) simply by dotted lines on the surface denoting the directions of the principal bending-moments for a uniformly-distributed load and three different conditions of edge support. In these directions, the bending moments (corresponding to distributions of principal stresses through the thickness like those sketched at (q) and in Figure 2.12 (a) to (c)) are either a maximum or a minimum and there is no twisting action. Thus, if these lines were drawn sufficiently closely together for adjacent lines to be virtually parallel to one another and cuts were made first along the lines of one set and then, separately, along those of the other, orthogonal, set, the curved strips left by each series of cuts could, together, carry the whole load as families of independent beams. Differently distributed loads, or, as illustrated, different supports, merely bring into action differently-aligned families of these 'beams'.

In the special case sketched at (t) – calling for continuous flexible edge supports of a particular stiffness – it will be seen that the strips run parallel to the edges and to one another. In general, however, if a slab were sliced as in (t), the resultant strips could not still carry the load in the same way, simply as families of independent beams. They would have to transmit some of the load to their neighbours by shear and to carry some of it by a twisting action. In this sense, the potentially greater strength and stiffness of a slab spanning in two directions, as compared with those of the same slab spanning one way only as a simple series of parallel beams (r), may be said to be realized by the development of the twisting resistances of these 'beams' as well as their bending resistances.

Despite the greater efficiency of the slab spanning in two directions as compared with the one-way slab and simple beam, it is still, like the beam of solid rectangular cross-section, an inefficient form in the sense that the stresses will vary greatly in magnitude over the depth: much of the material will be very lightly stressed. With a slight increase in overall depth, similar strengths and stiffnesses may be obtained with less material by changing to a ribbed form. This will, of course, take away much of the three-dimensional freedom of action of the simple slab that has just been considered, and impose, in its place, a particular pattern of action determined by the pattern of the ribs. The main bending actions will always be in the directions of the ribs, and variations in the distribution of the loads or in the conditions of support will have to be accommodated to a considerable extent by different twists of the ribs, bringing into play their torsional resistances. With the usual rectangular grid (two sets of straight ribs intersecting at right angles), the twists may have to be considerable except for the particular conditions of support corresponding to those assumed at (t) for the simple slab. In the slabs of some of Nervi's buildings [8.17], they were minimized by making the ribs follow the curves of the directions of the principal bending moments in a simple slab of the same size subjected to a uniformly distributed load.

Trussed and framed analogues

Figure 5.11 shows, alongside the simple arch and beam and a simple rectangular grid, typical examples of a much larger family of forms.[4] None of these forms is strictly elemental like those considered hitherto; but each is closely analogous to one of them. Each is an assembly of ties and struts, or columns and beams, which gains the overall stiffness necessary to act in a manner similar to that of its simpler analogue either from the triangulation of these members [(a), (c), (d), (f), (g), (i), (j), and (l)] or from their rigid interconnection (e).

Portal frames

The portal frame (e), consisting of a beam rigidly connected to two columns whose feet are supported on pins so that they are free to rotate, may be regarded as an arch whose thrust line will always pass outside its cross-section at most points. Wherever it does so, particularly at the centre of the beam and in the neighbourhood of the joints between the beam and the columns, bending will tend to predominate over simple compression, so that a greater depth of section is necessary in these regions than would be required if (as in a simple arch) the profile conformed more closely to the thrust line. On the other hand, the depth of section at the centre of the beam need not be as great as if the beam merely rested on the heads of two fixed columns as in 'post and lintel' construction [1.6, 3.5, etc.]. The other differences, as compared with 'post and lintel' construction, are that the form is capable of resisting considerable side loads acting in its own plane, and that, as a kind of arch, it will be subject to a net compression through-out, and will thrust outwards and call for inward reactions at the feet.

Plane trusses and space frames

All the triangulated, or trussed, analogues may be regarded as the result of taking material away from the truly elemental solid forms where the stresses would be relatively low even under critical conditions of loading, and concentrating it where they would be high. This is clearly most profitable where the maximum stresses in the solid form would vary most over the depth of the cross-section, as happens where bending action predominates over simple compression or tension. A trussed form would be less efficient than a solid member to resist pure tension. It is most efficient in the role of a beam or slab [(g), (i), (j), (l)]. It is also valuable where bending arises from or is associated with an inevitable or possible non-axial compression [(d), (f), (a), (c), and the straight latticed struts seen in Figure 3.17].

Unless the truss members are unusually wide in proportion to their depths, triangulation in one plane only [(a), (d), (g)] gives, of course, a form analogous to a very slender though relatively deep solid form and, therefore, one which will be equally prone to buckling out of this plane in a manner akin to that seen in Figure 2.16. To avoid such buckling, the plane truss usually calls for some kind of lateral support or bracing. Triangulation in two or more planes [(c), (f), (j), (l)] does not have this shortcoming. The resultant three-dimensional truss, or space frame, is stiff in all directions, particularly when, as is frequently the case, the number of members is such as to make it statically highly indeterminate.

In the fully triangulated form, all loads applied at the joints may be carried by simple compression or tension in each member of the truss. For a single pattern of loading, each member may also be stressed to the same proportion of its effective strength by appropriate choices of the truss geometry, the cross-sections of the members, and, in the case of the statically indeterminate truss, any initial lack of fit of the members. Any restriction of the overall depth, thickness or other critical transverse dimension should preferably lie for a given span or principal axial dimension will depend on the materials, the scale, the ratio of imposed load to self-weight, and the conditions of support. Economy in the use of materials tends, however, to favour solid, I-section or ribbed forms for materials like concrete and normal reinforced concrete, and I-section, ribbed, hollow or trussed forms for materials like steel and prestressed concrete. It also tends to favour the more solid forms for the lesser loads and smaller scales and the more attenuated

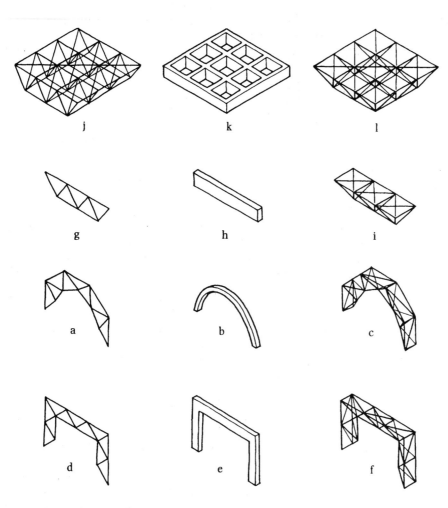

Figure 5.11 Framed and trussed analogues of the simpler elemental forms. (Source: author)

free relative rotations of the members at the joints will modify this simple behaviour; and varying distributions of the load will mean that all members cannot always work at the same proportion of their full capacities. A highly economical use of material will, nevertheless, be possible in most cases. Within reason, the greater the overall depth or critical transverse dimension in relation to the span or principal axial dimension, the greater will be the potential economy.

Proportions

For any elemental form – whether simple solid, I-section, ribbed, hollow, or trussed – that does not act purely in tension, the range within which forms for the greater loads and larger scales. As a result, optimum ratios of overall depth, thickness or equivalent dimension to span or principal axial dimension are fairly similar, in practice, for most variants on a particular basic form.

For the beam and the column or strut, the ratio is usually of the order of 1/20, increasing to about 1/10 for the trussed analogue. For the slab spanning in two directions and its trussed analogue, it decreases to about 1/40 and 1/20. (It has been exaggerated throughout in Figures 2.12, 5.5 and 5.11 for greater clarity.)

Corresponding ratios for the arch vary more widely according to how closely the profile conforms to the thrust line (which depends a good deal on the manner in which loads are transmitted to the arch by the structure of which it is a part) and the ability or inability of the material to resist tension. For materials able to resist tension, the ratio may vary within a range of about 1/40 to 1/80 if the profile conforms well to the thrust line, but at the other extreme it must increase to values more appropriate to the beam and column if the profile diverges as widely from any possible thrust line as it does in most portal frames.

For most vaults and shells, as already explained, much-reduced thicknesses are adequate. For the domical shell, the ratio need only be of

Figure 5.12 Simple demountable triangulated dome with fabric weather skin. (Source: author)

the order of 1/400 except, perhaps, near the edges, and for singly curved shells and folded plates only a little greater. Except for very large spans, it is therefore quite feasible to substitute a single-layer triangulation for a continuous surface, giving another family of triangulated analogues to the truly elemental forms. Figure 5.12 shows a simple example of such a form. For very large spans, the necessary overall thickness can, however, be obtained more economically by substituting a double-layer system of struts and ties, more like those of Figure 5.11(j) and (l).

For the membrane or catenary acting purely in tension, there would be virtually no lower limit to the ratio if only self-weight had to be carried and stiffness was not important. Actual reasonable thicknesses depend on the additional load and the required stiffness. They tend to lie between 1/10 000th and 1/100 000th of the span but may rise to the order of 1/1000th where there is a large additional load. Until recent developments of high-strength fabrics, networks of cables supporting an unstressed or lightly stressed covering had to take the place of continuous membranes for large spans [5.10].

Before turning to the complete structure we should note again that, if they are regarded merely as geometrical shapes, most of the elemental forms except those with the lowest ratios of depth or thickness to span or equivalent axial dimension are potentially capable of acting

in a variety of ways. The most versatile of the forms considered is the simple flat plate. Given a thickness ratio of about 1/20, this is potentially capable of acting as a wall [5.5(d)], a shear wall or panel [5.5(c)] or a tensile membrane [5.5(b)] if loaded in its own plane; as a slab spanning in one or two directions [5.5 (r) to (u)] if loaded transversely; or as two of these simultaneously if loaded both in its own plane and transversely. Since slab action is, itself, equivalent to arch action plus catenary action or to the action of a shallow shell plus that of a shallow inverted membrane, the potential action embraces these also and can, under certain circumstances, become either predominantly arch or shell action or predominantly catenary or membrane action. It can, for instance, become the former (as illustrated for the beam in Figures 2.12(d) and 2.13) if the material is weak in tension but the supports are capable of resisting the outward thrusts. Conversely, it can become the latter if the compressive strength is exhausted but there is enough tensile reinforcement and the supports can resist its inward pulls.

Complete structures

None of the elemental forms considered hitherto can be regarded as a complete structure, in the sense of being entirely self-sufficient, because each is dependent on something else to provide the reactions that finally balance the self-weight and any imposed loads. Ultimately the reactions must be provided by the ground. For most elements, though, they will be provided at least partly by adjacent elements, and will depend on the immobility or otherwise of these adjacent elements and usually on the ability of the interconnections to transmit tension as well as compression – either as a direct pull or as shear, bending, or torsion.. The immobility of adjacent elements will depend on the form of the structure as a whole, while the performance of the interconnections will depend more on their internal detailing.

The importance of the detailing of the interconnections is illustrated by the striking contrast between the two partial collapses of multi-storey buildings illustrated in Figures 5.13 and 5.14. The geometric forms were not very different. Each was a large multi-cellular box. The chief difference lay in the internal detailing. The first was a framed structure in which the main loads were meant to be carried by steel beams and columns. Each panel-like wall or floor element was set between four members of the frame; and these frame members had sufficient tensile continuity throughout the building (supplemented to some extent by direct tensile continuity of the reinforced-concrete slabs which constituted the floor elements) to tie all the elements together. As a result, when a bomb explosion deprived the entire end of support from below, this end was able to cantilever out over the void with the floor slabs and beams acting partly in direct tension and the side walls acting partly as shear walls or diagonal struts inclined against the unsupported weights [5.13].[5] The second had no separate frame. It was constructed entirely of large precast concrete panels whose interconnections were capable of developing relatively little tensile continuity beyond that developed by friction (and thus dependent on the compressions due to self-weight and other vertical loads). When a gas explosion blew out a large load-bearing wall panel four floors from the top of the building, the unsupported wall and floor panels above were, therefore, unable to cantilever out. They fell and, in falling, caused the further collapse of much of the structure below [5.14].[6]

From another point of view the very different internal detailing of these buildings is one illustration of the great freedom of choice now possessed by the designer. Indeed, even without departing significantly from the method of construction selected for the second one, its designer could have achieved much more tensile continuity between the panels by calling for additional reinforcement across the joints. The initial loss of one panel might not then have entailed such an extensive collapse.

The overriding requirements governing the choice of form for the complete structure are the geometrical one concerning the relative disposition of the elements in space that was discussed in the earlier part of Chapter 2, and the complementary one that all elements should individually be capable of the actions required of them and of being joined to one another in such a way as to develop these actions.

With the wide choices of materials, internal details and methods of construction available today, these are not usually very onerous requirements. Clearly they preclude the adoption of forms that are inherently unstable. Equally clearly the range of choice will be progressively narrowed as the scale or the anticipated imposed loading increases. Forms that can be simply and economically constructed to give simple, direct, and reposeful patterns of action will also, in general, be preferable to others that cannot be so constructed and that call for some kind of structural acrobatics. Elimination of the latter will narrow the choice further. But a choice will usually remain. Coupled with the very wide and growing diversity of the basic functional requirements to be met, it means that there is no apparent limit to the range of possible complete forms. With the much narrower choices of materials, internal details and methods of construction available in the past, there was less freedom to choose the complete form. But the range of forms actually chosen still greatly exceeded the range of elemental forms from which they were put together.

Because of the wide ranges of possible forms for complete structures, it seems less

Figure 5.13 Detail of a
seven storey block of
offices after a bomb
explosion had deprived
the end columns of direct
support. (Crown
copyright: Building
Research Establishment)

Figure 5.14 Partial
collapse of a block of flats
(Ronan Point, London) as
a result of local damage
caused by a gas explosion
on the eighteenth floor.
(Source: author)

appropriate than it was for the elemental forms to attempt here a systematic classification. The one basic distinction that might be usefully made is that between forms that are broadly analogous to one of the simple elemental forms plus the necessary support and those that are not.

The first category may be exemplified by Nervi's Roman Sports Palaces [1.2 and 13.6]. These are analogous to the simple domical shell [5.5(m)] plus direct support and restraint to the outward thrusts at the base. The second category is well represented by the two multi-storey buildings just discussed [5.13 and 5.14]. These might be likened to free-standing single walls or columns, and this analogy has some value when the complete form is unusually slender. It completely ignores, though, the dual role of the floors in the actual buildings – in which they span as slabs between the vertical supports to transmit to these supports the loads imposed by the occupants and simultaneously act as horizontal interconnections of the same vertical supports to help them to act in unison to resist horizontal loads. Today this second category also includes structures in which even the main spanning or space-enclosing elements are hybrids of more than one truly elemental form.

In the absence of a more detailed classification in purely structural terms, something rather different has seemed desirable as the principal basis for the later discussion of the developments of complete structural forms in Chapters 11 to 15. The more chronological basis that has been adopted there will be described at the beginning of Chapter 11.

Part 2: Structural elements

Chapter 6: Arches and catenaries

Figure 6.1 Bridge
between Tiryns and
Epidaurus.
(Source: author).

In turning to more detailed discussions of the developments of individual elemental forms, it is important to recall that, in most built structures, these elemental forms are merely parts of a larger whole. Particularly in the past, they have not developed in an independent isolation, but as essential parts of these larger wholes. To set this fact temporarily aside is largely a matter of convenience, as in the earlier discussions of such forms in Chapters 2 and 5. Yet it does also have some historical justification. As the development of each form proceeded, it gradually

enlarged the vocabulary on which future designers could draw.

We shall concentrate on developments that were structurally significant rather than on subsequent variations on these primary developments exploiting them for aesthetic or similar ends. Such developments, in the case of the arch, were ones that permitted wider gaps to be spanned, that led to improved strength and stability and a reduced dependence on heavy abutments, that facilitated construction, or that took advantage of the potentialities of new materials in one or more of these ways. That some of them were also, in themselves, aesthetically significant will be apparent from the illustrations. Most of them were also potentially so, simply by widening the range of choice and extending the limits below which fairly free modulations of the forms were possible. The actual subsequent exploitation of them for aesthetic and similar ends belongs more, though, to the histories of architecture and decoration, except where it is relevant to the developments of complete structural forms that are to be discussed later.

Origins

All developments must have a beginning, or perhaps more than one. It would, however, be futile today to seek to establish with certainty the origins of any of the more basic elemental forms considered in Chapter 5. The earliest attempts at building were, probably, no more than simple adaptations or copies of naturally occurring forms like a fallen log or hanging liana spanning a gap, or a boulder wedged between two others to give a rudimentary arch. They would be unlikely to survive for longer than these natural prototypes, and they are likely to have varied from place to place as widely as the prototypes themselves. It will be more profitable to start by looking at some of the surviving primitive forms that probably mark, or closely reflect, slightly later stages in the

development, and to do so more with a view to seeing how they may have provoked and stimulated this by their shortcomings or hints of other possibilities than to establishing any definite sequence of events. For the arch this is not too difficult.

The brick arch and the stone voussoir arch may be taken as the first fully developed forms. Neither exists as a direct natural prototype, nor do we find the monolithic natural arch [4.2] in any of the areas where they first appeared. Among the early man-made forms that may have contributed to their development are (1) a fairly straight copy of the accidentally-wedged boulder [6.1]; (2) the alternative simplest form consisting of two long blocks of stone inclined inwards to meet as an inverted V [6.2]; and (3) the equivalent of the latter consisting of two logs similarly inclined towards each other, or two lighter saplings or bundles of reeds set more vertically in the ground and then bent over until their free ends met and could be tied together, much as in the hut in Figure 1.1.

The illustrations of the first two should be self-explanatory, though Figure 6.2 hardly conveys the vast size of the blocks used to span an opening of only about 2 metres. Both could have been constructed without centring unless the practical difficulties of controlling the simultaneous lowering of the two blocks forming the lower of the two superimposed 'arches' of the second proved too daunting. Even if it did, a temporary wall across the opening would have served the purpose.

The third is known in early times only from a few surviving representations of huts so constructed; but it can still be seen today in various parts of the world. In the marshes of southern Iraq, for instance, roughly parabolic 'arches' up to 6 m in span have, at least until recently, been made by setting bundles of giant reeds in the ground in two rows spaced that far apart, then bending their heads to meet one another and tying them together. These 'arches' were then connected transversely by more slender bundles placed horizontally, and finally covered with reed matting.[1] Their action is not, of course, the purely compressive one characteristic of pure arch action. If, therefore, their earlier prototypes had any role in relation to the development of the masonry arch, it was probably to hint at a fourth possibility using bricks instead of saplings or reeds.

It is, indeed, quite possible that such reed 'arches', much more modest in scale, provided the model for the earliest known true arches of mud-brick, which have survived in Egypt from early in the third millennium B.C.[2] They were constructed of only a few bricks, very likely by the process observed by Somers Clarke in Upper Egypt in 1896 for constructing small window arches without using any centring:

The walls being raised to the necessary height from which the window arches should

Figure 6.2 North
entrance, Pyramid of
Cheops, Giza.
(Source: author)

Figure 6.4 (a) and (b)
Internal static equilibrium
of a false arch; (c) and (e)
alternative construction
details; (d) and (f)
possible modes of collapse
Principal compressions are
shown in (a) and (b) by
full lines, and principal
tensions by broken lines.
(Source: author)

spring, a little pile of four or five bricks was set up on either jamb, with plenty of tough mortar between like dough. One master craftsman manipulated one little pile, one the other. Each put a hand on the inner side of the pile and tilted it sedately and slowly, giving it a curved form as it moved until the two touched at the top. A cry, 'One more brick!' – in it went, and the little arch was complete.[3]

Alongside these forms, we should also note the early and widespread construction of 'false' arches and vaults, sometimes with stepped soffits and sometimes with the soffits dressed back to a smooth inverted V or a smooth curve [6.3]. Their distinguishing characteristic is that all the blocks or bricks are bedded more or less horizontally on one another. At each course they project slightly beyond those of the course below, so as to narrow progressively the gap to

ing extension of the walling of which it was a part. It must have been recognized – or soon discovered in the course of building – that no block could project more than half its length beyond the one below if it was not to tip over and fall, and that a considerably smaller projection was safer. It must similarly have been recognized that the total weight of masonry bearing down behind the edge of the opening must always exceed the total weight that projected in front of it. Hard experience must, however, have shown that, even if these requirements were satisfied, the independent stability of the two halves would not necessarily be ensured.

In the light of the more recent concepts of structural actions described in Chapter 2, it can be seen that each projecting half must act as a cantilever if it is not to be sustained partly by the horizontal thrust from the other half that is characteristic of true arch action. It must act in much the same way as the portions of the beams in Figures 2.12 (b) and (c) that would project outside the supports if these figures were inverted. The masonry must, therefore, be capable of developing tension as well as the compression that is characteristic of normal wall action. Herein lay the difficulty.

If each half were completely monolithic, the pattern of internal principal stresses would have to be roughly as sketched in Figure 6.4 (a). Since no tensile reinforcement was used, the necessary tension could be developed most effectively by using large blocks of stone, each projecting back behind the edge of the opening at the foot at least as far as it projected in front of it [6.4(c) and the left side of 6.3].

If shorter blocks or bricks were used as in Figure 6.4(e), the potential tensile resistance of the individual units could be developed only by frictional forces at the horizontal bed joints. There would be no resistance across the vertical joints, and the total resistance developed over several courses could never be more than about half that of a long continuous block of similar depth. Deficiencies in the tensile strengths of the individual blocks or bricks, and insufficient friction at the horizontal joints, would lead to cracking and slipping and tipping over as in Figure 6.4 (d) and (f); and thus to partial collapse unless this was forestalled by a thrust from the other side and the initiation of a kind of true arch action as sketched in Figure 6.4(b). The 'arch' at Selinunte [6.3] provides a good example of this on the right. The upper projecting block is relatively short and the longer block below has cracked in two under the excessive tension, allowing the upper one to tip forward. Collapse has been prevented only by a horizontal thrust from the left, leading to a very inefficient kind of true arch action in which the thrust acts at only a small angle to the bed joints between the upper and lower blocks. Its concentration on the tip of the lower block has led to local spalling here.

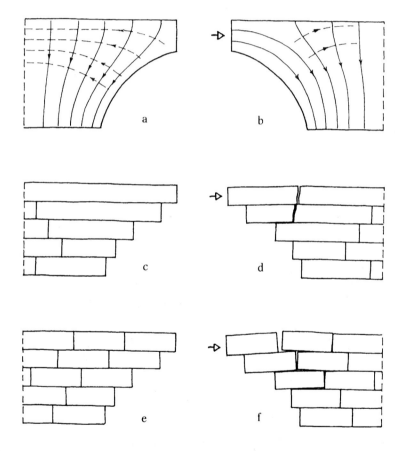

be spanned. Either the two sides meet eventually at the centre or they approach near enough to be spanned finally by a single block.

The intention behind this 'false' form (which, judging from its occurrence in early tombs and pyramids, is as old as the other early masonry forms) was probably that each half should be stable independently of the other, and that it should be possible to construct it, without any temporary support, as merely an overhang-

Figure 6.3 Small gateway
in the northern defence
walls, Selinunte.
(Source: author)

The behaviour seen at Selinunte is typical of that of most surviving false arches, with the exception that, where relatively small blocks or bricks were used throughout, the change to a kind of true arch action has taken place as in Figure 6.4(f) without any prior tensile failures of the blocks or bricks. Final stability without a horizontal thrust has probably been achieved only for very modest spans of about 2 to 3 m. and when enough long blocks have been incorporated and adequately anchored by friction to the mass of the wall behind the opening to develop the necessary tensions.

The chief merit of all these early forms – and one that would argue for their primacy over the true voussoir arch even if the archaeological evidence were less clear – was that they could be constructed with little or no centring or other temporary support beyond what might be desirable to facilitate the handling of the larger blocks. Their chief shortcomings were the very limited spans of which they were capable, plus the impermanence of the light timber and reed forms and the massiveness of most of the others. The wedged boulder had to be huge if it was to span a gap of any useful size, and its supports had to be correspondingly massive if it was not to push them aside by its thrusts. The inclined blocks of stone had to be almost as massive if they were not to break under the bending action to which they would be subjected in addition to direct thrust. They also called, therefore, for massive abutment. The false arch was the least efficient form of all unless it was merely an opening in a long continuous wall.

Masonry and concrete arches

The simple true arch of mud-brick, constructed probably as described by Somers Clarke, pointed the way to more efficient forms. Its direct further development seems to have been very closely linked with that of its longitudinal extension the barrel vault. In the type of vault illustrated in Figures 4.16 and 4.17 and described in Chapter 4, an end wall served in lieu of centring while the first few incomplete rings of bricks were set on edge leaning back against it. As the vault advanced away from this end wall, each fresh ring leant back against the previous one and temporarily gained such support as it needed from it. Where a greater depth than that of a single brick was required, further rings were built above the first one.

To adapt this procedure to the construction of a free-standing arch called only for the provision of light centring to provide temporary support for the first rings of bricks set on edge. When one or two superimposed concentric rings had been so constructed, additional stiffness could be imparted by adding further rings with the bricks set in what is now the normal way with their longer dimensions running radially and at right angles to the face of the arch. Figure 6.5 shows an arch with a span of about 4 m

Figure 6.5 Brick archway,
Tomb of Mentuemhet,
El Asasif, Thebes.
(Source: author)

Figure 6.7 Porta di Giove,
Falerii Novi (before
restoration of several
collapsed courses of the
adjacent wall).
(Source: author)

Figure 6.6 'Pseudo arch'
in the polygonal masonry
of the defence wall. Segni.
(Source: author)

constructed in this way in the seventh century B.C. Four rings of bricks have collapsed from inside leaving six more still standing. In a nearby arch with a span a little under 3 m, one ring of bricks on edge supports three rings with the bricks set like those remaining in the larger arch. The culminating achievement of this kind was probably the vault, still surviving in part, of the Taq-i Kisra of the Palace of Ctesiphon near Baghdad. It has a span of 25 m and may date from the third century A.D.

With numerous prototypes in brick available as models almost throughout the evolution of ancient Egyptian architecture, it is, at first sight, surprising that the true stone voussoir-arch was not developed sooner as a preferable alternative to the massive lintel, the inverted-V form, and the false arch. The reason cannot have been prejudice against the curved soffit, since curved soffits were often given to false arches and to those of inverted-V form, at least from the Middle Kingdom onwards, as in the later Greek example at Selinunte. The reason was probably that nothing would have been gained in the overall economy of construction. Handling of the much smaller blocks of stone would have been easier. But, against this gain, there would have had to be offset a very large increase in the labour of accurately pre-cutting the blocks, and the need to provide much heavier centring than was called for by the brick arches – in a country where good timber was very scarce and there was little skill in using it otherwise than as a column or beam. Monumental Egyptian stone architecture seems to have been geared

throughout to massiveness and to economy in the effort of cutting the stone. The quality of the stone was, moreover, good enough to permit the desired spans to be achieved without the voussoir arch.

It was in Greece, Rome and Etruria that true voussoir arches began to appear in stone in about the fourth century B.C.[4] There, they usually replaced not the massive lintel but the false arch constructed of numerous projecting courses. Apart from the possibility of a fairly direct copying of the brick arches of Egypt or farther east, there is the further possibility of the taking of a hint from the arch-like forms that tended to appear at intervals in the polygonal masonry often used for walls at the time [6.6]. Timber for centring would not now have been so hard to come by, and there was more experience in its use.

As a well-preserved example from the third century B.C. at Falerii Novi [6.7] shows, the early arches were semicircular in profile and had voussoirs of considerable depth even for very modest spans. The voussoirs were closely fitted together without mortar as in the somewhat later arches of the Pont du Gard [4.7]; and the semicircular profile was no doubt preferred for the same reasons. The depth of the voussoirs (without the outer moulding, which was formed from separate blocks) is close to two-thirds of the radius. This may be compared with the depth at which a similar semicircular arch, adequately supported at the base, free-standing, and subject only to gravity loading, would theoretically be just on the point of collapse through hinging

Figure 6.8
Pont Julien, Provence.
(Source: author)

Figure 6.9
Flavian Amphitheatre
(Colosseum), Rome.
Detail of the upper tiers
of the outer arcade
on the north east side.
(Source: author)

rotations at the base, the crown, and points on the haunches – a mere tenth of the radius. Even this depth could be reduced if the haunches were backed by substantial masonry like that of the wall at Falerii Novi.

As experience grew in the course of the next two centuries and the great cautiousness of these very deep arches – particularly those that were stabilized by masonry built up against the haunches – was recognized, designers became more daring. Arches of the first century B.C. may be typified by those of bridges like the Pons Fabricius in Rome and the Pont Julien in Provence [6.8]. The profiles of these are still circular arcs, but the depths of the voussoirs are not much greater than one-tenth of the radius for spans up to 25 m.

Apart from developing in this way the type of arch seen at Falerii Novi, the Romans introduced a number of other innovations. The first

three of these were probably linked, to some extent, to one another. They were (1) the arch of pentagonal voussoirs bonded into the spandrel masonry alongside and above, or of flat-topped voussoirs constituting both arch and spandrel [6.9]; (2) the flat or lintel arch; and (3) the arch of joggled voussoirs.[5]

Structurally the chief characteristic of the first was that the arch could not so readily accommodate itself to a spreading of its supports brought out by its outward thrusts simply by hinging rotations of the voussoirs. It was more likely to do so largely by relative slipping of the voussoirs. When hinging rotations did initially occur they usually led to local crushing or splitting and spalling of the corners acting as fulcra, which then precipitated slips. Both relative slipping and local crushing and spalling can be seen accompanying the gross deformations of some of the arches in the figure.

Figure 6.10 Flat arch of joggled voussoirs, Kalat Siman. (Source: author)

The second was little more than a variant of the first, with a horizontal instead of a curved soffit. Its depth had to be sufficient to contain a thrust line in equilibrium with the loads that was not so flat as to cause the supports to be pushed apart far enough to permit a collapse. This meant that accommodation to actual spreading was again most likely to occur largely by relative slipping of the voussoirs. Slipped voussoirs were, in fact, so common later in the ruins of ancient Rome that they appear frequently in Renaissance paintings and were copied as a deliberate conceit by Mannerist architects like Giulio Romano.

The use of joggled voussoirs was a fairly obvious expedient to reduce the likelihood of slipping. A fifth-century example is illustrated in Figure 6.10. A good earlier example is seen in the Temple of Jupiter at Baalbek [10.6] where the blocks above the architrave at the centre of each span are joggled to relieve it of load there. Apart from reducing the risk of slipping after construction was completed (especially in areas subject to earthquakes), joggling would have had the advantage of facilitating construction when the centring was not completely rigid. This may, for instance, have been the reason for its continued later adoption, for instance in the great flat arches of the fireplaces of the warming room of Fountains Abbey and of those of a number of roughly contemporary English manor houses. Its value in earthquakes (which may momentarily reduce the usual thrusts and associated frictional resistances to slipping) may, however, have been a more important reason for its widespread adoption in Turkish masonry arches of the Seljuk and Ottoman periods. The extreme elaboration of the interlocking of the voussoirs in many of the Ottoman examples must, however, be considered as purely decorative. In the eighteenth-century flat arches shown in Figures 3.15 and 8.10, slipping of the voussoirs was restrained by tie bars and cramps.

The remaining innovations were the concrete arch and barrel vault. The simple barrel vault appeared first, probably as a fairly direct copy of the false vault in cut stone finished with a smoothly curved soffit like the arch in Figure 6.3. A small vault over a well in Acrocorinth, cast in concrete bound with a lime mortar, appears to date from the third century B.C., and the extensive series of vaults of the Porticus Aemilia in Rome, with spans of about 8 m, from early in the second century B.C. By the beginning of the first century B.C. at the latest, such vaults were commonplace in and around Rome, even over non-rectangular spaces as at Palestrina [3.11]. They were cast on formwork and given a thickness comparable with that of the earlier false stone vaults. Sometimes, fairly large flat splinters of stone were laid, voussoir-fashion, immediately above the formwork, though not bearing directly on one another like the cut and fitted voussoirs of the arches just considered. More often all the aggregate was

Figure 6.11 Severan
substructures, Palatine,
Rome. (Source: author)

irregularly shaped and was randomly placed throughout in horizontal beds. In either case the mortar, placed separately and not pre-mixed with the aggregate as in modern concrete, was worked well into the gaps so as to prevent, when it hardened, any shearing deformations. Even when the soffit was coffered as at Palestrina, the rib-like projections were not at the outset structurally differentiated in any way. Only from about the middle of the first century A.D. were

brick ribs increasingly incorporated, for reasons to be discussed in Chapter 7. Where the ends of the vaults were not enclosed by walls, they were usually faced with stone voussoir arches or, later, with arches of the large flat Roman fired bricks. As mere facings, most of the bricks were then only half-bricks.

The concrete arch was slower in making its appearance, despite the fact that it was little more than a short open-ended barrel vault.

Partly, no doubt, this was due to the time that was taken to perfect the technique of making concrete to the point where confidence could be placed in its ability to stand up to the more concentrated loading of a primary supporting element. Well into the first century A.D., even in buildings in which concrete was extensively used for vaults, supporting arches were constructed of stone or brick over their full widths. In addition to their already proven strengths, such arches would have obviated the need for soffit formwork and been ready sooner to stand on their own and carry further loads. These advantages probably accounted for their continued use in structures like the Roman Colosseum [6.9] and the Pantheon when they had been superseded elsewhere.

From about the middle of the century, though, these arches were mostly superseded, under the stimulus of the very widespread adoption of concrete in walls as well as vaults, by brick-faced 'concrete' arches having a soffit as well as a core of concrete. The concrete was cast integrally with that of the spandrels or vault above, but it differed in one characteristic way from the solid mass of concrete of the earlier barrel vaults. This is typified by the flat lower arches over the doorways in Figure 3.10 and by all the arches of Figure 6.11. All their cut facing bricks have been robbed, exposing the cores. It can thus be seen that these were penetrated at fairly regular intervals by full bricks and thereby divided into voussoir-like sections. Once the concrete had fully hardened, this division would have served no useful purpose – as is clear today from the absence of distress where, in

other instances, the continuous concrete above such an arch has been left to take over its whole role by the complete removal of the arch itself. But when the arches were constructed it was presumably considered a prudent precaution to depart as little as possible from the proven voussoir form. This caution would have reduced the risk of spalling from the soffit before the concrete hardened.

The most radical post-Roman innovation was the 'chain arch'. This was a deliberately flexible stone arch, conceived as an answer to the problem of building bridges on very poor foundations which were liable to spread or tilt to a degree beyond that normally tolerated by a stiffer form of construction. Whether it originated in China on the Yangtze delta (where most examples are now to be found)[6] or in Europe is not clear: the example illustrated [6.12] is in Venice and is probably only a few hundred years old. The arch consists largely of long blocks of stone curved in the direction of the span. They are morticed or otherwise keyed into other blocks set transversely and acting as hinge pins. The whole arch can thus deform very easily by rotations at these 'hinges', as can be seen at several points in the photograph. Collapse is prevented by the stabilizing action of the spandrel fills. This action is really no different from that already noted in the discussion of the excessive depth of the earliest true voussoir arches. It arises because the fills are not mere inert dead weight – like a heap of sand – but have some stiffness and will weigh down more heavily where the arch tries to rise against them, and less heavily where it tends to

Figure 6.12 Canal Bridge (Ponte delle Colone), Venice. (Source: author)

drop. What was new was the way in which full advantage was taken of it in a situation where the deformation itself could be accepted.

More significant, however, in relation to the wider development of structural and architectural form, were various departures from the simple circular-arc profile – usually a full semicircle – preferred by the Romans. Structurally, the ideal profile will always be one that conforms exactly to a thrust line in equilibrium with all the loads [2.11(b), 5.3 and 5.4]. If it is desired to minimize the horizontal thrust, the rise should also be greater than it is in the semicircular or segmental arch. Some, but by no means all, the profiles actually adopted – notably the pointed profiles of many Gothic arches and arched ribs – came close to meeting these structurally desirable requirements. It is highly questionable, though, whether this was at all clearly understood by their builders. It is more likely that the profiles were chosen primarily for the advantages they offered in ease of construction (easy setting out plus the possibility of using somewhat lighter centring than was required for a semicircular or segmental arch) and in solving some of the aesthetic problems associated with groined and ribbed vaults having semicircular transverse profiles. Some of those which came less close to the ideal departed from it very widely. These more fanciful choices of profile – horseshoe, ogee, cusped, etc. – were possible only because the depth of the voussoirs was always sufficient to contain a continuous thrust line in equilibrium with the loads.

The idea that the best profile was that obtained by inverting a similarly loaded hanging chain appears to have been first arrived at in the latter part of the seventeenth century.[7] In the eighteenth and nineteenth centuries, considerable attention was given to a more scientifically based choice of form and the possibilities of masonry construction were pressed close to their limits.[8] In 1903, Séjourné completed a bridge in Luxembourg with twin arch ribs of 85 m span.[9] By then, however, the masonry arch had already lost its long ascendancy as the principal form available for spanning all but the shortest distances. Iron, steel, and reinforced concrete had all come into use, and, particularly with the last two, the arch was not necessarily the best form for one-way spanning except over fairly limited ranges of span. For the further development of the form we must therefore look to these materials, remembering that, when they were employed, other forms were developed in parallel to serve similar ends.

Timber, iron, and steel arches

While the masonry arch remained in the ascendant, the only alternatives over any but the smallest spans were forms constructed of timber or, from the latter part of the eighteenth century, cast iron. In terms of their structural potential, these two materials had a good deal in common: the higher strengths and stiffnesses of cast iron were largely offset by its much higher density [3.3]. In relation to stone, brick, and concrete, their chief merit was a better performance in tension. They could thus span considerably farther and carry higher imposed loads as simple beams than a stone lintel was able to do. The spans at which it became necessary to adopt some other form were therefore greater. When they were exceeded, moreover, some kind of truss was often a possible alternative to an arch and there was the further possibility of a hybrid form – a trussed arch or arched truss. When the simple arched form was adopted, the chief problem was to ensure its stability without resorting to the depth of solid cross-section characteristic of masonry arches. This would have been difficult to achieve on account of the limited girths of most trees and the limitations of the casting process.

In the third of the most primitive arch forms discussed at the outset, deficiency in the depth of cross-section was made good by the strength and stiffness in bending of the reeds or saplings. If the thrust lines did not follow the curves to which they were bent, such 'arches' acted partly as cantilevers well anchored in the ground. In medieval timber cruck building (in which pairs of curved single lengths of timber were inclined against each other in the same manner as the stone slabs of Figure 6.2) stability similarly depended partly on strength and stiffness in bending; though, as the crucks were not anchored in the ground, they could not act as cantilevers. Rotations of the crucks about their feet, and relative rotations at the top, were prevented only by their balanced opposition to one another.

Three is, however, the maximum number of joints at which free rotation can be allowed if the stability of an unbraced arch is to be ensured. If more joints were essential in a slender timber arch, most of them had to be incapable of rotation. They had to be as strong and stiff in bending as the individual lengths of timber. The only alternative was to provide some kind of bracing – analogous to that provided by the spandrel fills of the chain arch of Figure 6.12 – at most of the joints. Because it was difficult to make joints that were capable of efficiently transmitting direct tension and, thereby, bending, there was considerable practical difficulty in adopting the first of these procedures. Usually it was combined with the alternative of adding bracing, frequently in a way that left the actual mode of action highly dependent on the precise manner in which the members fitted together.

The first major timber arches of which it is now possible to form a reasonably clear picture were those of a bridge over the Danube built by Apollodorus of Damascus in A.D. 106. They are represented in one of the reliefs on Trajan's Column in Rome and, according to Dio Cassius, there were twenty-one spans each of considerably more than 30 m.[10] The highly simplified

Figure 6.13 Westminster Hall, London. (Source: author)

and stylized representation in the relief shows them as having three concentric timber arched ribs interconnected by radial struts which were continued upward to support the level roadway. It is impossible to infer any details of the construction because some that are shown are obviously incomplete or distorted. Probably they were not very dissimilar to those of Perronet's centring seen on the left of Figure 4.15 as far as the arch ribs themselves were concerned. Alternatively it is possible that each rib was built up from several sets of timbers bolted or clamped together alongside one another with the butt joints between the individual timbers staggered in such a way that each was lapped by a continuous member on at least one side. In either case there must have been more radial struts than are shown in the relief and more diagonal bracing.

Some of the finest surviving medieval timber arches are in Westminster Hall, London [6.13]. Here each arch was built up from three sets of timbers side by side, originally held together by oak pins. Those in the middle stopped short at each intersecting member of the roof structure, into which they were individually tenoned. The outer ones were joined roughly midway between these joints so as to lap them on both sides.[11] Some continuity in bending was thereby obtained, but stability was further ensured by the incorporation of the arch in what is known as a

hammer-beam truss, as described in chapter 9. The roof was completed in 1402 when the large timbers necessary for its span of 20.5 m were still easily obtainable.

When, by the sixteenth century if not sooner, large timber was no longer so plentiful, it became necessary to make do with shorter lengths of reduced thickness. One answer was the trussed arch, also discussed in Chapter 9. The others were various developments of the principle adopted at Westminster Hall, whereby the individual lengths were joined to give a more or less continuous member. Some (like that advocated by Philibert de l'Orme) were again based on setting the individual pieces side by side; others (like that illustrated by Faustus Verantius) on setting them above and below one another.[12] The joints were always staggered. The adjacent pieces were held together by bolts, pins, wedges, or nails. Transference of load along the arch was frequently further assisted, when the pieces were set above and below one another, by keying them together by means of a variant of the joint illustrated in Figure 3.13(c). The transept arches of the Crystal Palace [15.11] followed the pattern advocated by Philibert de l'Orme, and those of the bridge illustrated in Figure 6.14 the alternative pattern. This latter was also widely followed elsewhere on the Continent, in the British Isles and in North America in the first half of the nineteenth

Figure 6.14 Emme
Bridge, Schüpbach, near
Berne. (Source: author)

century, often in combination with various kinds of truss.[13] Multiple layers of quite thin planks were then found to be adequate for building up the arches if the planks were effectively bolted together.

Today gluing makes a much more effective joint and has given the glued laminated arch. With a glue as strong as the wood itself, such an arch can even be stronger than one cut from a single piece of solid timber – if a suitable piece could be found – since it becomes possible to cut out all natural defects that might have a weakening effect. A very free choice of profile is also possible, as is shown by the arched rib illustrated in Figure 3.14.

The first cast-iron arches of any size were those of the bridge at Coalbrookdale completed in 1779 [6.15] . They closely followed the timber form that was adopted long before by Apollodorus of Damascus and, as was noted earlier, the joints [3.16] were mostly copied from typical timber joints. An engraving of a previous design showed an even closer resemblance to the Roman timber form.[14]

Later cast-iron arches either followed almost equally closely existing timber forms (apart from a continuing liking for open circular members in the spandrels) or, as in a bridge over the Wear at Sunderland built between 1793 and 1796, came closer to masonry prototypes in using short open-work castings assembled as voussoirs for the arches themselves.[15] A noticeable difference, in comparison with most masonry arches, was that, after Coalbrookdale, much shallower profiles were nearly always adopted. Good examples are Telford's Craigellachie Bridge of 1810 [6.16] and his

somewhat later Mythe Bridge. Equally shallow profiles had already been adopted in masonry, as in Perronet's bridges at Neuilly and Pont-Sainte-Maxence designed in 1768 and 1771.[16] But for long masonry spans they had the disadvantages of exerting large horizontal thrusts and of making the settlement of the arch on removal of the centring [4.15] a much more serious threat to its stability. The great reduction in weight of the iron arch due to its greatly reduced cross-section compensated for this increase in thrust, and the close fitting of the castings reduced the settlement. Apart from their convenience in reducing the necessary height of the roadway for long-span bridges, the shallow profiles had the further advantage of permitting a reduction in the effective depth of the arch ribs without the likelihood of the thrust line passing outside it and inducing tension on the other side of the rib. Full semicircular or similar profiles exerting lesser horizontal thrusts continued to be preferred, though, for wide-span roofs.

Wrought iron and steel, though weaker in compression than cast iron, not only possessed higher tensile strengths but could also be formed more readily into long members of uniform cross-section – flats, angles, channels, Ts and Is – which could be joined together by rivets, bolts, or welds to develop fully these tensile strengths. The chief result of their introduction was, therefore, to extend considerably the range of potential application of the simple beam and the truss when full advantage was taken of these characteristics. This reduced still further the earlier dependence on the arch for spanning all but the shortest distances. Where the arched form was retained for longer spans, it became

Figure 6.15 Iron Bridge. Coalbrookdale. (Source: author)

Figure 6.16 Craigellachie Bridge (Telford). (Source: author)

possible to reduce the amount of material called for by taking advantage of the reduced liability to buckling that could be achieved through a greater strength and stiffness in bending. Excluding again, for the present, the fully triangulated trussed form that was adopted, for instance, by Eads and Eiffel [4.19], this was usually done in the later nineteenth century by adopting a built-up cross-section that was essentially an open-webbed I or box.

Two examples, both used for wide-span roofs, are illustrated in Figures 13.3 and 13.4. It will be seen that the arches carrying the roof of the Galerie des Machines were provided with pinned bearings between the two halves at the crown. They were also pinned at the feet, so that

they were, in this respect, similar to the medieval timber cruck. This made their action statically determinate, which was a considerable convenience from the point of view of design at this time as well as facilitating erection and eliminating the variations in stress that would otherwise have resulted from thermal expansion or contraction or slight relative movement of the feet.

More recently, I- or box-sections with continuous webs, or, following a precedent set by Brunel in the arched members of his Saltash Bridge,[17] continuous tubes, have usually been preferred, the third hinge at the crown usually being retained only as a temporary aid to construction. Elimination of this hinge in the completed structure adds to its strength and stiffness, and this has become more important in bridge construction as the continuously varying live load of the traffic has become more significant in relation to the unvarying dead load for which the arch profile is selected.

Simply as a demonstration of what is now possible, the Gateway Arch at St Louis [6.17] has been pre-eminent since its completion in 1966. Indeed, having been built solely as a monument and without any of the stabilizing additions that contribute to the stability of most other arches, it goes far beyond what would normally be considered the proper use of the form. It has a span of 192 m (measured between the outer faces of its legs at ground level) and is

192 m high. The profile was chosen to give fairly uniform compressive stress under dead load. The major problem in design was, however, the choice of the cross-section. Without the lateral and other bracing that large arches employed in roofs and bridges receive from other elements of the structure, this was required to provide the full bending stiffness necessary to resist the live load of the wind. When the wind blows at right angles to the plane of the arch, the whole arch must act as a great cantilevered curved beam subject to large bending moments likewise acting at right angles to this plane. The chosen cross-section was a double-walled triangle of welded steel plate reducing in dimensions towards the top, and, for increased stiffness in the lower half, the spaces there between the walls were filled with concrete tied to the rock below so that most of the weight is in this region. In a structure such as this, a great deal of effort had also to be devoted to devising a suitable erection procedure with suitable controls to ensure that the final form was as intended and had an acceptable distribution of stresses.[18]

Reinforced- and prestressed-concrete arches

The unreinforced Roman concrete arch was not widely used outside central Italy. Elsewhere its usual counterpart was the stone voussoir arch or the arch constructed throughout of fired brick. Even in Rome itself it was progressively supplanted by the latter as larger and larger quantities of brick came to be used as the coarse aggregate to give body to the concrete. Further development could not take place until the new strong concretes made from artificial cements became available in the nineteenth century. And, by the time that concrete arches of significant span were again constructed, the practice of reinforcing the concrete with embedded steel bars or rolled sections was so well established that these arches were almost invariably reinforced. Where bars were used as reinforcement, the forms usually closely resembled masonry forms. Where heavier rolled sections were used, as in the Melan system, the resultant form was sometimes virtually a steel arch, simply encased in concrete for protection and to give added lateral stiffness.

The development of new forms, based essentially on the use of relatively thin slabs as the most efficient elementary form for concrete reinforced with bars, was largely the work of three men towards the end of the nineteenth and in the early decades of the present century. They were Hennebique, Maillart, and Freyssinet. Examples of arches constructed by Maillart are illustrated in Figures 6.18 and 14.8;[19] and of arches constructed by Freyssinet in Figures 6.19 and 14.11.[20]

Maillart's Thur Bridge [6.18], though not built until 1933, has twin arches, side by side, of a kind first used in 1905 in a bridge at Tavanasa

Figure 6.17 Gateway Arch, St Louis (Saarinen, Severud). (Source: author)

Figure 6.18 Thur Bridge. Felsegg (Maillart). (Source: author)

Figure 6.19 Airship Hangars, Orly, under construction (Freyssinet). In the foreground is one of the centres that carried the formwork for casting the trough shaped sections of the vault. It was moved on rails from one section to the next as construction proceeded. (From *Eugène Freyssinet 1879-1962*, privately printed, Paris, 1963)

later swept away by a landslide. Each arch (72 m in span) is U-shaped in cross-section, being formed of a bottom slab only 180 mm thick over most of its length but somewhat thicker towards the springings, and two equally thin vertical slabs. The reinforcement binds the slabs integrally together, and these are further interconnected and stiffened by thinner vertical slabs set transversely at intervals. In the central portion they are also integral with the road slab to give an enclosed box section. As in the steel arches of the Galerie des Machines, hinged joints were incorporated both at the springings and at the crown – an added reason for the third hinge here being the inevitability, in a concrete arch, of progressive shrinkage of the concrete,

which could otherwise lead to cracking. The increase in total depth of the arch away from the hinges recognized a corresponding increase in the bending moments induced by concentrated live loads when near the quarter-span points. Such bending moments are more important in a relatively light arch than in a heavy masonry one because the thrust due to the dead load is smaller. The pointed profile of the soffit (not yet adopted at Tavanasa) could be justified, in turn, by this variation in depth.

In the Valtschiel Bridge [14.8] and others like it, the arch is so slender that there was no need for hinged articulations. It is a single slab only 270 to 290 mm thick for a span of 43.2 m. It is capable only of acting in direct compression

Figure 6.20 Esbly Bridge (Freyssinet). (Source: author)

in carrying the loads from the transverse vertical slabs to the abutments, and is the closest modern parallel to the masonry 'chain arch'. Bending stiffness to resist the bending actions due to concentrated live loads in different positions on the deck is provided entirely by the stiffened roadway slab.

Freyssinet's arches forming the continuous barrel vaults of the airship hangars at Orly [6.19] may be compared with the Gateway Arch at St Louis. The profile was similar, having been based again on the thrust line for dead load. Such difference as there was stemmed only from the selection of a more uniform cross-section throughout the span, leading to a more uniform distribution of dead load. The span was only about half that of the Gateway Arch, but a more significant difference from the point of view of the choice of cross-section was that the live load of the wind when blowing at right angles to the plane of the arches was here relatively unimportant because of their mutual support when subject to such load. Appreciable bending actions were induced only by the wind blowing in line with the arches. These were resisted by giving each arch a trough-shaped section. Because the material was now reinforced concrete and not steel, this was single- and not double-walled. The principal slabs forming the troughs were 80, 90, and 200 mm thick at the crown, increasing to almost twice these thicknesses at the feet.

In the bridge at Saint-Pierre-du-Vauvray [14.11], completed a year earlier in 1922, the arches had a span about midway between those at Orly and the one at St Louis and were more nearly free-standing. They were therefore given a rectangular-box section formed of slabs 200

and 330mm thick at the crown and somewhat thicker near the springings. There were no hinged articulations. Freyssinet preferred to have the greater stiffness of a continuous arch without hinges, and to overcome the awkward effects of shrinkage of the concrete by jacking the two halves of the arch apart at the crown before finally inserting strongly reinforced closing sections there. This jacking also lifted the arch slightly from the centring and thus freed it and greatly facilitated its removal.

Most subsequent wide-span reinforced-concrete arches have followed the precedents established at Saint-Pierre-du-Vauvray or Orly, with variations designed chiefly to economize further on the requirements for temporary centring and formwork. These have included the complete or partial precasting of the arch in voussoir-like sections, and the revival and adaptation of the Melan system of reinforcement to serve as permanent centring, as described in Chapter 4. The first has the further advantage of allowing most of the shrinkage of the concrete to occur before the 'voussoirs' are set in place. It also, as noted before, allows a freer choice of their individual forms. The desirability of making the mean profile closely follow a thrust line for dead load nevertheless remains a major constraint on the choice of form if economy is to be achieved without prestressing.

The use of tensioned reinforcement in Freyssinet's Esbly Bridge [4.22 and 6.20] permitted a relaxation of this constraint and the adoption of a very shallow profile and shallow depth at the crown. While leaving the concrete in compression throughout or very nearly so, the tensioned reinforcement allowed the cross-section to resist quite large bending moments where the thrust line passed outside it.

Without jacking at the supports referred to in the earlier description of the construction procedure in Chapter 4, the action would, however, have been essentially that of a portal frame. Bending would have predominated over true arching action; and the bending moment would have vanished at the hinged supports and at about the quarter points of the span, but would have been quite large at midspan. Shrinkage of the concrete and tensioning of the reinforcement would, by themselves, have further reduced the horizontal thrust and the true arching action. They would thereby have led, as noted previously, to a further increase in the moment at midspan – in other words to an action approaching more nearly that of a simply supported beam and without the shortening of the span due to shrinkage of the concrete and tensioning of the reinforcement.

By means of the jacking, the horizontal thrust was increased to a value greater even than that for a simple portal frame. A greater measure of true arching action was achieved; and the upward bending moment which the additional thrust created in the two-pinned statically indeterminate span reduced the net moment at midspan to almost zero and permitted a continuous reduction in depth to that point.

As compared with that of the simple masonry voussoir arch, the resultant behaviour of the prestressed-concrete arch can, therefore, be complex. Indeed, structurally speaking, the form is a hybrid one. Visually though, it can be one of great elegance.

Catenaries

The development of the catenary – if we use this term rather loosely to denote here any curved structural element which spans in the same way as a hanging chain or *catena* rather than the particular curve assumed by a freely hanging uniform chain loaded only by its own weight – is more easily described. Considered as an isolated element, the catenary is, structurally, the simplest possible form after the straight tie. The only problems of design and construction that it presents are those of choice of material, fabrication as a continuous element, and attachment of the two ends: it will automatically assume the curve necessary for equilibrium. More complex problems arise only when it is combined with other structural elements, usually in such a way as to provide support for these and, at the same time, to be stiffened by them. Discussion of these problems can, therefore, be deferred to Chapters 7, 13 and 14.

The first man-made catenaries were, almost certainly, slung ropes made from natural fibres and serving as very simple temporary bridges or as supports for tent-like shelters. In Roman times at least, the latter use was developed on quite a large scale for the temporary shade canopies (*velaria*) of theatres and amphitheatres. The Roman Colosseum, for instance, still bears around the top storey the attachments for masts to carry the ropes, and there are stone bollards around the base for their anchorage at ground level.[21]

The substitution of eye-bar chains of wrought iron for ropes offered considerable

Figure 6.21 Conway Suspension Bridge (Telford). The cables above the eye-bar chains are a subsequent addition. (Source: author)

increases in strength and durability. The first Western representation of such chains is that of Verantius published in 1620, but they were probably used much earlier in China.[22] Figure 6.21 shows a typical early nineteenth-century use by Telford. Before adopting them, he had initially intended to weld the bars together into single continuous lengths, but abandoned this idea. This was probably chiefly because the welded bars would have been too stiff to handle in the way that a chain can be handled, so that they would have had to be assembled *in situ* on temporary staging. Too much would also have depended on the strength of every weld.

The final development was the further substitution of cables made from iron wire and later steel wire. Cables of iron-wire rope had already been used in several places before the Conway Bridge and its neighbour over the Menai Straits [14.22] were built, but a further impetus was given to their development by the pioneering work of the Séguin brothers and a report on some French bridges by Vicat, which strongly favoured such cables as compared with eye-bar chains.[23]

Almost from the beginning, they took two forms: the parallel-wire cable and the twisted- or stranded-wire rope. Both had the advantage over solid bar chains that is associated with the intrinsically higher strength of a relatively thin wire, whether this is drawn from wrought iron or high-tensile steel. As first made, the stranded form was, however, much less compact, considerably more extensible, and more prone to corrosion. The parallel-wire form did not have these disadvantages, but it was much less flexible unless the individual wires were free to slide on one another. Vicat calculated that local

variations in tensions of up to 20 per cent had probably arisen between the wires of some such cables, which had been spun straight, as a result of subsequently bending them over supports.

The answer to uneven stressing of the wires in the parallel-wire cable was to spin the wires in such a way that, as the cable was built up, they already followed a curve very similar to that which they would finally have to assume. Vicat was himself the first to propose doing this by spinning *in situ*. It was, however, left largely to John A. Roebling and his son, working a little later in the United States, to develop the idea into a completely practical and economical procedure. With little change this procedure has now been used for well over a hundred years to spin the cables of most large suspension bridges in Europe [14.25], America [2.1 and 6.22], and elsewhere.[24] After all the wires have been spun, they are compacted and then bound together by circumferential wrapping.

The original drawback of the greater extensibility of twisted- or stranded-wire rope cannot be completely overcome. It arises because, as the rope or cable is stretched, the wires pack more closely together. But it has now been considerably reduced by the development of the locked-coil cable. In this all the outer wires are of specially shaped cross-sections designed to lock closely together even when unstressed [10.14(h)]. Extension under load is thereby reduced and much better protection against corrosion is possible. Since a fully prefabricated cable simplifies construction, provided that its diameter is not too great for easy handling, locked-coil cables are now increasingly used for both catenaries and straight ties of medium span or length, and occasionally over longer spans.

Figure 6.22 Parallel-wire cables of the Verrazano Narrows Bridge, New York Harbour, before compacting and binding. The travelling sheaves used to carry the wires over the span ran along the four single cables above. (Source: author)

Chapter 7: Vaults, domes, and curved membranes

It will be apparent, from the discussion of the development of the arch, that not all the forms actually built conform closely to the simple norms illustrated in Figure 5.5 or even to simple analogues of the kind illustrated in Figure 5.11. Particularly today, many forms are intermediate between, say, the arch and the beam or the arch and the portal frame. The basic norms for the forms whose development is to be considered next are those illustrated in Figure 5.5 (k) to (p) – doubly curved forms generated by rotating a simple arch or catenary about its vertical axis or by moving it sideways along another curve. It is convenient also to include the groined vault, whose norm is a pair of barrel vaults [5.5(i)] intersecting one another, with the parts inside the intersections (or groins) omitted. (The simple barrel vault continuously supported at the base has been considered already with the arch.) Ribbed variants in masonry or reinforced concrete will be included, since their development and structural action are closely related to those of the corresponding forms without ribs. Cable networks will similarly be included as

variants of the corresponding tensile membranes. Trussed or framed analogues of the vault and dome, and the barrel vault supported only at its ends to span as a beam [5.5(h)], will be considered later.

All the forms to be considered are, alone, capable of enclosing space. This tends to give them an aesthetic dominance over the interiors of buildings in which they are used, as well as, in some cases, over their external forms and silhouettes. Coupled with associations of various kinds and the fact that the precise geometry of a form like a dome is not a critical factor in determining its own potential strength and stability, this aesthetic dominance has led, as already noted in Chapter 1, to various non-structural objectives playing a much greater part in the pattern of development than was the case with the arch and catenary. Purely structural considerations have often made themselves felt more in the design of the total form in which a dome or series of vaults was used than in that of the dome or vault itself. With membranes and cable nets, this is less true. Their structural

Figure 7.1 Navajo Indian hut, near Monument Valley. Arizona. (Source: author)

behaviour must always be considered alongside aesthetics and other matters because they are much closer to minimal structures. But it is so dependent on suitable support that it is even more necessary to consider the adequacy of the total form. Fuller discussion of the whole structures of which these spanning elements are a part will, however, be left to Chapters 12 and 13.

Origins

The earliest forms were, almost certainly, very simple dome-shaped huts constructed with a framework of light saplings, reeds, or perhaps heavier timbers [1.1 and 7.1] and covered with some kind of thatch, matted material, turfs, or skins.[1] It was probably transposed first into solid rammed earth or mud, then, according to locality, into mud-brick or stone, though it is very difficult to say now when these changes first occurred. Low circular walls of stone, which may have been surmounted by more or less conical domes of mud or mud-brick, have survived from as early as the sixth or fifth millennium B.C.[2] The domes themselves have perished. Stone is more durable, and was definitely being extensively used for the dome itself by about the middle of the second millennium B.C. By 1300 B.C. it had already achieved monumental proportions in the so-called 'Treasury of Atreus' or 'Tomb of Agamemnon' at Mycenae, with a diameter at the base of 14.5 m [11.10]. By this time it was also being used quite widely, though on a lesser scale, in other parts of southern and western Europe.

It is quite possible that the earliest domes of rammed earth or mud were built up around a light framework of the kind that had previously been given a less durable covering. If so, this framework would have served as permanent centring of a kind. With that possible exception, all these early forms would, like the early forms of arch, have been constructed without centring.

The stone domes, in particular, were all false domes, constructed, like the false arches described in the previous chapter, by bedding each course more or less horizontally but projecting a little inward from the one below [7.2]. They were always capped by a single larger stone. As with the false arch, the profile had to be fairly sharply-pointed because of limitations on the projections of the individual blocks; but it was not necessary to make each course project back behind the springing as far as it projected forward over the void because it was prevented from falling forwards by its own true arch action as a complete horizontal ring. From this point of view, the form was also stable at every stage of construction provided that there was a complete ring at the top. This ring was capable of acting as the connection between the opposite sides that was needed to develop the horizontal radial thrust that was also necessary for equilibrium. On the other hand, the inclined resultant radial

Figure 7.2 Looking up
into the vault of a stone
beehive-hut near
Gordes, Provence.
(Source: author)

Figure 7.3 Dwellings with
corbelled domes, Haran,
Turkey. (Source: author)

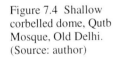

Figure 7.4 Shallow
corbelled dome, Qutb
Mosque, Old Delhi.
(Source: author)

thrust had, like the inclined resultant thrust in the false arch, to act obliquely to the bed joints. It was therefore similarly liable to cause the blocks to slip on one another if they tipped slightly as in Figure 6.4 (f). To prevent this, considerably greater thickness and weight were required than would have been necessary if the bed joints had been more or less radial in relation to the profile.

Some early brick domes were constructed in a slightly different way by arching up the otherwise horizontal brickwork a little at two opposite points on the circumference of the circular base wall and canting it forwards a little at the same time. Successive courses thus made two more-or-less conical fans which met each other above the transverse diameter.[3]

An incidental advantage of these freehand methods of construction was that they readily lent themselves to variations on the basic circular plan. It was not essential, for instance, for the horizontal projections of the individual blocks of a false dome to be the same at all points in each horizontal course right from the start. If it was desired to cover a room that was initially square, it was merely necessary to commence the forward projection at a lower level in the corners than elsewhere and then to extend it progressively towards the centres of the sides until a near-circular base was achieved for the completion of the dome. The early Etruscan Tomb of the Diavolino at Vetulonia, now in the Archaeological Museum in Florence, was one example of this technique. Figure 7.3 shows much later examples near the Turkish-Syrian border. In a similar manner, the early method of constructing brick domes described in the previous paragraph could be adapted to a square plan by commencing the arching up and canting forwards of the brickwork in the four corners.

Examples of this technique survive only from Classical times.[4] But it was then practised with such assurance to construct unusually flat saucer domes that it was probably a good deal older in origin.

The prototype membranes must have been the tensioned coverings of some nomads' tents.[5]

Concrete and masonry domes and related forms

The true dome of cut stone – with each block bedded more or less at right angles to the profile of the inner surface – was a late-comer compared with the voussoir arch, and did not develop directly from it. There were probably several reasons for this. One was the greater difficulty of cutting the blocks to fit closely if they had to be slightly wedge-shaped in two directions. Another was the fact that individual blocks in the upper part of the dome would tend to slide forwards until the horizontal ring in which they were set was complete. This could make some centring essential, whereas it was not called for by the false form. These reasons alone were probably largely responsible for the long-continued use of the false form in places where its construction was further simplified by the ready availability of a stone easily cut into long squared blocks [7.4].

In addition there does not seem to have been as pressing a demand for an improvement on this form in the third, second, and first centuries B.C. as there was then for the development of the voussoir arch for gateways and bridges. Despite the wide distribution of simple circular hut forms, large circular buildings or spaces within buildings were rare. For the few circular temples, timber roofs seem mostly to have been considered adequate. Occasionally, as in the

Figure 7.5 Temple of
Vesta, Tivoli. Etching of
1761 from the *Vedute
di Roma*, G B Piranesi,
Rome. (Source: author)

Tower of the Winds in Athens, a simple timber form (that of Figure 7.1) was directly copied in stone. Finally, when another alternative to a timber roof was required, the need was first met not in stone but in the new Roman concrete. This called for formwork as well as centring; but it eliminated the other purely technical difficulties.

The early development of the concrete dome is a little obscure. We know nothing of possible failures along the way and have to piece the story together from a very small number of survivals, resisting the temptation to assume, merely on the basis of an implausible Renaissance reconstruction,[6] that a structure like the long-roofless Temple of Vesta at Tivoli [7.5] originally had such a dome of perfect hemispherical form.

The earliest surviving domes are those of the *frigidaria* or cooling-rooms of the Stabian and Forum Baths at Pompeii, constructed in the late second or early first century B.C. They have the open eye at the top that was also typical of most later Roman concrete domes, but their conical internal and external profiles seem to have derived directly from the simple timber form which could, of course, have served to support the necessary formwork.[7] They are about 6 m in diameter.

A hundred years or so later, this conical form had given way to an internally hemispherical one, as is shown by the so-called Temple of Mercury at Baia. This was really another bath hall enclosing a pool fed by one of the thermal springs for which Baia was famous in Antiquity. Its dome again rose to an open central eye from a circular base punctuated by recesses and doorways, but its span of 21.5 m greatly exceeded those of the earlier domes at Pompeii and there were four windows at about mid-height. Being very close to Pozzuoli, it was built with a fine natural pozzolanic mortar, but its builders were still clearly unsure of the bonding capacities and strength of this. As in some early concrete barrel vaults, the larger *caementa* or pieces of aggregate consisted of flat splinters of stone (tufa) set almost radially, instead of smaller irregularly shaped lumps tipped on the formwork in more or less horizontal layers.[8]

For a much fuller realization of the vast potential, architectural as well as structural, of the basic domical form, it is necessary to jump forward to the highly productive period between the reconstruction under Nero that followed the great fire in Rome in A.D. 64 and the completion of the Pantheon as it now stands (the last of several reconstructions), under Hadrian in about A.D. 128. Though the development then from the octagonal room of the Domus Aurea through such structures as the Domus Flavia and Domus Augustana on the Palatine, Trajan's Market and Baths, and Hadrian's Villa at Tivoli and Palace in the Gardens of Sallust to its major culmination in the dome of the Pantheon can be traced in much more detail,[9] it will be necessary

to concentrate here on the first and last of these structures.

The octagonal room of Nero's Domus Aurea was 13.5 m wide from side to side and 14.7 m across the corners. Its dome began internally as an octagonal domical or cloister vault with straight sides gradually converging inwards towards one another and then, about halfway up towards the large open eye, merging into a more spherical form [7.6]. Externally it was given the form of an octagonal pyramid throughout, though this pyramid carried triangular buttress-like walls over each corner that connected it directly to the radial walls and vaults of the small rooms opening off the octagon.[10] It was entirely constructed of simple cast concrete apart from the bricks used in the lintel arches over the doorways to the small rooms and a brick ring around the eye. Since any original facing has now disappeared, the impressions of the boards of the internal formwork may be clearly seen. In its three-dimensional geometry it also exploited, as never before, the ability of wet concrete to assume any form that was desired, provided only that this form could first be reproduced in timber. Indeed it probably went too far in this direction from the point of view of ease and economy of construction. It must have called for considerable external formwork also, and this must have interfered a good deal with the placing of the concrete.

Figure 7.6 Vault of the octagonal room, Domus Aurea, Rome. (Source: author)

The dome of the Pantheon [1.7 and 7.7] was more than three times as large.[11] Its internal diameter of 43.3 m remained unequalled until the Renaissance. Internally, it rose from its circular base essentially as a hemisphere, though the mass was lightened in the lower parts and its surface broken up by five rows of coffers diminishing in height towards the top. Once the concrete had hardened sufficiently at a particular level, circumferential centres could have been strutted up from these coffers to support the formwork for the next lift, thereby obviating the need for complete radial centres and all the problems of constructing and later easing them. Externally, the profile was much flatter and the

weight further reduced by the incorporation of progressively lighter *caementa* in the concrete of the upper layers (the concrete being cast in horizontal layers throughout as indicated in the section) and by an open eye at the top. This, like the eye of the dome of Nero's octagon, was strengthened by a solid ring of brickwork.

This dome, together with a few lesser ones of Trajanic or Hadrianic date, served as the model not only for most subsequent Roman concrete domes but also, less directly, for many constructed of other materials. The lesser Roman concrete ones were notable chiefly for a bolder modelling of the inner surface to accommodate it more expressively and convincingly to

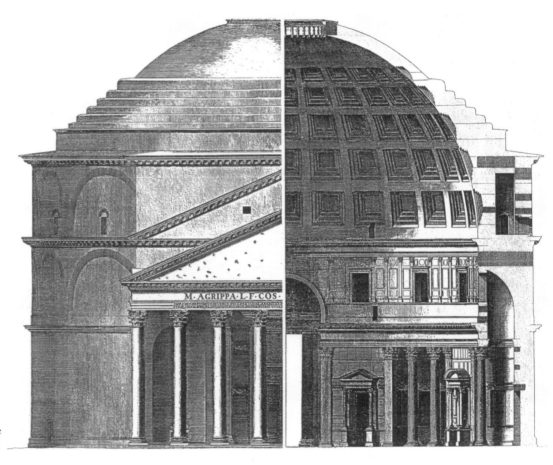

Figure 7.7 Part elevation and part section of the Pantheon, Rome, from A. Desgodetz, *Les édifices antiques de Rome*, Paris 1682, plates III and VII with details of the brick arches that penetrate the concrete mass added to the section. (Source: author)

apparent springing-level much higher. This profile approximated more to that of a shallow cone, but was made up of a series of steps which would have called only for vertical formwork, thereby simplifying the placing of the concrete and further simplifying the work of the carpenters.

The result of the different internal and external profiles was a much greater thickness in the lower part. This gave greater strength and weight where these might be most advantageous. But the total mass was hollowed out here at regular intervals around the circumference to leave a series of internal chambers spanned by arches that essentially repeated the arched construction of the base structure. The upper part, as well as being thinner, had its

a non-circular plan than was done in the early experiment in the Domus Aurea. Typically, this surface was scalloped and groined to follow smoothly the recesses and projections of the supporting walls at the springing level. No sign of this was visible externally, though. The external form remained more or less a flat cone and all the internal modulations of the basic hemispherical form were contained within the thickness of the concrete.

Architecturally the result was nothing less than a revolution. It would be an exaggeration to say that all earlier architecture had been primarily concerned with mass and sculptural form – with the external aspect rather than the internal space. There had been occasional tendencies to emphasize the interior in earlier assembly halls,

119

basilicas and throne rooms, but achievement was always severely limited by the spanning limitations of simple beam and truss types of roof. In larger structures it was never really possible to break away from the box-like space bounded by real or implied plane surfaces and encumbered by large numbers of internal columns. The Roman concrete dome did away with these limitations and permitted, for the first time, an architecture of large unencumbered interior spaces that could be experienced only from within; an architecture capable, moreover, of the subtlest and most intricate modulations of these spaces, their bounding surfaces, and the light falling on them.

in Chapters 2 and 5. If, moreover, the hoop tensions had been resisted in this way, there would, as we have seen, have been no outward thrusts on the supports.

It is fairly clear that the possibilities of drastic reductions in thickness were not appreciated. On the other hand, it is quite possible that hopes were initially entertained that, by virtue of the near-monolithic character of the set concrete, no more horizontal thrust would be exerted than by a truly monolithic lid cut into a domical form such as that of the Monument of Lysicrates in Athens or the much later and larger one capping the Tomb of Theodoric at Ravenna, and that it was only later experience that showed these

Figure 7.8 'Temple of Minerva Medica'. Rome, about 1790. Detail of a painting by J. C. Klengel in the Gemaldegalerie Alte Meister, Dresden. (Source: author)

Figure 7.9 Vault of the North Baths, Jerash. (Source: author)

Structurally the new forms were almost as revolutionary. They still fell short, nevertheless, of a full exploitation of the increased possibilities of structural action (as compared with those open to the arch) that were conferred on them by their double curvature. Their thickness at the crown or eye was usually between one-tenth and one-fifteenth of the radius, which was not much less than that of the voussoirs of the singly-curved masonry arch of a century or so earlier. Lower down it was always considerably greater than this, and corresponded more to the thickness of the voussoirs of the masonry arch plus the spandrel fill above its haunches. Full exploitation of the double curvature might have permitted a reduction of these thicknesses to about 1/200th of the radius (or 1/400th of the span) provided that the conditions of support at the base were appropriate and that the circumferential or 'hoop' tensions in the lower part could be resisted without cracking as explained

hopes to have been misplaced for most larger domes.

According to modern statical theory, the maximum hoop tensions to be resisted at the foot for ideal conditions of support and for self-weight loading only should indeed have been less than 0.5 MN/m^2 even in the Pantheon dome, and the fully hardened concrete was probably well able to take a stress of this magnitude. In fact, though, it usually seems to have suffered extensive radial cracking, as seen in Figure 2.15. Such cracking is particularly pronounced in the Pantheon dome, though patching of the exterior surface and re-rendering internally now make it necessary to enter the upper internal chambers to see it at first hand.[12] There were probably two contributory factors. One was that, unless the construction took place very slowly, the concrete would not have developed its full strength by the time that the full tension became effective. The other was that the conditions of support never were ideal. In particular, additional tensions would, almost certainly, have been developed as a result of restraints offered by the supports both to the drying shrinkage of the concrete and to contractions due to cooling after being heated by the sun. The result of the cracking was the transformation of the lower part of the monolithic non-thrusting 'lid' into a ring of thrusting arches with a common 'keystone' formed by the still unbroken ring of the upper part. This did call, after all, for a thickness more appropriate to an arch than to a domical shell, for additional thickness in the 'haunch' region, and for a base capable of resisting the thrusts. (It should be

noted, in passing, that these greater thicknesses could do less than might be expected to reduce the hoop tension that would have had to be resisted for ideal dome action. This is because, for a proportionate increase in thickness throughout, the tension would be increased to the same extent as the cross-sectional area available to resist it.[13])

In later Roman domes, efforts continued to be made to reduce the weight, and thereby the outward thrusts, by using lighter *caementa* for the concrete of the upper parts and by creating voids in the lower parts by setting empty amphorae in the concrete there. But the inevitability of some outward thrust seems to have been accepted. The designers were chiefly preoccupied, therefore, with providing supports capable of resisting the thrust and with facilitating the process of construction. The design of the supports will be considered in Chapter 12. The main emphasis in relation to the process of

construction was placed on subdividing the total mass of the concrete to make it easier to place, and on reducing the need for heavy centring and extensive formwork.

One outcome of the latter preoccupation was the incorporation of systems of embedded ribs best exemplified today in the surviving part of the dome of the so-called 'Temple of Minerva Medica' constructed in the mid-third century [7.8]. These ribs were constructed in very much the same way as the brick-faced concrete arches described in the previous chapter. Most of the bricks in them were broken ones set in parallel rows spaced slightly apart, and these rows were joined together at fairly regular intervals by full bricks set, rather like the rungs of a ladder, to create a series of voussoir-shaped voids that were filled with concrete. They were built up almost simultaneously with the placing of the concrete in the intervening compartments of the dome.[14] They could not, therefore, have carried the formwork for these intervening compartments. They would, together with continuous horizontal rings of brickwork spaced more widely to correspond with the successive layers of concrete and the steps in the external profile, have helped to stiffen the wet mass during construction and thereby to reduce the support required from the centring. They would also tend to concentrate the radial thrusts on themselves if they were slightly stiffer than the intervening concrete, and to concentrate the radial cracking in this concrete.

Numerous sixteenth- to nineteenth-century representations of the 'Temple' such as the one reproduced here show that, in this dome at least, both things did happen. Several ribs were then still standing in the part of the dome that has now collapsed, but the concrete between them had fallen away. The deliberate placing of the widest ribs between the windows and over the main supports also strongly suggests that it was intended that these ribs should carry the main thrusts.

A more radical outcome, though one of less relevance to later developments, may have grown out of the practice of setting empty amphorae in the concrete to lighten it. This was the technique of building up the dome with a core consisting of a continuous spiral of terracotta tubes so shaped as to fit into one another end to end. The result was a very light kind of dome requiring only minimal centring and formwork; but it lacked the strength required for the larger spans. It is best exemplified today by some of the fifth- and sixth-century domes of Ravenna.[15]

It marked, of course, a very considerable departure from the technique adopted in constructing the domes of the Domus Aurea octagon and the Pantheon, and it was not then alone in doing so. Once the new forms had been developed and the feasibility of constructing them on a vast scale amply demonstrated, they were widely copied in other parts of the Empire where the

Figure 7.10 Dome of
Hagia Sophia, Istanbul.
(Source: author)

Figure 7.11 Dome of the
Gunbad-i-Kharka,
Friday Mosque, Isfahan.
(Source: author)

natural ingredients for making Roman concrete were lacking. This usually meant reproducing them in brick or cut stone.

Good examples of brick domes from early in the third century are those of the Mausoleums of Galerius (later the Church of St George) in Salonika and of Diocletian (now the Cathedral) in Split. Presumably because of the expense of the cutting, domes of cut stone were never as large. Figure 7.9 shows one of the finest survivals, a slightly earlier Syrian example.

The immediate result of changing from concrete to brick or cut stone or even hollow tubes was that it was no longer possible, in effect, to carve the form freely from the solid. Since it had to be built up ring by ring, and layer by layer if more than a minimal thickness was required, it was much easier to keep to a simpler geometry and to vary the thickness, if at all, in discrete steps. This usually meant keeping, internally, to the hemispherical form or to another surface of revolution, either flatter or more pointed. It also precluded the free-flowing and geometrically rather indeterminate transitions from non-circular plans to essentially hemispherical forms that was noted at the Domus Aurea and is characteristic of many Roman concrete domes.

Over the relatively small spans of the four supporting arches at Jerash, the problem of the transition could be avoided, as seen in Figure 7.9, by making the radius of the dome itself considerably greater than that of its circular springing-line at the level of the arch crowns and simply carrying it down with the same radius to the points where the arches met. In terms of the construction sequence, this meant starting at the four low points with a radius equal to half their diagonal spacing and maintaining this radius throughout. When this procedure is followed, the parts below the circular springing-line of the dome proper are referred to as merging pendentives. Its drawbacks were that the increase in radius and flatter profile of the dome proper made construction more difficult and led to increased thrust.

One way of avoiding these drawbacks was to build up to a circular base in the same way by constructing similar part-spherical triangular pendentives, but afterwards to resume construction with more or less vertical springings and a reduced radius of curvature. This was the expedient adopted for the sixth-century early Byzantine dome of Hagia Sophia in Istanbul [7.10]. The alternative, usually favoured by Islamic architects but pre-Islamic in origin, was to first span the corners of the ground plan by means of secondary arches referred to as squinches. If necessary, a further tier of smaller arches could be superimposed to give eventually a many-sided close approximation to a circle as in the fine eleventh-century Seljuk dome of the Gunbad-i-Kharka in the Friday Mosque in Isfahan [7.11]. Both alternatives led, at the base of the dome proper, to a pronounced discontinuity in the diagonal cross-sectional profile of

the complete vault. Stability was ensured by circumferential compression at this level. Usually it was further assisted by loading the pendentives or squinches externally with additional material, coursed horizontally.[16]

In the construction of these brick and stone domes themselves, the necessity of building up any thickness greater than that of a single brick or block of stone by adding further layers or shells had the advantage that centring, if any was used, was not required to support more than the innermost shell. The support that this called for was, moreover, again reduced by the fact that each completed ring would be self-supporting once any mortar had hardened sufficiently to take, without undue deformation, the compression to which it would initially be subjected. It seems likely that in Hagia Sophia, as in the Pantheon, only circumferential centres were used for the requisite temporary support of successive rings. Here there were no coffers, but these centres could have been tied back to the outside of what had already been built through circular holes left at regular intervals through the thickness. Sometimes the need even for circumferential centres was reduced or eliminated by the adoption of special bonds. One possibility, seen in the dome of the Mausoleum of Diocletian, was to set the bricks in numerous small radiating fans disposed around the circumference and superimposed on one another like the scales of a fish. Another was to set them in a herringbone pattern in which spiralling ribs of bricks set on end interrupted each ring at intervals or in some other simplified variant of the largely decorative bond seen in Figure 7.11.[17]

In overall design there were two important innovations. The first was the incorporation, in brick domes at least, of circumferential ties around the base. Where these were of timber, the practice probably stemmed partly from a long tradition of incorporating 'bonding' timbers in masonry walls. But it must also have been intended to limit the radial cracking seen in earlier domes and to reduce the outward thrusts on the supports. In the domes like that of Hagia Sophia, rising from marble cornices above the pendentives, iron cramps that tied together the marble blocks were probably intended to serve a similar purpose.

The second was the dome formed of two distinct shells, usually interconnected at intervals but elsewhere separated by a void. This was, in retrospect, a fairly natural development from the practice of building up the total required thickness by adding superimposed layers of masonry. It had the advantages of divorcing the weathering surface from the inner shell and thereby giving improved weather protection, and of reducing the weight for a given overall breadth of construction. It also permitted an increase in the external size and height of the dome to make it more imposing without necessarily increasing the height internally. But it came about only slowly. The almost invariable practice up to about the eleventh century was simply to rest a timber roof, with a covering of tiles or lead, directly on the working masonry of the dome, or to protect this masonry by a conical roof of stone slabs supported on it by a rubble fill with some internal voids to reduce the weight as in most Seljuk examples [7.12].

The earliest known masonry double domes crown a pair of eleventh century Iranian tomb towers [7.13]. Both appear to have had inner and outer brick shells of similar thicknesses and profiles (though, of course, of greater diameter and height for the outer ones). These shells were completely independent of one another apart from a few interconnecting timber struts.[18] They may well have been modelled, in part, on earlier timber double domes such as that of the Dome of the Rock in Jerusalem (which, in the early eleventh-century reconstruction that now survives only in part, consisted of two independent sets of radial ribs covered, one externally and the other internally, with planking) or the similar and slightly later one of the Great Mosque of Damascus. Indeed the possibility that the upper parts of the outer shells of the Iranian domes were constructed partly of timber cannot be completely excluded since they are now lost.

There is no such doubt about the much larger early fourteenth-century brick dome of the Mausoleum of Oljeitu at Sultaniya in a much more accessible location some 75 km to the north of the tomb towers [7.14]. When restored in the mid-1970s, this dome still retained much of its original complete outer shell of brick, somewhat thinner than the inner shell but here connected to it and partly carried on it by numerous shallow radial and circumferential ribs. It has an internal diameter of 26 m and a slightly pointed profile. Close inspection during the restoration left no doubt that the two shells

Figure 7.12 Tomb tower, Ahlât, Lake Van. (Source: author)

Figure 7.13 Tomb towers with double-shell domes, Kharraqan, Iran. (Source: author)

were constructed together entirely without centring. This was made possible not only by the circular plan form and pointed profile but also by setting the bricks at a flatter-than-radial angle and by the use of a quick-setting gypsum mortar. Lengths of poplar built into the rising masonry

probably, that part of the outer shell which projected beyond the springings was also stiffened and partly carried by walls built up vertically from the supporting drum.

The outstanding double dome in the West, in terms of both the achievement that it represented

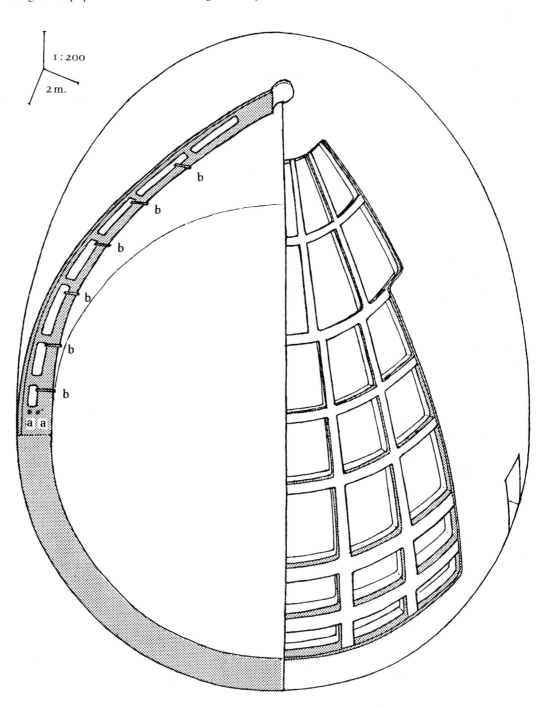

Figure 7.14 Dome of the Mausoleum of Oljeitu, Sultaniya, Iran. (Source: author)

and subsequently sawn off almost flush with the inner surface (b,b) supported working platforms. Vestiges remained of twin circumferential ties of poplar built in at the base (a,a).[19]

In later Iranian double domes, the outer shell was often given a bulbous form as seen in Figure 1.10. When this was done, its stability was partly ensured by means of circumferential and radial ties – the latter in the void above the now much lower inner shell. Of greater importance

the time of its construction and its subsequent influence, was Brunelleschi's octagonal domical vault over the crossing of Florence Cathedral, Santa Maria del Fiore, commenced in 1420 and completed, save for the lantern over the central eye, in 1436 [7.15-7.17]. Well before much detailed thought was given to the design of this dome, the tall octagonal drum from which it rises had been completed. Internally this had a diameter of 42 m between faces and 45.5 m between corners, so that the mean diameter marginally exceeded and the critical corner diameter substantially exceeded that of the Pantheon dome. It was 4.2 m thick. For the sake of appearance, if nothing else, the dome almost

fairly widely spaced walls built up from this vault. Only at the top were its slabs set directly on the lower vault. With this precedent alongside, it seems to have been a fairly natural choice to construct the cathedral dome also with

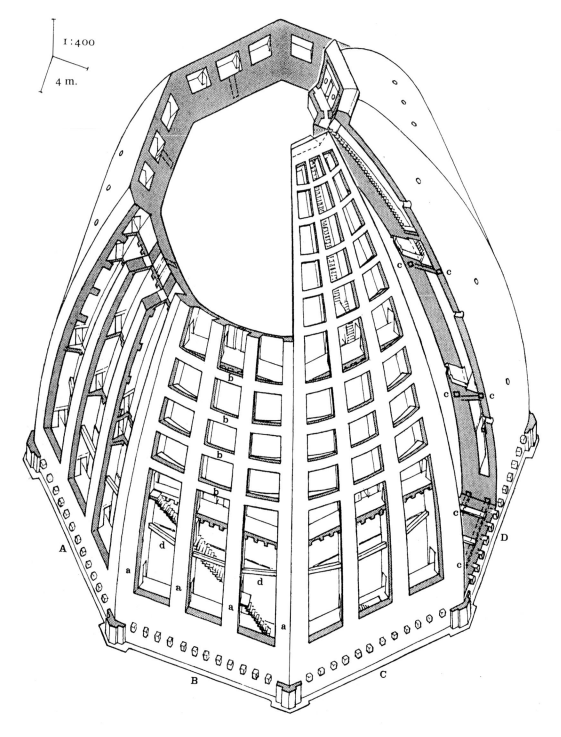

1:400

4 m.

A

B

C

D

Figure 7.15 Partly cut-away view of the dome, Florence Cathedral. In the sections above A, B and C the outer shell has been cut away between the radial ribs (a,a). The circumferential arched ribs (b,b) have also been cut away above A. (It should be noted that, in the actual dome, both sets of ribs are bonded integrally into the outer shell.) Above D is a complete cross-section through both shells in which three pairs of masonry and iron ties (c,c) are seen in section at the base and at the levels of the two internal walkways. A single timber tie-ring runs between the two shells at (d). (Source: author)

had to conform closely to it, both internally and externally, at its springing. To have given the dome a solid thickness of 4.2 m, even if this was reduced progressively towards the top, would, however, have made it excessively massive by any standards. The earlier octagonal Baptistery already had a double dome, its outer pyramidal roof being supported on the vault below by means of stone barrel vaults spanning between

two shells for most of its height.

The written specification of 1420,[20] in recording the decision to adopt the double-shell form, referred particularly to the first and last of its advantages already noted – namely improved weather protection and a more imposing external aspect. In a dome of this size and octagonal rather than circular in plan, the implications of the choice were, nevertheless, far from being as simple as in the smaller domes so far considered. A further complication was the decision to dispense, if at all possible, with the temporary support of centring – the design and erection of which probably presented as daunting a prospect as the construction of the dome itself.

The form, as finally evolved, was much too complex to be described in detail here.[21] Its most essential features are shown in the partly cut-away sketches in Figures 7.15 and 7.16. It will be noted first that the inner shell was made much thicker than the outer one. It seems virtually certain that it was given this thickness in order that, at every stage of construction, the topmost section should contain within itself a complete circular ring as a guarantee of its stability without extraneous support. Without referring directly to this dome, Alberti seems to

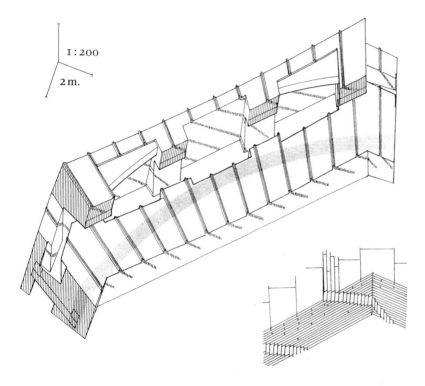

1:200

2 m.

Figure 7.16 Detail of the construction of the dome of Florence Cathedral. The outer shell is cut at a higher level than the inner one to show the first ring of circumferential arches. Stippled shading on the cut surface of the inner shell denotes the circular dome that exists within the octagonal one. (Source: author)

say as much when he writes that 'You may likewise turn the Angular Cupolas without a Centre, if you make a perfect one in the Middle of the Thickness of the Work'.[22]

Above the level up to which (by virtue of the small inclination of its bed joints to the horizontal) the masonry would be prevented from slipping during construction by friction alone, it is composed almost throughout of large flat purpose-made bricks. These were laid circumferentially in true conical beds interrupted, in the manner shown in Figure 7.16, by others set vertically in radial bands never more than about 1 m. apart. These bands spiralled upwards as construction proceeded, projecting like teeth from the conical beds. They thus served three purposes: they keyed successive conical beds to what had already been constructed; they divided each bed into voussoir-shaped sections; and they limited to two or three the number of bricks that

had to be laid together before they were prevented from slipping by adjacent teeth. At most it should therefore have been necessary to provide some light temporary support to each of these short runs of normally bedded bricks while they were laid. Taking into account also the relatively slow rate of construction (sixteen years to complete the dome proper), the keying together of successive courses should also have allowed the development of sufficient hoop compression at each working level in the inner shell to make centring unnecessary.

The outer shell was linked to, and partly supported by, the inner one by means of the radial ribs (a,a). They were constructed integrally with the shells, even to the extent of continuing into them the spiralling bands of bricks set on edge. In order, though, to make the outer shell more self-supporting during construction, the circumferential arched ribs (b,b) were also introduced. These merge with the outer shell in the central part of each face where the diameter is least (and probably have no independent existence there though they have been sketched as if they do) but approach the inner shell near each corner so that, in the words of an amendment of 1426 to the original specification that called for the first of them, 'the live arch should be complete and not broken' – in other words so that there should be a complete circular ring formed by each set of ribs, as there was, throughout its height, within the inner shell.

The records agree that construction was indeed completed without the use of any substantial supporting centring; and analysis of the stresses throughout construction (in terms of modern theory) confirms that, with the slow rate of construction, this should have presented no difficulties.[23] Brunelleschi, of course, lacked this assurance; so it does nothing to belittle his achievement, which was, and still must be, counted as one of the greatest in all structural history. He seems, moreover, to have anticipated in an almost uncanny way just what sort of provision needed to be made to resist the tensile hoop stresses that slowly developed in the lower parts of both shells as construction advanced higher up. At three levels he built in continuous 'stone chains' (c,c). These consisted of large sandstone blocks with superimposed iron chains. In addition, the entire lower part of the dome was built of large well-fitted blocks of stone, which, merely by virtue of the lapping of their vertical joints and the large frictional forces due to the weight above, should have been well able to resist the tensions without further assistance after the completion of construction. (It may also be noted that these tensions were roughly halved by the adoption of a fairly steep pointed profile rather than a semicircular one, though this choice was more likely to have been based on its obvious advantage in easing the problems of doing without centring.) The radial cracks that do now run down three faces of the dome have probably

Figure 7.17
Florence Cathedral.
(Source: author)

arisen more from the effects of repeated expansion and contraction due to changes in temperature than from excessive tensile hoop stresses due to dead load.

The double dome of St Peter's in Rome was modelled fairly closely on the Florentine dome, except that its circular plan rendered unnecessary the expedients adopted by Brunelleschi on account of the polygonal plan of the latter. Thus, with an internal diameter only slightly less than the mean internal diameter of the Florentine dome, the inner shell was not much more than half as thick, there were no circumferential arched ribs to help stiffen the outer shell, and the radial ribs were reduced in number (from twenty-four to sixteen) and relative thickness.[24] Why they were not omitted is not entirely clear, since, with a circular plan, there was no more need for them in the finished structure than in the early Iranian double domes and many later ones.

Apart from the influence of the Florentine precedent, the reason probably lay in their value in allowing much faster completion by first rapidly constructing the ribs on centring and then adding the intervening sections. In the earliest account of the construction, Fontana both states that this was the procedure adopted and reproduces a design for the centring. Confirmation is provided by the fact that construction was completed in not many more months than the Florentine dome took years.

Some advantage was taken of the double construction to make the outer dome rise more imposingly above the other roofs of the church. To increase its height still further, the springing-line was raised well above the cornice of the already high and slender drum. This meant that when, in spite of the incorporation of circumferential iron chains, it nevertheless cracked radially, its outward thrusts proved to be excessive for the limited buttressing capacity of the drum. Equilibrium was more precariously re-established only by much more widespread cracking and deformation [16.6]. Extensive remedial works had to be undertaken in the latter part of the eighteenth century.[25]

The majority of later double or multiple masonry domes were constructed with more widely spaced shells that were, structurally, largely or wholly independent of one another. This gave much greater freedom in pursuing the typically Baroque desire for impressive height externally. At most, usually, the shells were braced against one another by struts and ties hidden from view in the intermediate voids in much the same way as the high stilted outer timber domes were, in the thirteenth century, carried over the low inner brick domes of St Mark's in Venice.

The individual shells, although they never approached the spans of the Florentine dome or that of St Peter's, were notable chiefly for their increasing daring in approximating to true shell forms with thicknesses an order of magnitude less than those required for stability as a ring of arches independently spanning the diameter. To do so with safety, they relied more and more on circumferential iron ties to resist the hoop tensions in the lower parts or to prevent spreading of the base under the thrusts generated by inclined springings like those of the intermediate brick cone at St Paul's, London [12.32]. The reductions in thickness also made this easier by reducing the weights and, thereby, the tensions or thrusts, and it seems likely that this was becoming recognized. On the other hand, the full potential of the true domical shell was clearly still not recognized.

For the final stage in the development of the masonry dome, it is necessary to turn to the work of the Guastavinos, father and son, on the eastern seaboard of the United States in the late nineteenth and early twentieth centuries.[26] They perfected a technique of light centreless vault construction that had already been traditional in Spain and Italy for some centuries and had been introduced into the south of France somewhat later. It used flat tiles that were set tangentially to the surface of the vault rather than at right angles to it and joined by a fast-setting mortar where they were butted against one another. The requisite stiffness was obtained by adding additional layers or plies, taking care to break joints in both directions in successive plies. Continuity was also obtained by virtue of the cohesiveness of the mortar and by spreading a substantial layer of this between the successive plies. The final appearance was very similar to that of the vault shown in Figure 3.8, though it is not known whether the visible surface here typifies the underlying construction or is merely a weathering surface. The chief contribution of the elder Guastavino to the technique was to experiment with and introduce the use of Portland cement mortar in place of the earlier lime and gypsum mortars. It also now became possible to use mild steel bar as a much more effective reinforcement against tension within the vault and against spreading of its supports.

The major achievement of the younger Guastavino was the construction of the temporary dome over the crossing of the Cathedral of St John the Divine in New York City in 1909. This was done using templates and measuring wires to control the form but no supporting centring.[27] Its method of construction apart, this dome really looked forward to modern shell construction in reinforced concrete. It was a flattish part-spherical dome with merging pendentives, with a radius of curvature of 20.2 m and a diameter, at the top of the pendentives, of 30 m At that level, six 19-mm steel rods were embedded in concrete as a circumferential tie, and similar rods were embedded at intervals in the pendentives. The dome itself was six tiles thick (190 mm) at the foot, but soon diminished to five, four, and three tiles (95 mm). Its average thickness was, therefore, only about 1/250th of the span.

Concrete and masonry groined and ribbed vaults

In the simple groined vault, the support roles of the missing triangular sections of the two intersecting barrels are taken over by diagonal arches effectively created at the groins within the thickness of the vault. These 'arches', of course, carry a much more concentrated load than the rest of the vault as they approach the springings. If the thickness is uniform throughout, the maximum compressive stress in them will, very approximately, exceed the maximum

of the domes already considered, with the concrete immensely thick in the lower parts. Problems of excessive stress in the groins never arose, therefore.

While it was difficult enough to cut the blocks for the groins of a stone vault, it was even more difficult to form adequate groins in a brick vault of the same surface geometry as the Roman concrete vaults unless they were constructed as virtually independent diagonal ribs. Many Romanesque groined vaults were, therefore, constructed in a manner not very different from early Roman concrete construction. The

Figure 7.18 Small vault of the library, Friday Mosque, Isfahan. (Source: author)

stress elsewhere in the vault in the ratio of the span to the thickness. On the other hand, the groins are naturally stiff like any other crease in a surface, so that there is little risk of their buckling.

The early development of such vaults closely paralleled that of the dome. To construct them in cut stone called for very complex cutting of the blocks forming the groins if these were to have the requisite strength. This is well exemplified in the vault of the lower storey of the Tomb of Theodoric at Ravenna. Casting throughout in concrete circumvented this difficulty. Groined vaults accordingly became commonplace in and around Rome during the first century A.D. Later, embedded brick ribs were introduced at the groins in much the same way as they had been introduced into the continuous surfaces of concrete domes and normal barrel vaults, but with the added justification here that they did follow lines of maximum stress. Good examples may be seen today in the Severan substructures to the southeast of the Palatine [6.11] and in the Baths of Diocletian. Spans exceeding 20 m were achieved; but the proportions remained similar to those

blocks were laid voussoir-fashion, but were only roughly cut to shape before being bedded in large quantities of mortar. Though they never approached in size the larger Roman vaults, they were, in proportion, equally massive; and they must, in the same way, have demanded substantial centring and formwork.

One way of reducing the difficulties and the need for centring and formwork was to depart from the surface geometry of two interpenetrating horizontal barrels in favour of a more domical form. This is illustrated by the early brick vaults in Figure 4.18. Where the supporting arches had a greater rise, the rise of the crown of the vault above them was correspondingly greater. In this form, the groins projected noticeably only near the springings and fairly soon died out above into the doubly curved surface of the whole vault. Where they did project, they were near enough vertical to be constructed without centring. Above this level, only the individual courses of brick or stone called for temporary support as in a true dome. The structural disadvantage of this expedient for applications such as the vaulting of church naves was that the arching along the axes of the

130

two interpenetrating vaults led to outward thrusts around the entire perimeter of the vault. Aesthetically it had the further disadvantage, for such applications, of over-emphasizing the individuality of each bay (though Angevin Gothic builders seem to have thought differently, influenced probably by a local domical tradition).

The alternative, whose general adoption by Gothic builders had much more far-reaching consequences, was to construct diagonal ribs at the groins before filling in the intervening webs or severies. Ribs constructed in this way, rather than *pari passu* with the rest of the vault and embedded in it as were the Roman ribs, had been used in the Islamic world since the tenth century if not earlier, but only in connection with domes and domical vaults [1.11 and 7.18] in which there was no real need for them structurally. As already noted in Chapter 1, they were probably regarded primarily as decoration – though, as present practice clearly shows, they would also have been valuable in a largely tree-less country like Iran as a means of economizing on centring and formwork. Once the ribs had been constructed on light centring, the webs could easily have been filled in freehand as they frequently are today.[28]

When applied to a groined vault, such ribs had a much larger role to play. Initially, they were probably still thought of primarily as cover strips for any unsightly irregularities in the groins, and only secondarily as aids to construction and as potential stiffeners in the event of any weakness arising in the vault proper. As fully developed in the latter part of the twelfth and early thirteenth centuries, they became an indispensable constructional aid as well as being of major importance both structurally and aesthetically. Indeed, as was also pointed out in Chapter 1, these aspects are not really separable except in retrospect.

Together with the adoption of much more adaptable pointed as well as semicircular profiles for the boundary arches and the ribs, the ribs facilitated construction in three ways. First, they continued to serve as cover strips to the groins and to simplify the cutting of the web stones there. Second, they simplified setting out (and simultaneously increased the freedom of planning) by transferring the emphasis from the cross-sectional profiles of the intersecting barrels to the profiles of the bounding arches and the ribs, leaving the webs to accommodate themselves to these latter profiles in any way that was convenient. Third, they again greatly reduced the need for centring. Full centring was required only for the bounding arches and ribs, and then only above the *tas-de-charge*, which consisted of blocks built out horizontally for some height above the springings.[29] The webs, which were usually appreciably thinner than in earlier groined vaults, were normally arched up sufficiently in the direction of spanning between the arches and ribs to make each course self-supporting when complete and to call, at most, for a light temporary support moved from course to course as the work proceeded.

In the completed vault, the potential structural role of the diagonal ribs in strengthening and stiffening the groins will be obvious from what has already been said. Their actual role has, nevertheless, been much debated, particularly in the face of the apparently conflicting evidence of some vaults from which the webs have largely fallen leaving the ribs standing [7.19] and of others, seemingly very similar, from which the ribs have fallen leaving the webs standing.[30] The simple explanation is that the vault as a whole was a statically indeterminate system. The ribs and arches had to be capable of standing alone to allow construction to proceed; the webs, provided that they formed continuous groins over the tops of the ribs, would also have had this capability as simple groined vaults. (Stresses in the lower parts of the groins would usually have been reduced to safe levels by the practice of filling the narrow space above the

Figure 7.19 South aisle vault, Ourscamp. (Source: author)

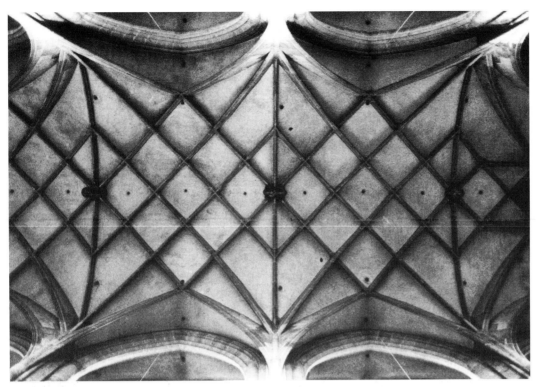

Figure 7.20 Nave vault, St George, Dinkelsbühl. (Source: author)

tas-de-charge with mortared rubble to assist in transferring the vault thrust to the wall or buttress.) As in any other statically-indeterminate system, they would, while both remained in place, have shared the load in proportion to their respective stiffnesses. But they would have done this to an extent determined also by the precise sequence and manner of construction – and therefore impossible to determine today merely by inspection of any particular vault unless very pronounced cracking of the webs or separation of them from the ribs shows only one mode of action to be possible. It is nevertheless a reasonable assumption, on the basis of likely sequences of construction, that the diagonal ribs would usually have taken much of the load. The transverse arches dividing bay from bay can, on the other hand, have played little structural role after completion of the webs on both sides unless the vaults were highly domical. There would otherwise have been no significant groins for them to support. Their presence and the great emphasis often given to them [7.19] could then be justified on non-aesthetic grounds solely by their value during construction in making it easier to complete one bay at a time.

The subsequent elaboration of the basic diagonal rib system by adding more and more ribs would similarly have helped during construction, in that it reduced the sizes of the compartments between the ribs and thereby simplified the construction of the webs. It would also have served to stiffen the webs without calling for them to be arched up in the direction of spanning.

In the fourteenth, fifteenth, and sixteenth centuries, particularly in central Europe but initially in the west of England, the elaboration took a new turn in which the identity of the individual bay tended to be lost. In its place, the continuity of the whole vaulting-system was emphasized by ribs spreading over its surface as a continuous net [7.20] or even as a net that was almost randomly disrupted [7.21].[31] Leonardo, when sketching a vault like that in Figure 7.20, thought that the outward thrust would be reduced by the reduction in the angle at which the ribs approached the lines of supports.[32] This was a mistake. The thrust exerted could be significantly reduced only by reducing the weight of the vault, by concentrating it nearer the supports, or by changing the whole surface geometry. Decorative effect seems, in fact, to have become the major concern, though one must admire the assured technical mastery that made it possible to achieve it in these ways.

The Baroque vaults of Guarini and Neumann similarly pushed technical mastery to its limits

Figure 7.21 Stairway vault, Vladislav Hall, Prague. (Source: author)

for the sake of visual effect. In the design of their doubly-curved ribs, the main interest was in the three-dimensional geometries of the interpenetrating surfaces making up the vaults. Neumann's drawings for the vaults at Neresheim [1.15] show that he intended them to be constructed of tufa reinforced by iron tie bars. A lighter form of construction was adopted after his death.

Masonry spires and fan vaults

Before we turn to the modern successors of these concrete and masonry forms, brief mention should be made of the spire and the fan vault. Structurally the spire may be regarded as one limiting case of the domical vault, and the fan vault as a set of splayed-out inverted spires, each usually sliced in half and each half carried by one of the supports.[33]

If the typical spire – a slender eight-sided pyramid – is compared with the octagonal domical vault over the crossing of Florence Cathedral, the structurally most significant difference is that the sides never curve in towards one another. Because of this there is no tendency for them to burst outwards in their lower parts if they are adequately held at the base, and thus no hoop tension. If horizontal rings can be inscribed within their thickness, these will be in compression at all levels under dead load. For the same reason there is no need for centring during construction – an advantage which may have led to the nave of St Ours at

Figure 7.22 Salisbury Cathedral. (Source: author)

Loches being vaulted, in the twelfth century, with two squat octagonal masonry spires and to the use of spires as the principal vault of some later Russian churches referred to in Chapter 12. The price to be paid for the absence of tension was an inevitable horizontal outward thrust on the supports. For tall slender spires the thrust was, however, kept within manageable limits by supports by the usual *tas-de-charge*. The vault of Henry VII's chapel in Westminster Abbey [7.23] differs in that the individual fans are differently sliced and the vertical loads of the principal ones are carried to the walls at a lower level by largely hidden transverse arches. Large pendant voussoirs at the points of support create the illusion that the vault simply hangs in the air.

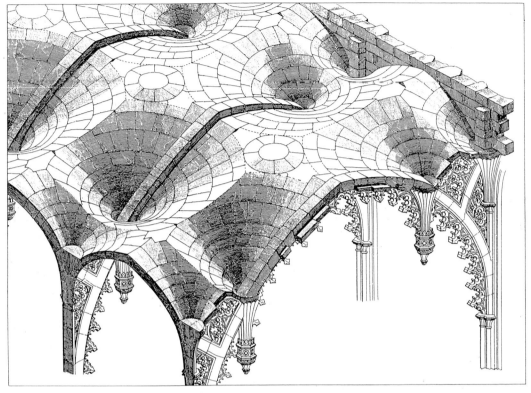

Figure 7.23 Drawing by Willis of the vault of Henry VII chapel, Westminster Abbey. (Source: author)

the near-verticality of the sides and the thinness of the masonry. Over the 53 m height of the crossing spire of Salisbury Cathedral [7.22] this averages less than 30 cm. The thrusts have been contained by iron ties at the foot.

If a spire were simply inverted and balanced on its point, the downward-acting compression would increase there to a very high value and the hoop compression would become a hoop tension throughout. The effect of splaying it out at the top would simply be to call for a larger hoop tension there giving way to a hoop compression lower down. Since normal masonry cannot develop such a tension, equilibrium can be established only by applying inward forces around the top.

When, in the fourteenth century, one development of the basic ribbed vault led, by multiplication of the ribs emanating from each support, to the fan vault, these inward forces had to be provided by adjacent fans leaning, in effect, against one another and by the supporting walls where there were no adjacent fans. A statical balance was achieved there by forces exerted inwards on the lower horizontal rings by the walls and the opposing action of the rings constituted the outward thrust of the vault. Excessive vertical compression at the foot was avoided by channelling the compression into the

In passing, it might also be noted that the trumpet-shaped fans of the fan vault were the first doubly curved structural elements with anticlastic curvatures – convex one way and concave the other.

Thin reinforced-concrete shells

Although there were a few earlier experiments, the real further development of the thin domical shell beyond the point to which it was taken by the Guastavinos began in a rather fortuitous manner with the construction of a planetarium dome and then of a larger and flatter dome at Jena in the early 1920s.[34] These were constructed by first erecting triangulated nets of light steel bars, then suspending interior formwork from these and spraying on a thin layer of concrete – just sufficient to envelop the bars and a much finer wire mesh stretched between them. Some circumferential reinforcement was incorporated near the base. On completion, the ribs projected slightly externally, but the rest of the concrete was only 30 mm thick in the first dome and 60 mm in the second for spans of 16 m and 40 m, respectively.

The chief difficulty that was encountered arose during the earlier stages of construction of the triangulated network of the second, much

134

flatter, dome. This proceeded upwards and inwards from the base, and the free edge showed a marked tendency to undulate as the work advanced so that temporary props had to be placed below it until, with the narrowing of the opening, the tendency disappeared. Its cause was identified as the incompatibility between the contractions due to the circumferential compressions of the network and the extension of the tie ring at the base as the outward thrusts developed the tension in it.

The principal reason for adopting the triangulated net of bars for the planetarium dome had been the way in which, with very accurate cutting of the bars, it allowed the exact hemispherical form to be achieved with relative ease. It was, on the other hand, a wasteful kind of reinforcement for the concrete. Once the possibility of constructing such shells had been demonstrated, the logical step was taken of substituting normal centring to carry the formwork, and normal reinforcement, aligned with the principal stresses, to take the tensions. An excellent example of this phase of development, still standing and marred only by its architectural detailing, is Torroja's slightly larger

also being constructed.[36] Thus, by the mid-1930s, most of the basic possibilities of using reinforced concrete to construct true shell forms that fully exploited the inherent stiffnesses of double curvature had been explored, including the use of prestressed ties where necessary. It remained only to exploit these possibilities more fully architecturally, since the development so far had been wholly the work of engineers and, with few exceptions, had been limited to utilitarian structures like factories and market halls.

Architectural interest was really awakened about twenty years later. (Gaudi's much earlier proposals to use forms very similar to the hyperbolic paraboloid for the vaults of the Church of the Holy Family [5.4] remained unexecuted and stimulated no wider interest at the time.) This awakening may be represented by Saarinen's Kresge Auditorium, Yamasaki's St Louis Air Terminal, and Candela's Church of the Miraculous Virgin, all constructed in 1954-5.[37]

The part-spherical form of the shell that roofed the first was chosen almost entirely for aesthetic reasons. It was exactly one-eighth of a full sphere and was supported at only three

Figure 7.24 Market Hall, Algeciras (Torroja). (Source: author)

(47.6 m. span) and still flatter dome constructed in 1933 for the market at Algeciras [7.24].[35] This was carried on eight isolated supports. Between these, narrow sections of flat cylindrical vaults projected slightly from the dome proper to stiffen the unsupported edges, and, for the first time, the continuous tension-ring just below them was fitted with turnbuckles so that it could be prestressed to do away with the incompatibility of deformations that had caused trouble previously.

Soon afterwards, the theoretical and practical advantages of the hyperbolic paraboloid having been realized (ideally a uniform state of stress throughout the shell, coupled with the possibility of generating the surface by straight lines only [4.4]), thin shells of this form were

points [7.25]. This entailed a considerable departure from the ideal state of stress for a simple membrane [2.14(b)]. The cutting away of large parts of the shell between the supports and the absence of direct support along the cut edges resulted in considerable bending actions even at some distance from the edges. These called for a greater thickness of concrete than would normally be required for the span, as well as for fairly deep edge beams and very heavy reinforcement near the supports. The outward thrusts were resisted by large concrete buttresses below ground level. To allow the shell to 'breathe', there was no direct contact between it and either the external glazing or the internal partitions.

The St Louis Air Terminal was roofed by means of three structurally independent thin

Figure 7.25 Kresge
Auditorium, MIT,
Cambridge, Massachusetts
(Saarinen, Ammann and
Whitney).
(Source: author)

groined vaults set side by side and separated by triangular panels of glazing [7.26]. Both the free edges and the groins were stiffened and strengthened by external ribs. Since the diagonal ribs here naturally collected most of the loads, the four-point support of each shell was quite logical. The outward thrusts were resisted by ties running between the supports below floor level.

In the church [7.27], each main column carried four hyperbolic paraboloid shells, two of them spanning half the width of the nave and the other two spanning the side aisle, and each pair meeting in a sharp groin on the transverse axis of the column. Adjacent pairs similarly met in

sharp groins midway between the transverse axes of the columns. Structurally, the compressions in the groins on the axis of each column must lead to considerable departures from the ideal state of uniform stress [5.5(o)]. But the spans were much less than in the other two structures (the width of the nave was 11 m. whereas the spans of the Kresge Auditorium dome and the St Louis vaults were 48 m and 36.6 m, respectively) so that it was possible to absorb the stresses within the minimum thicknesses of concrete that were convenient from the point of view of construction.

Two years later, Nervi's smaller Sports Palace [1.2 and 13.6] and Esquillan's CNIT

Figure 7.26 Upper
concourse, Air Terminal,
St Louis, Missouri
(Yamasaki, Roberts and
Schaefer, Becker).
(Source: author)

Figure 7.27 Church of the
Miraculous Virgin,
Mexico City (Candela).
(Source: author)

Hall [7.28] broke new ground structurally in two
ways.[38] In the first, and in the larger Sports
Palace constructed a year later, Nervi adapted to
the requirements of large flat domes the
methods adopted earlier in Turin [4.8] to do
away with the need for soffit formwork. The
entire soffits were formed from precast units
which also served as stiffening ribs for the thin
in-situ shells subsequently cast on top. (Somewhat similar methods were also used at about

this time for the construction of more utilitarian
domes in Algeria and in Russia.)

In the CNIT Hall, a complete double shell
was constructed to obtain the necessary additional stiffness required on account of the great
span (206 m between each pair of corner supports), the relatively low rise, and the basic
choice of form. This last approximated to a triangular groined vault with three horizontal
ridges and very slight circumferential curvatures

Figure 7.28 CNIT Hall under construction, Paris (Camelot, de Mailly and Zehrfuss, Esquillan). (Source: Bahr)

below the ridges so that the loads would be transmitted as directly as possible to the three supports instead of having to be transmitted partly by edge beams as in the Kresge Auditorium shell. The action was therefore largely that of fans of arches converging on the supports. To make the two shells (mostly only 60 mm thick) act in unison, they were connected at intervals by precast concrete webs and transverse diaphragms of the same thickness. They were also given additional stiffness by corrugating them between each pair of webs, i.e. transversely to the main arching action. The construction had to be very carefully phased to maintain symmetrical distributions of load and to control the deformations. For the latter purpose, the ties set between the supports and below ground level to resist the outward thrusts were also progressively prestressed as the temporary support of the centring was removed.

Other examples, to be discussed in Chapter 13, show a little more of the characteristic versatility of the doubly curved shell form. In particular, Saarinen's TWA Terminal at Kennedy Airport illustrates the extent to which, given a sufficiently skilful engineer alongside the architect, a free sculptural modelling of the form is possible without any need for the massiveness of Roman concrete vaults. On the other hand, Utzon's original proposals for Sydney Opera House serve as a reminder that there are limits, set by the inviolable laws of static equilibrium, beyond which it is impossible to go without calling for more than a thin shell.

More recently, other limits have increasingly

been set by the rising cost of labour and hence of conventional formwork. Attempts have been made to circumvent them by using inflated forms, though these impose their own limits on feasible shapes. Isler, in Switzerland, has taken another approach. He has continued to use timber forms and centring, and by rationalizing the construction procedure and making repeated use of the supporting centres has been able to built a large number of very elegant shells of more modest size. These have been designed with the help of simple funicular physical models to give little or no bending under self-weight load.[39] But in general since the 1970s, shells have tended to be supplanted by their purely tensile counterparts or by space frames.

Timber counterparts

Before leaving them for the tensile counterparts, some note should, however, be taken of timber shells. Passing reference was made earlier to framed domes of timber such as that originally covering the Dome of the Rock in Jerusalem. Such domes must have been far more numerous than the present survivals, which are mostly only the outermost elements of double or triple domes otherwise built of masonry. They all seem to have consisted of radial and circumferential members arranged as in the simple hut in Figure 1.1 and carrying an outer weathering surface, though they were sometimes braced or partly carried by the masonry dome below.

Only since the 1950s have more truly shelllike timber vaults – vaults not relying on ribs as their main supports – been built. To construct such vaults it was necessary to laminate two or more continuous layers of planks with the planks in the different layers set roughly at right angles to one another so that stress could be resisted in all directions. Surfaces that could be generated by straight lines – such as hyperbolic paraboloids or conoids – were easiest to construct, and most timber shells were of these forms. Since it was easier to bend a plank to a slight curve than to twist it and hold it twisted until finally fixed, it was usual, however, to set the planks of a hyperbolic-paraboloid shell to follow the principal curves rather than the straight-line generators.

A further possibility is to deform a timber lattice to a shell-like form. This does not call for centring. Deformation can be imposed by lifting part of it in such a way that the laths are bent to the desired curves. If the mesh is originally square, the squares then become lozenges. After deformation, the deformed shape is maintained by attaching all the laths to suitable boundary members. Stresses are then carried by the laths acting as individual arches. There will be little or no bending to be resisted if the deformed shape is a funicular one, like the inversion of Gaudi's hanging chain model [5.3]. The chief hazard, as in all compression structures, is buckling, especially under varying distributions of imposed load.

After Otto had built some relatively small domes in this way in the 1960s at Essen,

Berkeley and Montreal, the real demonstration of what could be done came in 1973-5 with the construction of a large pavilion for a federal garden show in Mannheim [7.29]. This pavilion consists of two large halls and connecting walkways on an irregular curving plan, all covered by a single continuously curved roof. A hanging-chain model gave the essential configuration. Much more was required to obtain the precise geometry, confirm the structural behaviour under wind and snow loads as well as under self-weight, achieve adequate stiffness to avoid a buckling failure and devise all the necessary details of connections and boundary members. To achieve the necessary freedom to deform from a flat regular lattice required flexible laths of relatively small cross-section and pinned joints at all intersections. After erection, the joints had to be bolted tight, and sprung washers were interposed to ensure that they would remain tight. But this alone would not have given sufficient stiffness. That called for a second lattice over most of the surface to give a greater overall depth, and for ingenious detailing to allow the two lattices to slide in relation to one another during erection but not afterwards. Lifting the lattice to its desired final configuration was performed using spreader beams, scaffold towers and fork-lift trucks. Bolting-up and connection to the boundary elements followed in a controlled sequence, followed by covering with a translucent skin. [40]

Figure 7.29 Timber lattice roof for the Mannheim Bundesgartenschau (Otto, Happold). (Source: author)

Slung membranes and cable nets

The first tensile counterpart of the dome or vault– the slung membrane of the first tent – may not have made its appearance until some time after the first simple domical huts. It is a more complex form whose construction would have called for a somewhat wider-ranging technical skill on the part of the builder. It was also much slower in developing further. In spite of widespread experience with sails and some further experience with shade awnings such as the Roman *velaria*, it had hardly developed beyond the large circus-tent until the early 1950s. In several exhibition halls designed by the Russian V. G. Suchov [13.22], the canvas of the tent was replaced by nets of iron bars riveted together at their principal intersections. But that was all.

There were three main obstacles to be overcome. The first was the relatively low tensile strength of most fabrics and the considerably greater difficulty of making effective joints to transmit tension than of making joints subject primarily to compression. The second was the inherent flexibility of thin, singly curved tensile forms and their consequent liability to flutter and billow in the wind. The third, associated with the first, was the difficulty of overcoming the flexibility by adopting forms with marked double curvature without, thereby, introducing excessive tensions at some points.

The much greater importance of fairly uniform stresses to utilize fully the available tensile strengths of the thin tensile membrane is, from the point of view of design, its main difference from the compression structure typified by the Gothic cathedral or Gaudi's design for the church of the Sagrada Familia. In these structures, thicknesses are sufficient to accommodate a satisfactory system of internal thrusts even if there are considerable departures from an ideal geometry. Stresses are also low enough for it to matter little if they are far from uniform. Even in the reinforced-concrete shell, quite large variations can be accommodated by varying the thickness or reinforcement. For the membrane, they do matter and choice of surface geometry is therefore more restricted. Moreover, the only doubly curved membrane that can match the ability of the singly curved membrane to assume automatically an ideal configuration is the soap bubble or soap film, which stretches or contracts until the surface tension is uniform.

Only since the 1970s has the obstacle of limited fabric strengths been largely overcome. In all large slung roofs prior to this, a continuous fabric had to be replaced as the principal load-bearing medium by an equivalent network of wire cables, by single parallel strands in singly-curved roofs, or by double-layer arrangements of cables held apart by vertical struts and tensioned against them.

The last of these approaches involves, of course, a considerable departure from the single membrane and really leads to a hybrid form.

One set of cables is convex, giving a dome-like exterior over a circular or oval plan when covered. But these cables are held up and kept in tension by the struts, which are supported, lower down, by the other principal cables through which the tensile prestress is applied. The geometric similarity to the dome (though not reflected in the structural action) is greatest in the system originally patented by Fuller in 1964 as the 'aspension' dome, but developed largely by others including Geiger and Levy.[41] In this form, only the upper cables are continuous. Circumferential and diagonal cables complete the structure, giving it some of the character of a space frame. In an alternative form, continuous upper and lower radial cables span, like the spokes of bicycle wheel, between an outer compression ring and a deeper central tension ring and are held apart by this latter ring and the vertical struts. This form was adopted by Zetlin in 1957 for an auditorium in Utica and shortly afterwards for the US pavilion at the 1958 Brussels exhibition. [42]

When only single parallel strands have been used, stiffness has been ensured by means of auxiliary guys, by the dead weight of a heavy covering (sufficient to make varying wind loads relatively unimportant), by encasing the strands in concrete to give them the stiffness of inverted arches, or by concreting the whole roof to give it the stiffness of an inverted barrel vault. The roof of Saarinen's Dulles Air Terminal [7.30] illustrates two of these approaches. Here Ammann and Whitney, the engineers, first encased each pair of suspension cables (slung at 3 m centres between the horizontal beams carried by the inclined piers) in reinforced-concrete ribs. To the stiffness of these ribs was then added the further stiffening effect of the dead weight of precast panels spanning between the cables to constitute the roof surface.[43]

More frequently, cable networks have been used and their stiffness has been ensured by double curvature and prestressing. One set of cables has been slung between peripheral arches, cables or trusses to carry the whole weight and the additional forces resulting from prestressing, and the other set, intersecting the first one at right angles, has been tensioned against it to hold it steady [5.5(n)]. The avoidance of excessive local tensions calls for fairly constant curvatures. As with a simple catenary, the flatter the curvature, the greater must be the tension to support a given load. It also calls for very careful attention to the geometry of the peripheral elements between which the roof is slung, particularly if these are themselves flexible cables whose tension will similarly depend on their curvature.

Before computers came to the designer's assistance, it was possible, in roofs of moderate span, to achieve a satisfactory geometry by adjustments of cable lengths during erection. To help in choosing and defining suitable configurations that were also visually satisfying, Otto,

Figure 7.30 Terminal
Building, Dulles Airport
(Ammann and Whitney,
Saarinen).
(Source: author)

the most adventurous pioneer designer of such
roofs, also made considerable use of preliminary
models such as soap-films stretched directly
over wire frames or between flexible threads.[44]
For similar roofs of greater span, the large
numbers of intersections make in-situ adjustment impracticable, and scale models cannot
give a sufficiently precise definition of an
acceptable configuration. More accurate theoretical analyses must be made, for which powerful
computers are essential.

The development of doubly-curved nets
largely began with the construction in 1952, by
Nowicki and Severud, of the large saddle-
shaped roof of the Raleigh Arena, North
Carolina [7.31, 7.32].[45] It spanned almost 100 m
between two reinforced concrete arches inclined
in opposite directions at a little more than 20° to
the horizontal so as to intersect one another
scissor-fashion at the lowest points of the roof.
These arches resisted the tensions of the roof
cables partly by normal arch action, but also
partly by their own dead weight. Although
stimulated by it, Otto concentrated on freer tent-
like forms that were well exemplified by the
German Pavilion at Expo 67, Montreal, for
which the tent shown in Figure 5.10 was a
prototype. In these structures, the cable nets
were stretched between heavier cables attached
at their high points to masts and at their low
points to ground anchorages. As in the Arena, it
was, of course, necessary to suspend a separate
membrane as the weather shield. The further
development of both these types of structure
will be discussed in Chapter 13.

Once suitably strong fabrics that could be
bonded together at seams, and that were capable
of keeping out the weather as well as serving as
the structure, became available, they offered an
attractive alternative. They understandably came
to be preferred to cable nets, though they did not
completely supersede the hybrid forms referred
to earlier. The fabric must likewise be doubly
curved – convex one way and concave the other
way – and prestressed by tightening it in one
direction against the other to maintain the desired shape. There is the same need for precise
shaping to avoid significant variations in stress
over the surface and a further need for precise
cutting of the fabric to give not only the desired
form but also satisfactory alignments of the
individual panels in relation to the weave and
overall geometry. Moreover, since the fabric has
different stiffnesses in the directions of warp
and weft, it is necessary to allow in the cutting
pattern for the different stretches that will occur
in these two directions when the prestress is
applied. Physical models, using stretch fabric
for instance, have been valuable in exploring
possible forms. But, here again, computer analysis has been essential for final design and confirmation of behaviour.

Continuous support for the fabric is required
at boundaries and at any sharp changes of
direction, such as can occur at ridges and
valleys. Direct support by a post is not possible
because it would entail very high local stresses.
Usually the requisite support has been provided
partly by cables slung between posts and/or
from posts to ground anchorages and partly by

Figures 7.31, 7.32 Dorton
Arena, Raleigh
(Nowicki, Severud).
(Source: author)

boundary cables. The stunningly elegant roof the new passenger terminal at Denver International Airport [7.33] is slung between concave ridge cables and prestressed convex valley cables, the former carried by pairs of vertical posts. The fabric is Telfon-coated fibreglass. The necessary prestress was applied by jacking

had a limited use for exhibitions and the like but will not be considered further here. Major structures have been all of the air-supported type.[47]

Ideally, in still air, the air-supported structure carrying nothing but itself can be an almost vanishingly thin separation between the air

Figure 7.33 Terminal Building, Jeppesen International Airport, Denver (Fentress Bradburn, Berger Severud, Birdair).
(Source: Robert Beck)

down the connectors to which the valley cables, boundary cables and diagonal anchor cables were attached.[46]

Air-supported and pneumatic membranes

As compared with the tent, the stretched membrane supported only by internal air pressure is a complete newcomer for purposes of space enclosure. After early successful experience with Radomes in the 1950s, it next became a frequent inexpensive choice for small temporary structures and for covering swimming baths and recreation halls [5.9]. Use for much larger structures has hitherto had a more chequered history. Two possible forms exist. One is the single-skin air-supported structure held up by a pressure difference between the whole enclosed space and the exterior. The other is the double-skin pneumatic structure held up, like a pneumatic tyre, by the stiffening effect of a higher pressure in the space between the two skins. Readily demountable pneumatic structures have

inside and that outside, and only a small excess of internal over external pressure will be called for. In practice, it must be inflated (and kept inflated) by an internal pressure at least marginally higher than the maximum fluctuating external pressure due to the wind and any other loads such as accumulated snow and must resist the maximum pressure difference. Since, also, the fabric of which it must be made is far less extensible than that of a child's balloon, it must be tailored to very nearly the desired final form. The tensions to which it is subjected by the pressure difference will depend, like those in a slung membrane, on this form, chiefly on the local curvatures. The flatter the curvature, the greater they will be.

The ideal form is that of a floating soap-bubble, continuously tied down at the base (i.e. similar to Figure 5.5 (k) but inverted). Other synclastic forms are also possible but will entail varying tensions in the membrane as the curvature varies. The anticlastic curvatures that are essential to give stiffness to light prestressed slung membranes are possible only as transitions between zones of synclastic curvature, just as waists can occur in a child's balloon only between bulbous zones.

These are the basic limitations to the choice of form; and they become more restrictive as the scale increases, the curvatures are thereby made flatter, and the tensions are in turn increased. In most large structures the flattening of the curvatures has been circumvented by associating with the membrane a cable net or ribs tied down to the ground. The internal pressure then stretches the membrane against the net or ribs. The major tensions are taken by the net or ribs, and the tensions in the membrane are greatly reduced because the latter bulges slightly between the cables or ribs and thus has much-increased local curvatures. A wider choice of form thus becomes possible. But this choice, as with the choice of form for thin slung membranes, can never be as free as that of heavier forms acting primarily in compression and not approaching as close to the strength limitations of the materials.

The internal pressure must, of course be maintained by a constant input of energy - a fundamental difference from all the other structures considered in this book. To reduce costs, that input must also be controlled so that it varies according to the outside pressure. Herein lies a considerable difficulty in both snow and storm conditions. Particularly in storm conditions, the outside pressures will then not only fluctuate widely at any place but will do so in very different ways around the structure and sensors cannot be placed everywhere. Numerous collapses have occurred as a result. Part of the answer has been to adopt designs that will minimize wind pressures (as distinct from wind suctions) and reduce the likelihood of accumulations of snow.

The most successful design hitherto has probably been that of the American pavilion for Expo 70 in Osaka [7.34]. This was constructed over a partly excavated elliptical area, 140 m long and 80 m wide, with the excavated material distributed around it to form a berm and deflect the wind smoothly over its shallow profile. The vinyl-coated fibreglass fabric was stiffened by a diagonal cable net made from bridge strand and anchored in a concrete compression ring at the top of the berm. Six 5.5 KW blowers were provided to ensure continued inflation in all circumstances. Revolving doors at entries and exits served as partial air locks.[48]

Figure 7.34 US Pavilion, Expo 70, Osaka (Davis Brody, Geiger Berger, Taiyo Kogyo). (Source: Horst Berger, from *Light Structures*)

Chapter 8: Beams and slabs

Geometrically, the simple beam and slab seen at (q) to (u) in Figure 5.5 are, together with the straight tie, column, and wall, the simplest of all elemental forms. Their primary structural actions, considered in terms of the principal internal tensions and compressions, are, on the other hand, among the most complex. The beam is capable of spanning between supports without exerting any horizontal thrusts only by virtue of containing within itself the actions of both arches and catenaries [2.12]. The slab similarly must contain within itself the actions of both shells and membranes. But all these actions take place beneath the surface and are belied by the simple exteriors.

Architecturally this amounts to a lack of the kind of natural structural expressiveness that is characteristic of all the forms considered in the previous two chapters. The beam, for instance, can never be exploited visually in quite the same way as the pointed arch was by Islamic and Gothic architects. The architects of Classical Greece were content to decorate its surface and allow its horizontality to complement the more dominant verticality of the supporting columns. The spanning capacities of the available stone were fully exploited but no attempt was made to go beyond the natural limits. Only when their neo-Classical successors wished to emulate their achievements without such excellent stone was some structural development called for. More recently, straight horizontal spans and cantilevers and the seemingly floating horizontal planes that they can create have been more directly exploited for architectural ends [1.16]. Nevertheless, the related structural development that has made them possible has been spurred on more by practical needs and a desire to make the most efficient use of new materials and techniques.

The major developments are, in fact, nearly all fairly recent ones. Another reason for this late start was the need to visualize correctly the internal actions brought into play in order to see what was needed and potentially feasible. Thus it was only by drawing upon the deeper structural insights of the past two hundred years or so that it was possible to make the most efficient use of the new tensile materials to resist the catenary-like tensions that always arise.

The earlier developments can therefore be traced very briefly.

Early forms

The widespread use, even today, of logs of whatever timber is available locally to span between two supports and thereby form part of a floor or roof or carry a wall over an entrance no doubt closely reflects very early imitations of the fallen log accidentally bridging a gap. Such logs may be merely cut to length and otherwise left as found. To make a floor or roof, numbers of them are set side by side or a short distance apart, and they are then covered with turfs, reed matting, mud, or some other handy material. In the case of a roof, they need not be set horizontally. A moderate inclination will be more suited to throw off any rain. But it will usually result, structurally speaking, only in adding some axial compression to the primary bending action, and some horizontal thrust to the vertical reactions at the supports.

It is possible that the undressed or very roughly-dressed stone slabs which today still roof large numbers of megalithic tombs similarly reflect very early direct imitations of slabs found naturally in suggestively similar situations. On balance though, a prior use of timber, leading only later to the use of stone as a much more durable material, seems more likely. Certainly the known megalithic remains are much later than the highly-accomplished Egyptian use of dressed stone [1.6], which originally imitated, at Saqqara[1], a prior use of timber logs or palm trunks.

Be this as it may, the earliest forms in both timber and stone would have been true beams, and not merely approximations to the true form as were some of the roughly contemporary forms from which the true voussoir arch later developed. Their chief shortcomings were the inevitable limitations on span that resulted from the limited strength of the available timber or stone and the maximum length in which either could be procured.

A careful squaring of the faces, which was the first and for long the only variation on the presumed original timber form, did nothing directly to permit any increase in span. Indeed, in reducing the original cross-section of a log, it could easily call for a reduction, particularly if, as sometimes happened, the squared timber was then set with its wider faces horizontal to give, perhaps, a better bearing surface at the end.

145

Spans had, therefore, mostly to remain small. This gave rise both to the long narrow rooms typical of many surviving early palace structures and to the close spacing of the columns typical of columnar temple structures [8.1]. Where wide roof-spans were required, it was necessary to import the only timber capable of providing them – the magnificent cedar of the Lebanon and the mountains a little to the north, which was used not only for Solomon's Temple but as far afield as Persepolis. Where greater carrying strength was required, this could, to some extent, be achieved by setting two or more beams side by side [8.1 again], provided that the span was not too close to that at which each beam could only just support itself. Where massive stone lintels were used over spans close to this limit, it became common in the New Kingdom in Egypt [11.13] and in Mycenean Greece [8.2] to reduce the load from the structure above by arranging for much of it to be carried down directly to each side of the opening rather than weighing heavily upon the central part of the lintel. A similar procedure was also widely adopted elsewhere, in India for instance, where the available stone was suitable.

Later timber and iron and steel beams

By the end of the Roman Republic, the spanning limitations of readily-available timber had led to the development of the simple truss for roofing purposes as will be discussed in Chapter 9.

However, its overall height in relation to its span made it unsuitable for floors. To what extent the Romans overcame the spanning limitations by other means is not known. Probably the usual solution, when either the clear span or the load was excessive for the available single lengths of timber, was still the introduction of additional supports, as, indeed, it long continued to be.

Most of the ingenious expedients illustrated in Figure 8.3 are known only from various later records. The first (a) from the sketchbook of Villard de Honnecourt was introduced with the words 'How to work on a house or tower if the timbers are too short'.[2] It would serve the purpose provided that the loads were not too great; but only at the expense of greater bending-moments in the individual beams than if they had spanned the whole way from wall to wall, and of a tendency for the beams to pull apart where one rested on another. The resulting floor would be appreciably more flexible than one with continuous beams from wall to wall. It nevertheless recurs with variations in numerous later texts up to the late-nineteenth century (b), and was used, for instance, by Wren in the tower of the Divinity Schools at Oxford and for framing the first floor and ceilings of Independence Hall in Philadelphia.[3]

The second, (c) and (d), was essentially a means of building up longer single beams than the individual lengths of timber used. This – like the Renaissance techniques for building up timber arches from short planks that have already been noted in Chapter 6 – may be expected to

Figure 8.1 Underside of architraves, Parthenon, Athens. (Source: author)

Figure 8.2 Lion Gate, Mycenae. The false arch above the great lintel relieves it of much of the load that would otherwise bear down upon it. (Source: author)

have become increasingly attractive as supplies of the better West European hardwoods became as depleted as supplies of cedar had been earlier. It was first described by Alberti and first illustrated by Francesco di Giorgio.[4] The illustrations should be self-explanatory, but Alberti's description of (d) is worth quoting for the growing insight that it shows into the nature of the internal arching and catenary actions:

> But if your Timber is so short, that you cannot make a Beam of one Piece, you must join two or more together, in such a Manner as to give them the Strength of an Arch; that is to say, so that the upper Line of the compacted Beam, cannot possibly by any Pressure become shorter; and on the contrary, that the lower Line cannot grow longer: And there must be a Sort of Cord to bind the two Beams together, which shove one another with their Heads, with a strong Ligature.

Its strength could never have equalled that of a beam of the same size cut from a single piece of the same timber because of the inefficient way in which the midspan tension in the lower 'Ligature' was transferred to the ends of the beam.

The remaining expedients gave a beam no longer than the longest piece of timber employed, but they could considerably increase the bending strength and stiffness, which would, in effect, increase the possible span whenever either of these was the limiting factor. The

8.3 Timber beams and floors of the thirteenth to eighteenth centuries: (a) 13th century Villard de Honnecourt; (b) 16th century Serlio; (c) and (d) 15th century Francesco di Giorgio and Alberti; (e) and (g) 18th century and earlier after Leupold; (f) 17th century, Wren. (Source: author, based on drawings in MS. Français 19093, Bibliothèque Nationale; S. Serlio, *Il primo libro d'architettura*, Paris, 1545; Codice Torinese Saluzziano 148; L. B. Alberti, *L'architettura di Leon Batista Alberti*, ed. Cosimo Bartoli, Florence, 1550; J. Leupold, *Theatrum pontificiale oder Schau-Platz der Brücken und Brücken-Baues*, Leipzig, 1726; and the paper by Fletcher cited in note 5).

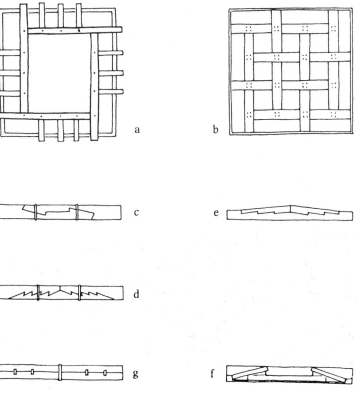

trussed beam (e) was a hybrid – part beam and part truss, but much shallower than the normal truss. An unusually elaborate version of it (f) was devised by Wren to take the heavy weights of the bookcases in the library of Trinity College, Cambridge.[5] The last (g) was a simple means of making two similar beams placed one above the other act almost as one by preventing any relative slipping when they were under load. Again it was probably not completely effective in doing this, but the principle was sound. Ideally the strength should have been quadrupled instead of merely doubled as it would have been without the interconnection.

As with the timber arch, the final development had to await the recent introduction of effective gluing techniques. These have made it possible to build up beams of almost any desired size and cross-section from quite short planks and sheets. But the forms actually adopted have been largely based on those developed in the meantime in iron and steel.

The iron beam had remained, until the end of the eighteenth century, of little importance structurally. Long before then it had been known as a bar of roughly square cross-section forged from pieces of soft blacksmith's iron welded together to build up the desired length and cross-section. Well-preserved examples even up to 10.6 m long and up to 280 mm square have survived from as early as the thirteenth century at Konarak on the north-east coast of India.[6] They were used to assist in bearing the weight of heavy masonry corbelled (or false-arch) ceilings. But they represented a prodigal waste of material and labour since the multiple welds were so defective that the finished beams can have had little useful strength. It has also been suggested that cuttings in the stone lintels of some Classical Greek temples originally housed similar, though smaller, beams.[7] But nothing has survived of these despite the good preservation of iron cramps and dowels in some of the same temples. It seems more probable that the insertions in the cuttings were of timber, perhaps sheathed in iron or sandwiched between thin iron plates.[8]

Towards the end of the eighteenth century, two factors led to the substitution of iron beams for the timber beams that had initially been used to carry the floors of the large new textile mills built to exploit the machines invented by men like Jedediah Strutt and Richard Arkwright. These were a desire to put an end to a series of disastrous fires in the original timber-framed buildings, and the general availability, by this time, of good cast iron as an obvious incombustible substitute for timber. The next fifty years saw an intensive development of 'fire-proof' floors and of the iron beam as the major element of these floors.[9] 'Incombustible' would have been a more accurate description, since collapse could still result from the action of the heat of a severe fire, particularly in the case of the earlier floors in which shallow brick vaults

spanned between the parallel beams and had their thrusts resisted by exposed or partly exposed wrought-iron tie rods.

Figure 8.4 (a) to (d) shows some typical cross-sections. In profile the depth was commonly increased towards the centre as in (e), correctly recognizing the increase in bending moment towards the centre when the ends were simply supported. The first cross-section (a), though much more slender, was otherwise closely modelled on a slightly earlier timber form in which triangular wooden skewbacks were attached to each side at the bottom to receive the shallow brick vaults spanning from beam to beam. It was soon altered to the (structurally) slightly more efficient form shown at (b). As yet, there was no top flange. But this was not a serious defect if the beam was simply supported, since the cast iron used was so much stronger in compression than in tension. It is, on the other hand, equally stiff under both. Therefore, when Tredgold first attempted to apply *a-priori* theoretical reasoning to the devising of the most efficient cross-section, he was correct in concluding that it was one with two equal flanges as at (c) if maximum stiffness (or least deflection at a given load) was taken as the criterion.[10] In practice though, there was usually ample stiffness whatever the cross-section; and maximum strength was more important, particularly since the brittleness of the material meant that any failure took place suddenly and without warning. Adopting, therefore, a criterion of maximum strength, Hodgkinson finally conducted a series of tests in which he progressively increased the width of the bottom flange of Tredgold's section in relation to that of the top flange until failure occurred in compression of the latter rather than in tension of the former. This gave the section shown at (d).[11]

A number of attempts followed to make good the tensile deficiencies of cast iron by the use of wrought-iron reinforcement. One expedient was the trussed beam shown in Figures 8.4 (f) and 8.5. The idea was very promising, being one of several early anticipations of modern prestressing techniques with the wrought-iron tie rods usually prestressed against the cast iron. But it fell into disrepute as a result of some serious collapses of beams that had been designed without adequate understanding of their behaviour, and was soon superseded by beams made wholly of wrought iron and then of steel.[12]

The wrought-iron and steel forms corresponding to those shown in figure 8.4 (a) to (e) are parallel flanged ones with cross-sections such as those shown in figure 8.4 (g) to (j). With no significant difference between the strengths in tension and compression, equal flanges at top and bottom become desirable for both maximum strength and maximum stiffness. The first two of these sections were early experimental ones; the other two are more recent ones produced in a wide variety of standard sizes for a variety of uses. The last section (j) has flanges of more

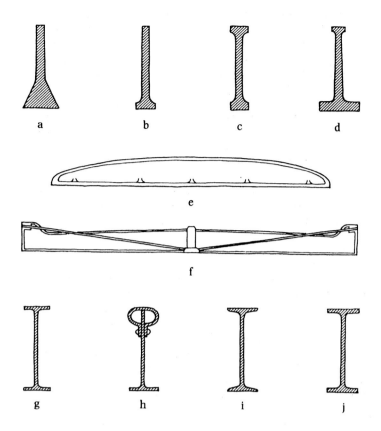

Figure 8.4 Iron and steel beams: (a) cast iron, Bage, 1796: (b) cast iron, Boulton and Watt, 1799 (c) cast iron, Tredgold, 1824: (d) cast iron, Hodgkinson, 1830: (e) cast iron, typical profile: (f) cast iron with wrought-iron trussing, typical profile: (g) and (h) wrought iron, Fairbairn and Zores, c.1850 (i) mild steel, c.1885 (j) mild steel, c.1930s onwards. (Source: author, with earlier examples based on drawings in the works cited in note 10).

uniform thickness which can be much broader if desired. It is therefore more efficient than (i), particularly where there may be bending about both axes of the section. All these sections have (or had) to be produced by passing white hot billets of the metal between a series of shaped rollers. A succession of opposed pairs of rollers is sufficient to produce (i). To produce (j) calls, on the other hand, for separate rollers for each face, suitably geared together. Development of suitable mills began around 1900 and sections of this form became more and more widely available from the 1930s onwards.[13]

The alternative to rolling single sections of the required shape and size in either wrought iron or steel was to build them up from simpler sections – usually flat plates and angles or Ts. These were riveted or bolted together and are now more frequently welded. All early large wrought-iron beams were, in fact, fabricated in this way, and the largest steel beams have continued to be so fabricated despite a progressive widening of the range of large sections available directly from the rolling mills.

Initially a hollow or tubular cross-section was preferred for large built-up wrought-iron

Figure 8.5 Trussed cast-iron beams, Boat store, Portsmouth Dockyard. (Source: author)

Figure 8.6 Conway
Tubular Bridge (Robert
Stephenson).
(Source: author)

beams. This was the outcome of an extensive series of tests and analyses undertaken by William Fairbairn and Eaton Hodgkinson for Robert Stephenson in connection with the design of the Conway and Britannia Railway Bridges [8.6 and 14.14].[14] The cross-section finally adopted for these bridges, as well as having the inherent lateral and vertical stiffnesses of the overall tubular form, had a cellular top flange for increased resistance to buckling under the compression to which it was subjected [5.7]. It also incorporated, for the first time, vertical stiffeners to the side plates to prevent the buckling which had been observed in one test on a large model of the tube having no such stiffeners. These took the form of pairs of Ts riveted to the sides at frequent intervals.

In due course it was realized that, with such stiffeners on the webs and other forms of lateral bracing where necessary, a built-up I-section could often be as strong as a box. The box was thus largely superseded by I-section 'plate girders', which were used almost universally by the time that steel took the place of wrought iron.

Now, the built-up steel box beam has staged a dramatic comeback for large bridges on account of its clean lines and inherent stiffnesses. Its torsional stiffness is particularly valuable where the traffic load (which constitutes an increasing proportion of the total load as the structure itself becomes lighter with stronger materials and better design) may sometimes act largely to one side, or where the span itself is curved. Often it is used in association with a ribbed-steel [8.7] or reinforced concrete [8.8]

deck that projects beyond it to carry the roadway but acts in association with it as part of the whole composite section. For a relatively narrow roadway a single box may be so used [8.8], whereas for a wider roadway two or more boxes side by side may be more economical [8.7]. On account of the thinner plates now used, such box beams have an even greater need than earlier tubular beams for local stiffeners to prevent local buckling. These usually take the form of light steel sections welded internally to the plates plus diaphragms (large transverse plates) at much wider intervals.[15]

Figure 8.9 shows a special form of steel box beam developed originally for the Severn Suspension Bridge, but subsequently used for a number of other bridges. One transverse diaphragm can be clearly seen closing off the last section to be erected. In addition to its torsional stiffness, it had the great advantage for a long-span suspension bridge of presenting little resistance to the wind. It also offered an easy method of erection: complete sections of the box terminated by pairs of diaphragms were prefabricated on one bank of the river, then floated out and hoisted into position.[16]

Reinforced-masonry beams

The simplest way of circumventing the spanning limitations of stone as a single-block lintel was always to turn instead to a false or true arch once these forms had been invented. If desired, the latter could even be given the flat profile characteristic of a beam [6.10]. But this led to

Figure 8.7 Zoo Bridge. Cologne, under construction (Rheinstahl Union Brückenbau). (Source: author)

greater horizontal thrusts than for an arch with a greater rise. Normally these thrusts called for substantial abutments at the supports. The alternative was to provide a tie or ties to take all the thrust by developing a matching tension. The flat arch then became a reinforced masonry beam, the horizontal thrust being resisted internally and no longer transmitted to the supports.

No real attempt seems to have been made to construct such reinforced beams until the deliberate return in eighteenth-century France, under the promptings of men like the Abbé de Cordemoy and the Abbé Laugier, to the

Figure 8.8 Road bridge near Innsbruck. (Source: author)

'primitive' forms of straight entablatures carried on free-standing columns as in the Classical Greek temple.[17] Since no stone was available to construct the entablatures from single blocks as in Greece, they had to be constructed as flat arches reinforced by iron bars to take most, if not all, the thrusts. One example, from the portico of the Paris Panthéon, is seen at the upper right of Figure 3.15. The first examples are shown more fully in Figure 8.10. The provision also of diagonal ties between the ends of the spans (equipped with means of adjusting the tensions) suggests, however, that the original intention was simply to resist the arch thrusts rather than to create non-thrusting reinforced beams.[18]

This form was relatively short-lived, partly because the fashion gave way to others and partly because later similar desires could be satisfied more economically in reinforced concrete or steel. A rather different but related form was, however, tested early in the nineteenth century and developed further a century later.[19] This is the beam of reinforced brickwork. In the more developed form, steel reinforcement − sometimes bars and sometimes a finer mesh − is incorporated into the mortar joints. It is valuable chiefly in permitting openings to be introduced into brick walls without calling for steel or reinforced-concrete beams to support the brickwork above. The result is usually a 'beam' whose ratio of depth to span is much greater than that of most other beams. Its structural action therefore resembles even more closely that of a normal arch tied across its foot (a series of 'arches' of substantial rise being formed, in effect, within the depth of the brickwork).

Before turning to the reinforced-concrete beam as the more direct successor of the beam of reinforced masonry, it is also worth looking briefly at a 'chain bar arc' patented by Thomas Pope and published by him in 1811 [8.11(a)].[20] This shows, in one way, an even clearer anticipation of the modern prestressed-concrete beam than did the trussed iron beam because, without the tensioned rods passing through the blocks of stone, these blocks would have immediately fallen apart if set up as a beam. The tensioning of the rods − by turning the nuts at the threaded ends − would both have held the blocks together and enabled the rods to participate immediately in supporting any transverse load. The placing of the rods shows, on the other hand, little evidence of an understanding of where the tension would be highest under a vertical load.

Reinforced- and prestressed-concrete beams and related forms

When concrete is used in place of masonry, its initial plasticity allows a very much freer choice of the disposition of the reinforcement. It is not surprising, therefore, that the earlier phases of development of the reinforced-concrete beam

Figure 8.9 Deck of the Severn Suspension Bridge under construction (Freeman Fox). (Source: author)

were marked not only by the use of many different types of reinforcement, as already noted in Chapter 3, but also by the adoption of many different dispositions.

In the light of our present understanding of beam action, there were really two feasible alternatives if the concrete was to be left to resist all the compression. One was to align the reinforcement so that it followed roughly the directions of the principal tensions under the expected loading − as shown by the broken lines at (a), (b), and (c) in Figure 2.12. The other was to run all or most of the principal reinforcement close to and parallel to the soffit over the whole length, and to introduce secondary reinforcement to prevent shearing failures near the ends due to the inclined tensions acting there. If the beam was to be continuous over one or more supports, the principal tensions would, of course, be at the top over these supports, as may be seen by looking at Figure 2.12(c) upside down again and considering the central load as the reaction from a support. The principal reinforcement would then be required at the top in these regions, but still at the bottom nearer to the centres of the spans. There was nothing to

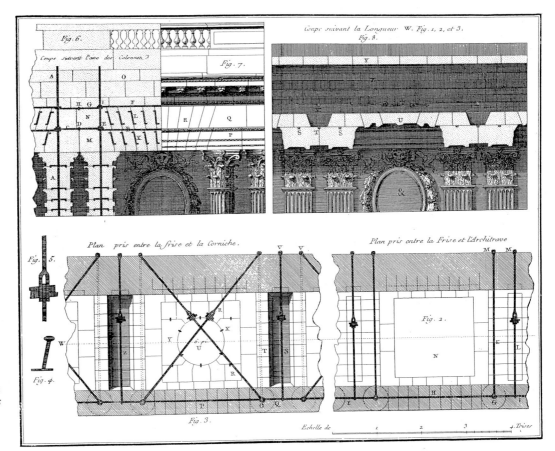

Figure 8.10 Details of the flat-arched ceiling of the east colonnade of the Louvre, Paris, as shown in Patte, P., *Mémoires sur les objets les plus importans de l'architecture*, Paris, 1769. (Source: author)

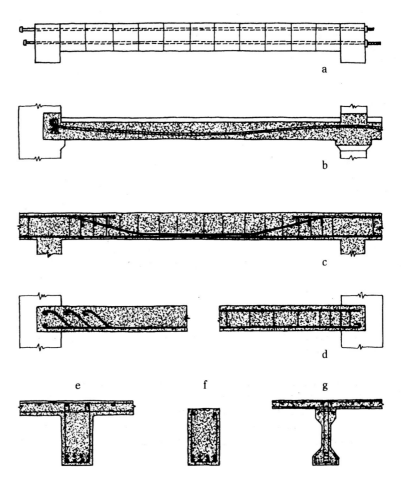

Figure 8.11 Reinforced masonry, reinforced concrete, and prestressed concrete beams: (a) Pope, 1811; (b) Wilkinson, 1854 patent; (c) Hennebique, 1897 patent; (d) to (f) modern reinforced concrete; (g) modern prestressed concrete. (Source: author, the earlier examples based on plate 3 of Pope, T., *A treatise on bridge architecture in which the superior advantages of the flying pendent lever bridge are fully proved*, New York, 1811, and the relevant patent drawings).

preclude the use of further reinforcement to assist in carrying the compression, but it could be regarded as an optional means of increasing the strength or providing for possible wide variations in the nature of the loading and hence in the pattern of principal stresses. Whether the first or second alternative or some combination of the two was adopted, it would also be necessary to arrange for all the reinforcement to be adequately gripped by the concrete or mechanically anchored to it so that it could develop the requisite stress.

The first alternative was adopted in Wilkinson's early patent of 1854 [8.11(b)]. Though the patent drawing actually showed a series of one-way spanning fireproof floors, he adopted a similar disposition of the reinforcement when constructing a supporting beam. Iron-wire rope was used as reinforcement, and its ends were flayed and formed into loops to grip the concrete.[21]

The second alternative was adopted by Hennebique towards the end of the century [8.11(c)]. His patent drawing showed part of a beam continuous over several spans. The bending up of one set of bars over each support was intended primarily to provide the necessary top reinforcement at the supports. Vertical 'brace pieces or stirrups' provided the secondary reinforcement against failure in shear. The stress was developed in the principal reinforcement by splaying out the end of each bar into a fish tail and by lapping the bars and interconnecting them where they were not continuous.[22]

Modern equivalents, designed with the benefit of a much fuller understanding of the structural actions, are shown, for a single simply supported span, in the left- and right-hand halves of Figure 8.11(d) and in the cross-section of the latter at (f).

A plain rectangular cross-section of uniform width and depth is not, however, the most efficient form for any but the simplest applications, such as a lintel over a doorway or window. When the beam is used to support a floor slab also of concrete, it is more efficient and easier to cast it integrally with the slab and to design the two as one unit so that the effective cross-section of the beam is a T as in Figure 8.10(e). When its length and the load upon it are sufficient to justify the slight extra expense of the formwork, it is also more efficient to vary the depth according to the variation in the bending moment, particularly in the case of cantilevers like those carrying the projecting canopy in Figure 8.12 and those carrying the roadway beams in Figure 8.13. (The hipped profile to the left of the cantilever in Figure 8.12 could, on the other hand, have no justification other than visual effect.)

For large bridge beams, and sometimes also for large beams in buildings, a hollow box-section is often, as in steel, the most efficient form. Its walls, as compared with those of an equivalent steel box, are necessarily thicker.

They are not, therefore, prone in the same way to buckling. But there is none of the saving on maintenance that arises with a fully enclosed steel box and almost twice as much formwork is required as for a solid section.. For really large spans the need for additional formwork is not of great importance. For somewhat smaller ones it calls for more than usual attention to rationalizing the process of construction to keep costs down unless, for instance, a sharp curvature in plan makes the torsional stiffness of the box form particularly valuable.

In the structure shown in Figure 5.2, some advantage could be taken of this torsional stiffness, but the box form would still not have been economical with the high cost of labour in California if the system of construction had not been perfected to the last detail on many similar structures by the Californian Division of Highways. In these structures there were usually both longitudinal and transverse diaphragms – the latter occurring particularly over the supporting columns in order to distribute the support reactions. Service pipes and conduits were carried, as required, within the boxes and threaded through the transverse diaphragms. The different loading

Figure 8.12 (top) Entrance and booking hall, Termini Station, Rome (Nervi). (Source: author)

Figure 8.13 (bottom) Corso Francia Viaduct, Rome (Nervi). (Source: author)

Figure 8.14 Prototype element for the roof of a grandstand. Madrid Racecourse (Torroja). (From E. Torroja, *The structures of Eduardo Torroja*, F. W. Dodge, New York, 1958)

requirements of particular structures were catered for by appropriate choices of reinforcement. All the concrete was cast *in situ*.

An alternative is to precast a sufficiently large number of more or less identical beams to allow a very free choice of their form. Nervi chose to do this for the roadway beams of the viaduct in Figure 8.13. Here each span was precast as six identical V-shaped troughs with transverse diaphragms plus separate flat-slab components that were placed on top to close the open troughs and provide a base for the roadway.[23] The troughs and slabs were connected to one another by concrete cast *in situ* around projecting reinforcement.

Though a closed box is stiffer than an open trough, the top closure is not essential. Both the open trough and the corresponding inverted form are capable of spanning independently as beams as shown in Figure 5.5(h). A fairly recent example of a continued series of such forms has already been illustrated in Figure 5.8. Figure 8.14 shows a pioneering example designed by Torroja in 1935 for the canopies of a grandstand at the Madrid racecourse. Again a continued series was used to form each canopy.[24] The photograph shows a single experimental element before casting of the concrete. Each element was cantilevered outwards from a support at its lowest point. The principal internal stresses were thus compressions converging radially on this point of support and tensions tracing a series of roughly concentric arcs at right angles to the compressions. For convenience the principal tensile reinforcement was straight for the greater part of its length; otherwise it followed the directions of the principal tensions. Greater stiffness was obtained by increasing the

depth of concrete in the curved cross-section towards the support, and the percentage of reinforcement was similarly increased. At the free ends the thickness was a mere 50 mm. The overall span was almost 13 m.

Some of the reinforcement at the bottom of the precast troughs in Nervi's viaduct was prestressed to make it easier to handle these components and to give the desired distribution of stresses in the completed deck. None of Torroja's reinforcement was prestressed, though. His canopies thus provide an excellent illustration of the relatively free choice of form that is possible even with normal reinforced concrete if one is willing to take enough trouble over the design and accept some additional costs for the construction of the formwork and placing of the reinforcement.

Prestressing is valuable in permitting the adoption of construction methods that would otherwise be impracticable, such as progressive cantilevering from opposite supports and the building-up of longer beams from short precast sections. It can also be used to modify the internal distribution of stress over the cross-section or the distribution of bending moments along the length of the beam. Indirectly thereby it extends the practicable range of some of the forms just discussed, and it permits freer choices of cross section or profile without loss of structural efficiency.

Examples of its use for constructing long spans by progressive cantilevering have already been illustrated and discussed [4.19, 4.20]. An example of its use both for building up longer members from short precast sections and for modifying the final distributions of stress and bending moment has also been illustrated and

Figure 8.15 Prestressed concrete viaduct. Bingen. (Source: author)

discussed [4.22, 6.20], though here the final form was a hybrid between a portal frame (with the central beam as its major element) and an arch. The principle of building up from short sections is, of course, no more than that of Pope's 'chain bar arc'. Figure 8.15 shows a well-detailed example, with particularly clean lines, of a kind of continuous-span prestressed-concrete box beam that has been much used for motorway viaducts. Prestressing here permits construction to advance progressively from one end of the structure and makes it possible to maintain a constant external cross-section of a slenderness that comes only from making full use of the strengths of both concrete and steel.

For spans that permit the precasting of full length beams, an I–section is usually preferred. This may have a wider flange at the top which serves directly as a base for the roadway. Or it may be given a relatively narrow and shallow top flange with a roughened top surface from which some reinforcement is left projecting. In the latter case, an *in-situ* slab is later cast on this flange to act compositely with it [8.11(g)]. Such beams have usually been set side by side at about 2 m from centre to centre.

In all but the shortest precast beams, the prestressed reinforcement takes the form of bars or wire strands. These are initially threaded loosely through ducts, and tightened only after the concrete has hardened around the ducts and construction has proceeded to a predetermined stage. Only some of the reinforcement is pre-stressed. The rest consists of normal mild-steel bars which serve to hold the ducts in position before and during the casting of the concrete. Both to reduce the friction of the prestressing bars or strands against the sides of the ducts during the stressing operation and to make the fullest use of these bars or strands, the ducts

follow smooth curves conforming as closely as possible to the expected principal tensile stresses [8.16]. This often gives them an aesthetic quality which is inevitably lost in the finished form,

Floor and deck systems and slabs

Apart from ingenious systems such as those illustrated in Figure 8.4 (a) and (b), floor systems remained until the late eighteenth century either shallow vaults of concrete, brick, or tile, or simple arrangements of large beams carrying lesser beams and, finally, some kind of wearing surface. Early fireproof floors with shallow brick or concrete vaults or 'jack arches' spanning between the lower flanges of iron beams were a partial break-away from the latter. They led, in the 1840s in France, to another kind of fireproof floor in which light I-section iron beams carrying lighter ironwork spanning from

Figure 8.16 Prestressed concrete beam for the approach viaduct. Medway Bridge, prior to casting (Freeman Fox). (Source: author)

one beam to another were largely embedded in gypsum plaster topped by a timber floor, and, in turn, to floors in which the iron was fully embedded in concrete.[25] Wilkinson, as we have seen, took the next significant step in substituting wire rope, capable of resisting tension only, for the iron beams [8.11(b)].

With the full development of the reinforced concrete beam as it is now known [8.11(d)], came also that of the reinforced-concrete slab treated simply as a very shallow laterally extended beam spanning between other beams of normal depth or directly between walls. Apart from being much broader than it was deep, such a slab differed from a beam chiefly in that the vertical shear along the edges where it was supported was rarely large enough to call for any special reinforcement. When it was supported on all four edges, it would, inevitably, span in both directions, and the exact distribution of bending moments would depend not only on the relative spans but also on the relative stiffnesses of the beams or other edge supports. When, moreover, it was continuous with edge beams or otherwise keyed to them to prevent any relative movement at the interface, and these beams projected below it as in Figure 8.11(e), it would also participate directly in their primary bending actions. The resulting action would be highly complex, and would change further at loads high enough to produce large deflections. But the slab could still be safely designed by making a number of cautious simplifying assumptions about the bands which would be effective in different ways. These led to a simple distribution of reinforcement running in the two main directions of spanning, and rising from the bottom to the top of the slab when this was continuous over a support. For long, the 'T-beam' contribution of the slab to the primary bending action of the supporting

beams was ignored when the latter were of steel and not of concrete cast monolithically with the slab. Later, in recognition of growing evidence that this was unrealistic, allowances for the interaction came to be made, particularly when the steel beams were mechanically connected to the concrete by 'shear connectors' (on the principle of Figure 8.4(g)) to prevent any relative movement.

For the moderate spacings of walls and columns that are encountered in most buildings, such slabs can easily span directly between the walls or main beams with uniform thicknesses of around 150 mm. For longer spans, approaching or exceeding say 10 m, intermediate secondary beams are desirable, and the alternative then presents itself of breaking up the whole soffit of the slab into coffers with a grid of beams or stiffening ribs below the slab itself. These coffers have usually been made square or at least rectangular, giving what is often called a waffle slab. A greater freedom in the placing of the supports when the space below the slab is to be completely open is, however, obtained by running the ribs in three directions giving triangular coffers [13.12]. Yet another possibility is to align the ribs with the directions of the principal bending moments for a particular predominant condition of loading as was first done by Arcangeli and Nervi [8.17].[27]

Experiments in doing away with beams completely in floors supported on regularly spaced free-standing columns began in the United States and Switzerland around 1900. It is fairly obvious that, with no beams and a uniform depth of slab, the principal bending moments will act radially in all directions in the vicinity of the column heads. Turner, the principal (though not the first) American exponent of the form, therefore distributed his reinforcement in four bands intersecting at 45° at each column head. He also used an enlarged column head and a square 'drop panel' above this to distribute the column reaction and avoid the excessive local vertical shearing-action that would otherwise have arisen there in the slab [8.18 (a) and (b)].[28] The four-way reinforcement was strongly criticized by Maillart, who was responsible for the independent development in Switzerland.[29] He felt that it led to an unnecessary concentration of bars at the column head and meant that some bands had to be set at too great a depth below the top of the slab to be very useful in resisting the reverse bending-action there. In his own designs he recognized that it was not really necessary to align the reinforcement with all the changing directions of the principal moments. If it ran in any two directions at right angles, it would also be capable of resisting tensions at its level in any intermediate direction. Also, in only two rather than four layers, it could all be placed at more nearly the correct levels to resist the bending actions. Thus it was more efficient to run it all in the directions of the column axes. Maillart also adopted a

Figure 8.17 Ribbed ceiling slabs of the Gatti Wool Factory, Rome. (Nervi). (Source: Cement and Concrete Association)

smoothly-flared profile for the column heads [8.18 (c) and (d)] to avoid any sudden changes in the intensity of the vertical shearing action: the depth reduced as the circumference over which the shear acted increased.

Initially these flat slabs were used chiefly for buildings like warehouses and factories which called for large open floor-spaces. Their chief merit was that they permitted a given minimum clear height to be maintained with a reduced overall storey height – though at the expense of calling for more material in the floors because of their reduced overall depth. Today, by eliminating the mushroom column head, two further advantages have been gained which have made the flat slab as attractive for office buildings as it is for multi-story parking garages. These are extreme simplicity of formwork – even its virtual elimination if the lift-slab technique [4.12] is adopted – and the complete freedom with which internal partitions can be placed beneath an unbroken level soffit. To permit the elimination of the head without running the risk of the column punching through the slab because of the highly concentrated shearing-action, a head composed of structural steel sections is often set within the thickness of the slab.

Nothing strictly comparable with the flat slab is possible in steel, since a solid flat plate of steel would have to be far too heavy in order to be stiff enough. There has, however, been one important parallel development in steel. This is the stiffened plate commonly known as a battledeck on account of the similarity of its construction to that of the deck of a warship. It is a much thinner plate stiffened by fairly closely spaced ribs running, like Maillart's slab reinforcement, in two directions at right angles to one another. It acts in very much the same way as a reinforced-concrete slab. In its usual application as a bridge deck, the ribs are much more closely spaced in the longitudinal direction than they are transversely.[30] Considered as an equivalent slab, it therefore has different stiffnesses in the two directions. Technically speaking, it is said to be orthogonally anisotropic, or, for short, orthotropic. Used in conjunction with much deeper I-section main beams, or as the top of one or more steel box beams, it is lighter than the equivalent reinforced-concrete slab even when this also acts compositely with the beams. This has made it preferable for long spans in spite of the greater cost of maintaining steelwork.

Figure 8.18 Early forms of reinforced-concrete flat slab: (a) and (b) Turner system; (c) and (d) Maillart system. (From Maillart, R., 'Zur Entwicklung der unterzuglosen Decke in der Schweiz und in Amerika', *Schweizerische Bauzeitung*, vol. 87, 1926)

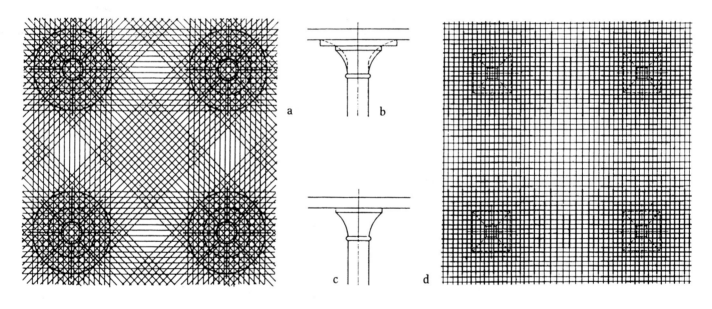

a   b   c   d

Chapter 9: Trusses, portal frames, and space frames

The wide range of forms to be considered in this chapter is that illustrated in Figure 5.11, excluding the most elemental ones, (b) and (h). None, as was pointed out in Chapter 5, is strictly elemental. Each is composed of a number of individual struts and ties or columns and beams; and only by acting in unison do these become analogous to the forms considered in the previous three chapters.

The primary structural actions are sometimes more and sometimes less complex than those of these latter forms. Mostly though, they are more assertively legible – even to the extent that any lack of clarity in the basic concept of the form is usually painfully obvious to a trained modern eye. Such lack of clarity tended to be the rule rather than the exception up to about the mid-nineteenth century, since it was only then that a clear understanding of the actions of any but the simplest kinds of truss began to be acquired. For this reason also, as well as on account again of the need for materials strong in tension and for joints effective in developing the tensions, the major developments, like those of the beam and slab, have mostly been fairly recent ones.

From a practical point of view, the trussed or triangulated forms have been employed chiefly as the most efficient – or the only – ones capable of spanning distances considerably greater than the individual members from which they were put together could have spanned by themselves. To a large extent, this practical advantage, and attempts to exploit it to the full by improving the efficiency, have dominated the development. Often, indeed, such forms have been boarded or plated over or kept out of sight in roof spaces or between floors when appearance was considered to be important. In recent decades, though, a greater architectural interest has been shown in the aesthetic potentials of both the portal frame and the triangulated space frame and in the freedom of planning offered by the latter.

Origins

Neither of the two principles of assembly of the individual members that confer on these forms their strengths and stiffnesses – triangulation and the use of rigid joints that restrain relative rotations – exists in nature in a sufficiently obvious way to have permitted simple copying of a natural prototype. Both must have been arrived at more through repeated attempts to make good the manifest lacks of stiffness, and the consequent instability, of unbraced rectilinear assemblies of posts and beams, and to prevent, for instance, the feet of coupled rafters from spreading or thrusting outwards.

The successful exploitation of rigid joints was, however, long frustrated by the difficulties of making joints capable of transmitting bending moments in the kinds of timber construction where the problems of instability would chiefly have arisen. It was further held back by the common adoption of other means of stabilizing the post-and-beam type of structure, either by infilling parts of it with relatively substantial walls or by setting the whole within substantial free-standing walls or tying it back to such walls. Initially, the joints between post and beam would have been made, as they still are in many parts of the world, by lashing them together. Later, joints more like those in Figure 3.13 (i) to (l) were probably first developed for shipbuilding and furniture construction. These joints have more capacity to transmit bending moments, but only after some rotation has occurred to take up any initial slackness.

The development of effective triangulation must also have been hampered by difficulties of jointing, but not to quite the same extent. Here there was no need to transmit bending moments. It was necessary only to develop the direct tensions in the tie members. The basic idea behind the principle was that of diagonal bracing. Two posts and a beam connected only by joints that allowed fairly free relative rotation constituted a mechanism like that of Figure 2.8(d'). But the addition of a diagonal brace as in Figure 2.8(e') locked this mechanism in one particular configuration and gave it structural stability in its own plane. Similarly, the whole assembly could be made stable in any plane by the addition of further braces. This idea also may well have been first systematically applied in shipbuilding and furniture construction, where the alternative of bracing by means of the equivalent of infilling or free-standing walls was not always available. It is in Egyptian furniture of the eighteenth dynasty (sixteenth to fourteenth centuries B.C.) and in representations of it that the earliest extant applications are now to be found. But the connections here were of such a kind that it is unlikely that the diagonal bracing members were capable of acting other than as struts. Probably only in ships' rigging were effective tensile braces to be found. The simplest true truss capable of spanning as a beam without exerting any horizontal thrust on its supports – a couple of inclined rafters connected together at their feet by an effective tie – does depend on joints capable of developing the tension in the tie; and there is no evidence for its appearance before the time of the Roman Republic. With the possible exception of some Sicilian examples, Classical Greek temples and auditoria seem to have been roofed without such trusses.[1]

Roof trusses

Vitruvius gives the first description of what appears to be a roof of coupled rafters connected by a horizontal tie.[2] The oldest surviving roof of this form is that of the sixth-century church of the monastery of St Catherine on Mount Sinai [9.1].[3] But there is every reason to suppose that most or all of the large Roman and Early Christian basilicas that were not vaulted were similarly roofed. Originally the roof was an open one: the ceiling panels between the horizontal ties are an eighteenth-century addition. At each end of each tie, the foot of the rafter is tenoned into it almost as in Figure 3.11(h). In addition to the basic triangular truss, it will be seen that there is a secondary system consisting of a central post and inclined struts notched and tenoned into it and into lengths of timber bearing against the undersides of the rafters. The central posts significantly do not rest on the horizontal ties. They are terminated well short of the ties and do, in fact, themselves act as ties supporting the feet of the inclined struts. These provide intermediate support to the rafters against any tendency to sag.[4] Roughly every second post now serves further as a support against sagging of the horizontal tie, being connected to it by an iron strap.

The only indubitably earlier trusses that survived long enough to have been fairly fully recorded (before being taken down by Urban VIII in the early seventeenth century to make cannon) were those of the Pantheon portico constructed in the early second century A.D. Drawings by Palladio and others show them to have been of the general form indicated in Figure 9.2(a).[5] The individual rafters, horizontal ties, and secondary struts and ties all appear to have been fabricated from plates of bronze riveted together at their intersections and splices. Two plates separated by a distance corresponding to their depth were used for each member, but it is not clear how they were interconnected. Presumably the interconnection was a substantial one in the case, at least, of those members subject to compression: they would otherwise have buckled out of the plane of the trusses. Probably it took the form of a continuous top plate giving them an inverted-U

section, like that shown for the purlins in some of the drawings.

The chief criticisms that can be levelled against these Pantheon trusses are that the central horizontal ties were too short and too high to absorb all the horizontal thrust of the central sections of the rafters without inducing some bending action in them, and that not enough differentiation appears to have been made between the cross-sections of those members subject to compression and those subject to tension. To prevent sagging of the central sections of the rafters, additional inclined struts were introduced; but there was nothing to prevent these from thrusting outwards. There is

thus evidence of lack of clarity in the basic concept as compared with that of the trusses at St Catherine's. But the faults are minor ones, if we bear in mind that there was probably little or no prior experience of similar construction in metal.

Later roof trusses in Italy and the Byzantine East long continued to follow closely the pattern of those at St Catherine's. Thus, for instance, an almost identical form, or a simple elaboration of it [9.2(b)], was consistently used by Palladio.[6]

The pattern was, however, less obviously appropriate and less immediately applicable to the design of the more steeply pitched roofs of the great churches and halls constructed in north-west Europe from the eleventh century onwards – if indeed it was known there at that date.[7] Early surviving roofs that did not serve as coverings for a masonry vault employed the lower horizontal members essentially as beams on which props were placed to support the rafters. The resulting forms bore a superficial resemblance to the earlier true truss forms, and they would to some extent have been capable of acting as trusses. But they were clearly not so

Figure 9.1 Roof structure of the nave, Church of St Catherine, Mount Sinai. (Courtesy of the Michigan-Princeton-Alexandria Expedition to Mount Sinai)

conceived. When the roof did cover a vault, it either rested on the vault or was carried by little more than pairs of rafters braced, along their lengths, by a certain amount of strutting to prevent sagging. Such a roof is illustrated (with some distortion of its proportions) in the sketch by Villard de Honnecourt reproduced in Figure 9.2(c). It would act more as an arch than as a truss. But it permitted the wall plates (the horizontal timbers along the tops of the walls on which the roof was carried) to be placed some distance below the crown of the vault. An early modification of this form which both retained the latter advantage and went some way towards eliminating the outward thrust of the rafters was the scissor type of truss, also illustrated by Villard [9.2(e)].

The true equivalent of the St Catherine's type of truss inevitably entailed raising the wall plates above the level of the crown of the vault to permit the placing of a continuous horizontal tie between the feet of the principal rafters. By the thirteenth century this had been accepted in the Île de France, as shown by the roofs of the choir and nave of Notre Dame in Paris. The trusses of the nave roof [9.2(d)] have the more fully developed form, in which the lower horizontals were unmistakably ties supported against sagging by the three central verticals.

A peculiarly English innovation was the hammerbeam 'truss'. Associated with arched ribs, it served, at the beginning of the fifteenth century, to span the whole width of Westminster Hall without any continuous tie member [6.13].[8] This span (which resulted from the elimination of two rows of columns that previously divided the hall into a central nave and two aisles) was well in excess of that of any of the contemporary French roofs, and approached that of the widest Roman and Early Christian roofs. Obviously [9.3 (a) and (b)], it made extensive use of triangulation. But to what extent was it really a truss?

If it were reconstructed today to the same design, but using members of steel, pinned or more rigidly connected together at all their intersections, it would certainly be a truss of a highly redundant or statically indeterminate kind. This means that it would be capable of acting in many different ways, according to the precise conditions of support and the individual fits or lacks of fit of the members. Two possibilities, both for self-weight vertical load only and assuming that all the members are initially of the correct size and are pinned together at all intersections, are shown at (c) and (d) in Figure 9.3.[9] The thickness of the lines denotes the magnitude of the direct tension (broken line) or

Figure 9.2 Roof trusses of the second to nineteenth centuries: (a) Pantheon portico, Rome, 2nd century; (b) Teatro Olympico, Vicenza, 16th century; (c) and (e) after Villard de Honnecourt, 13th century; (d) nave of Notre Dame, Paris, 13th century; (f) Sheldonian Theatre, Oxford, 17th century; (g) church at Steinach, 18th century; (h) Royal Hospital, Greenwich, 18th century; (i) Euston Station, London, 1837; (j) Lime Street Station, Liverpool, 1849. (Source: author, earlier examples based on drawings in MS Français 19093, Bibliothèque Nationale, and in or reproduced in the works cited in notes 5, 6, 7 and 10)

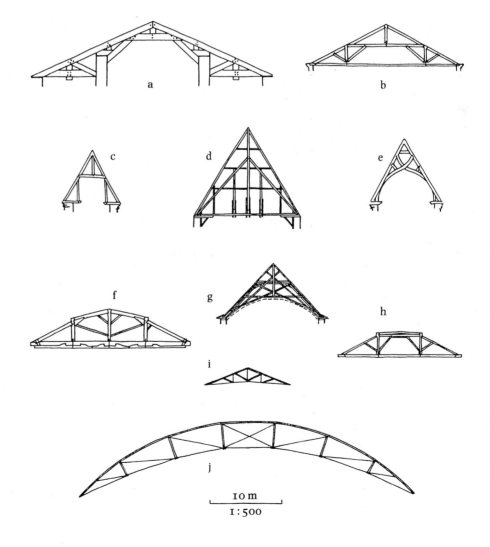

10 m

1:500

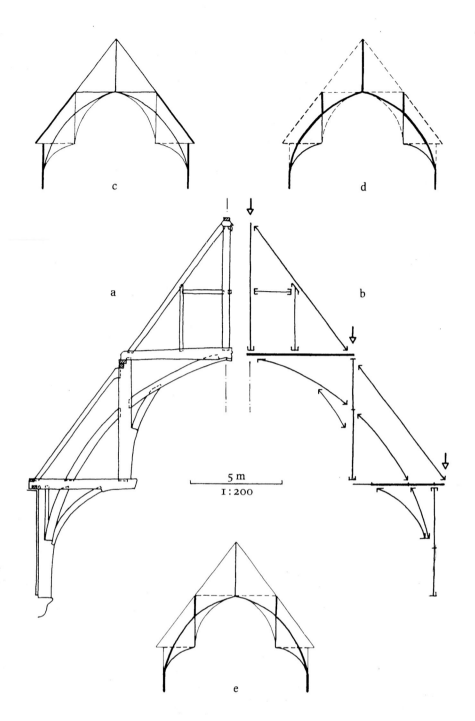

Figure 9.3 Westminster
Hall roof; (a) and (b)
details of construction of a
hammerbeam 'truss'
showing, diagrammatic-
ally at (b), the types of
restraint originally
provided by the joints and,
by means of heavier lines,
the only members or parts
of members in which these
restraints could have
developed appreciable
tensions; (c) to (e)
possible actions of trusses
with the same geometry
but different types of joint
and support (see text).
(Source: author).

compression (full line) in the members; and the very great differences between the two possibilities are wholly accounted for by the fact that the first calls for horizontal thrusts to be resisted at the level of the wall plates, whereas the second calls for them to be resisted lower down at the feet of the posts set against the walls. Which of these alternatives would be closer to the true behaviour would depend on the extent to which the walls moved outwards under the thrusts. In either case the thrusts would be there, so that the action would be more akin to that of an arch than that of a really efficient truss.

In fact, the members were all of timber, and none of the joints was pinned. At some intersections one member was continuous and able not only to transmit a tension past the joint but also to resist a bending moment there. But only in two places could a tension have been developed near the end of a member [9.3 (a) and (b)]. In these circumstances the action shown at (d) was clearly impossible. Taking into account also the likely relative deflections of the 'truss' and the wall, the absorption of the entire horizontal thrusts at the level of the wall plates and the large compressive loads in the lower lengths of the rafters shown at (c) were almost equally impossible. It is much more probable that the action approximated to that shown at (e). If so, it was more that of a stiffened arch; and was probably so envisaged by the designer of the roof, Hugh Herland.

Also illustrated in Figure 9.2 are three timber trusses of medium or large span constructed in the seventeenth and eighteenth centuries.[10] The first (f) is Wren's design for roofing the

Sheldonian Theatre in Oxford, which presented an almost identical problem to that solved by Palladio for the Teatro Olympico in Vicenza (b). Wren's design is markedly heavier, partly because he apparently persisted in regarding the horizontal tie member also as a beam. At (g) is one of a series of designs by Jacob Grubenmann for roofing Rococo churches. It is a considerable refinement of the early scissor truss (e) – a form that was particularly suitable for the purpose over a medium span, since a curved

This fault is corrected in the second design (j) whose unprecedented span of 46 m was but one of many indications of the great progress made in structural engineering in the mid-nineteenth century as a result of the introduction of new materials and construction techniques allied with a rapidly growing understanding of structural behaviour. Its crescent form may be compared with the forms of the earlier scissor-type trusses (e) and (g). Though it resembles an arch, it could span without exerting horizontal

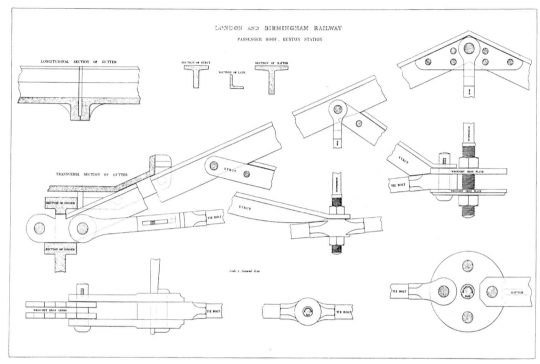

Figure 9.4 Details, from F. W. Simms, *Public works of Great Britain*, London, 1838, of the wrought-iron truss at Euston Station, London, illustrated above in figure 9.2(i). (Source: author)

plaster ceiling could be hung directly from the lowermost members. At (h) is Samuel Wyatt's design for the roof of the Royal Hospital at Greenwich. It is generally similar to Wren's design, but suggests a clearer concept of the essential role of each member.

Finally, at (i) and (j), are two trusses of wrought iron constructed in about 1837 and in 1849, respectively. Though the material had been used before for tie rods and straps in association with timber or cast iron, the earlier of these designs was probably the first to employ it throughout. As compared with previous designs, it is particularly notable for the simple clarity of its triangulation and the equally clear differentiation between the cross-sections of the members subject to tension and of those subject to compression [9.4]. Another notable detail is the provision made for adjusting the tensions in both the main 'tie bolt' and the vertical 'suspension rods'. It can be criticized chiefly on the grounds that the latter are shorter than the inclined struts that work in association with them. Since the strength of a tie is independent of its length, but that of a strut is reduced by an increase in length (due to the increased liability to buckling), it is preferable in any truss to make the shorter members carry the compressions wherever possible.

thrusts providing that the tension in the lower tie was fully developed – either by allowing the feet to spread or by pre-tensioning it. The crossed diagonal ties in the centre panel allowed for some asymmetry of the live load.

The chief subsequent development for roofing purposes has been the introduction of three-dimensionally triangulated space frames, which will be considered at the end of this chapter.

Trussed equivalents of the beam and arch

The roof trusses considered in the previous section are all, of course, trussed equivalents of the beam (if non-thrusting) or arch (if thrusting). For bridging purposes in particular, there was, up to the mid-nineteenth century, a largely independent development of trussed forms that much more closely resembled the simple beam and arch. Only from then onwards were they also used for the longer spans in buildings.

Early hints of such forms are to be found in the very frequent Roman use of crossed diagonal bracing for timber parapets, if not actually for fully load-bearing trusses,[11] and, once more, in the notebook of Villard de Honnecourt.

In the late fifteenth century, Leonardo sketched several designs for arched trusses similarly braced.[12] It is not quite clear how, in detail, he intended that these should be constructed, beyond the fact that he seems to have envisaged the crossed diagonals as being slender ties of, perhaps, iron wire. The shorter radial interconnections of the parallel arched ribs are shown with the greater thickness appropriate to the role they would then have as struts. Under any particular imposed loading, one tie in each pair would go slack and thus be redundant. The trusses seem, though, to have been intended as centering for masonry arches. They would thus have been subjected to large and changing imposed loads as construction proceeded, putting some diagonals in tension in the earlier stages and others when the load was differently distributed later on. The redundancy could have been eliminated only by giving a single, one-way, set of diagonals the capacity to act either as struts or as ties. Contemporary techniques for jointing timber hardly allowed this.

The first surviving more fully detailed designs for bridge trusses are those published by Palladio in 1570,[13] two of which are illustrated in Figure 9.5. These are of timber throughout, except for some iron straps; and the jointing indicates which members were envisaged as acting in compression, which in tension, and which were continuous beyond a node so that they could, if necessary, also act as beams.

The lower design (d) closely resembles Leonardo's arched trusses, but with the shorter radial interconnections of the main ribs now acting as the ties because of the simpler jointing details that this allowed, and the crossed diagonals as the struts. One of these in each pair would now go slack for each position of the imposed load unless the initial assembly was tight enough to precompress all of them and pretension each radial tie. But the design was well in advance of that of any known contemporary roof truss in that it showed how, using only quite short timbers subject only to direct tension or compression, it was possible to span a much greater distance. In a beam-like truss it is more difficult to do this because it is not possible to rely on self-weight, acting against the inward thrusts of the abutments, to bind together all the individual timbers of both chords (as the top and bottom members of a truss are called).

Understandably, the beam-like Cismone truss (a) was less ambitious. Its top chord was continuous past every other node, and its bottom chord was shown as continuous throughout the span, though it is doubtful whether single timbers of the requisite length could have been found. If this truss is analysed as depicted, it is seen that it was really a beam – the bottom chord lettered DD – stiffened by two largely independent systems of trussing. Two of the uprights and the main upper chord constituted one system (b); the remaining diagonals and uprights constituted the other (c). The absence of crossed diagonals in the centre panels and,

apparently, of any positive interconnection of the central upright and upper chord, would have meant that severe distortion under the action of a heavy moving load could have been avoided only by bringing into play the stiffnesses of the chords in bending.

During the next two centuries or so, many timber road bridges were built with a measure of diagonal bracing to give them the necessary stiffness. Almost without exception, though, they were inferior in overall clarity of concept to the designs just considered. Among the more notable achievements were several bridges constructed by the Grubenmann brothers in the second half of the eighteenth century.[14] Figure 9.6 shows the highly redundant framing, part arch and part truss, for the two spans of one of them. To an indeterminate degree, these spans

Figure 9.5 Two bridge trusses designed by Palladio: the upper one built over the Cismone River near Bassano; the lower one a project only. They are shown at (a) and (d) as illustrated in Palladio, A., *I quattro libri dell'architettura*, Venice, 1570, and the component trusses of the Cismone Bridge are added at (b) and (c). (Source: author)

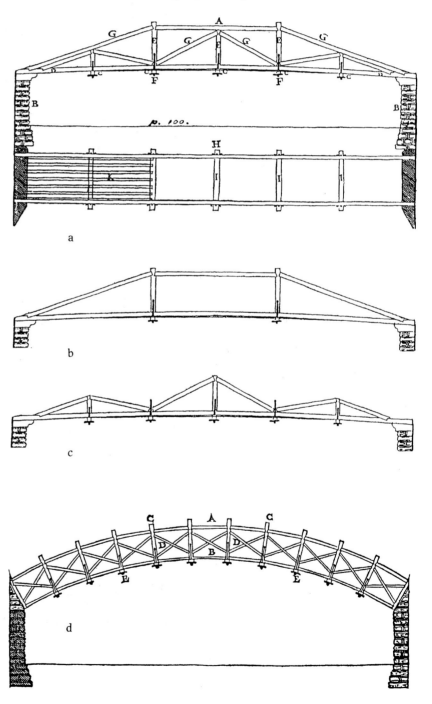

164

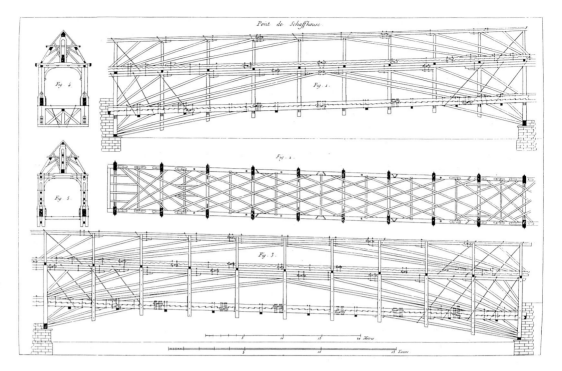

Figure 9.6 Side elevations (top and bottom) of the timber framing of the two spans of the Grubenmann brothers' covered bridge at Schaffhausen, together with typical cross sections and a plan of the lateral bracing at floor level of the shorter span, from J. Rondelet, *Traité théorique et pratique de l'art de bâtir*, vol.4, part 1, Paris, 1814. (Source: author)

were further stiffened by boarding applied to the sides and roof. They marked an advance on the Cismone truss to the extent, at least, that the lower chord was no longer required to act partly as a beam but was frankly treated as just a horizontal tie.

It was in the first half of the nineteenth century that the beam-like truss as now known really emerged, particularly in North America under the stimulus of an almost insatiable demand first for road and then for railway bridges. Figure 9.7 shows a few of the most significant from among the very large number of designs proposed and constructed up to 1848.[15] The first three were intended for construction wholly or primarily in timber; the others marked the transition to iron. This is reflected in the most consistent difference between them. In the earlier examples iron tie rods, where employed, were set vertically as was then most convenient,

Figure 9.7 Bridge trusses of the first half of the nineteenth century: (a) Burr, 1804; (b) Town, 1820; (c) Howe, 1841; (d) Pratt, 1844; (e) Whipple, 1846; (f) Warren, 1848. Single lines represent the ties in (c) to (e). (Source: author, based on patent drawings and drawings of actual bridges reproduced in some of the works cited in notes 15 and 17).

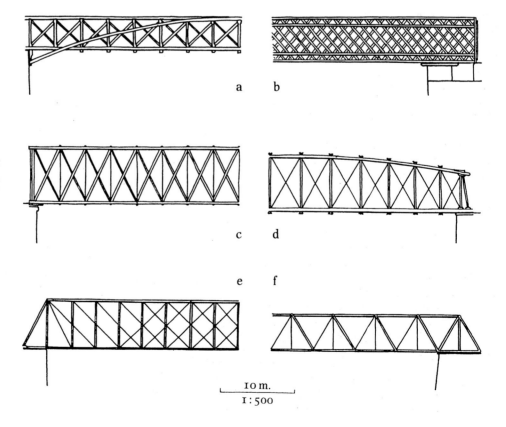

a   b

c   d

e   f

10 m.
1 : 500

Figure 9.8 Bollman truss railway bridge. Savage, Maryland. (Source: author)

whereas in the corresponding later examples they were used as diagonals to keep the struts as short as possible.

The first two designs may be regarded as transitional. The Burr truss (a) was highly redundant in rather the same way as the framing of the Grubenmann bridge. The load was shared by a truss proper and a continuous arch rib; and every panel of the truss was braced by two diagonals. The Town lattice truss (b) did away with the arch but substituted such a close mesh of light crossed diagonals that they became almost the equivalent of the continuous web plates of a deep I-beam or box. Only in the Howe truss as usually constructed (c) was the basic concept of Palladio's arched truss first realized in a non-thrusting beam-like form with the lower chord as a tie.

The Pratt truss (d) introduced the significant change of making the diagonals into the ties. It also incorporated, almost for the first time, a polygonal top chord; giving a total depth that varied with the variation in maximum bending moment. The fact that it still had crossed diagonals in all panels showed, however, that its designer still lacked sufficient understanding of the variations in stress in the diagonals as a heavy moving load passed over the bridge to realize that, towards the ends of the span, it would always be the diagonals sloping down catenary-wise towards the centre of the span that would be in tension. Since those sloping down towards the supports could only be subject to compressions that they were too slender to resist, they might as well have been eliminated.

Whipple was the first to arrive at a sufficiently clear under-standing of the action to take this step.[16] His truss (e) was provided with crossed diagonals only near the centre of the span, where the shearing action that tended to

deform the individual panels was likely to change its direction as the load passed. Shortly afterwards Warren, in England, proposed an alternative highly-efficient form (f), which was probably the first of these forms to have been actually constructed wholly of cast and wrought iron.[17] As originally proposed, it had only diagonal members interconnecting the two chords. Those near the centre might be subject alternately to tension and compression as a load passed over; but those near the ends would always be subject only to one or the other. As actually constructed, vertical posts were usually introduced, either as shown or in every panel.

During the next few decades, yet other configurations were proposed and built. None really succeeded, though, in improving on the Warren truss and the Pratt truss as later modified by the elimination of its superfluous diagonals. The Bollman truss [9.8] marked, for instance, a reversion to a highly redundant, or statically indeterminate, system.[18] Its intention was clear enough – the trussing of the upper (compression) chord by multiple V-shaped ties like those of the trussed beam in Figure 8.4(f), and much more effectively placed with their low points held well below this upper chord by the vertical struts. But a proper sharing of the trussing action between the shorter and longer ties anchored at the foot of each strut could have been achieved only by repeated careful 'tuning' of the ties by means of the threaded ends or turnbuckles that seem to have been provided for the purpose.

In comparison with forms like the Bollman truss and its derivative the Fink truss (with a different arrangement of the V-shaped ties), the more nearly statically determinate forms had the advantages of permitting easier analysis of the maximum forces in the members, and easier assembly. They largely avoided the problems caused by redundant members not fitting exactly or by the need to keep them 'in tune'. In the latter part of the century, these advantages were felt particularly in North America on account of the slower progress made there in understanding the behaviour of statically indeterminate systems and the usual practice, in both iron and steel construction, of reproducing as closely as possible the assumed freedom of rotation at the joints by actually pinning the members together. A pinned joint, however, must fit closely if the whole structure is not to be unacceptably loose, and some of these joints were of considerably complexity where members of very different size met one another from several directions, as in the arches of the Eads Bridge.[19]

In Europe the position was a little different. First, familiarity with some of the advantages of statical indeterminacy and a more general appreciation of its implications were acquired sooner. In particular, the advantage of making a beam continuous over several supports (all but two of which are statically redundant), and thereby reducing considerably the maximum bending moments, had been exploited as early as 1847 in the Britannia Bridge over the Menai Straits. Second, joints were usually made by means of gusset plates and rivets rather than the pins that were assumed in design. A measure of statical indeterminacy was inevitably introduced by the restraints offered by the gusset plates to any relative rotations But the rivets filled the holes only after being driven, so that some initial tolerance could be allowed in the lengths of redundant members. Experience also showed that, though the restraints induced some bending, it was not harmful if the members did not join at too acute an angle. The net result was that a simple triangulation of the Warren or modified Pratt type was again preferred, but that the further measure of statical indeterminacy introduced by adding one or two extra members

to make adjacent spans continuous over a support was more readily accepted.

The forms that prevailed were mostly developments of the Warren and modified Pratt trusses, but with riveted (or, more recently, bolted or welded) joints in place of pins. The further developments were chiefly the addition of some secondary shorter members in very deep trusses to provide intermediate support to the main struts and compression chord and thereby reduce the lengths liable to buckle. In all these forms, the material that is continuously distributed throughout the web of an I- or box-beam is concentrated, instead, in a relatively small number of struts and ties closely proportioned to the forces they must carry. This considerably increases the economical ratio of overall depth to span and has, thereby, considerably extended the range of spans that can be bridged with only vertical loads exerted on the supports.

For steel arches (in which both upper and lower chords should always be in compression and horizontal thrusts are inescapable), a modified Pratt or Warren type of bracing similarly jointed has likewise prevailed over systems with crossed diagonals resembling those proposed by Leonardo and Palladio. Given adequate abutment, this has made it possible to construct arches of even greater spans such as those crossing Sydney Harbour and the Kill van Kull at Bayonne [9.9].[20]

Portal frames

The true rigid-jointed portal frame is essentially a recent form, whether made of rolled steel, reinforced or prestressed concrete, or built-up sections of timber bolted or glued together. It is well illustrated by the 36.5 m-span welded steel main frames of the Illinois Institute of Technology's Crown Hall [9.10].

Such frames have two advantages over beams sitting without any rotational restraint on

Figure 9.9 Bayonne Bridge over the Kill van Kull between New Jersey and Staten Island (Ammann and Dana). (Source: author)

Figure 9.10 Crown Hall, Illinois Institute of Technology, Chicago, under construction (Mies van der Rohe, Kornacker) (Source: Bill Engdahl, Hedrich-Blessing)

the heads of the supporting columns. First, they can easily be given ample stiffness in their own plane to resist any tendency to sway sideways under the action of a horizontal load, such as that of the wind, acting in this plane. This stiffness can be achieved even with the feet of the columns freely pinned. Second, the restraint to rotation of the ends of the beam reduces the bending moment at the centre due to vertical load. This permits a reduction in the maximum depth of the beam at the expense of calling for some extra resistance to bending in the columns and inward horizontal reactions at their feet. The depth of the beams of the Crown Hall frames is only 1/20th of the span. Some further reduction is possible if the beams are given a hipped or polygonal profile, in which case the frame as a whole approximates more closely to an arch. From the planning point of view, these characteristics make the form particularly appropriate for wide unobstructed spans in assembly halls and the like or in factories. Aesthetically it is notable for its very clean lines.

Prior to the development of sufficiently rigid joints in steel and reinforced concrete and then in prestressed concrete and timber, this true rigid-jointed form could only be approximately realized. Even the best traditional beam-to-column timber joints [3.13 (k) and more especially (l)] were not, as already noted, completely rigid. At least they could not be relied upon to remain so as the timber of the posts shrank.

Figure 9.11 Barley Barn, Cressing Temple (main frame probably late twelfth century). (Source: author)

They were, moreover, incapable of developing the bending capacity of the full depth of the beam. While the latter joint was used with very deep beams to build frames having considerable resistance to sidesway, it was therefore still considered necessary in many surviving examples to provide further stiffness in the form of short diagonal braces.[21] Such braces between the beam and columns could not, by themselves, make the main joints completely rigid if their own joints were fully effective only in compression.

An alternative to the simple rigid-jointed portal frame is its trussed analogue. Either the entire frame may be trussed as in Figure 5.11 (d), or the beam only. Modern examples of both possibilities have adopted forms of trussing similar to those used for simple beam-like spans. Again (in aircraft hangers, for instance) they permit very wide spans to be constructed more economically than if the beams are made of I or box or similar form.

In the nearest frame carrying the roof in Figure 9.11 may be seen an early example of the second possibility. The beam here is really a fairly deep primitive truss with top and bottom chords braced by two sets of crossed diagonals. One diagonal of each pair is continuous past the lower chord to a point lower down on the column to increase the rigidity of the whole assembly. At the points where this diagonal intersects the other diagonal and the lower chord, both members are halved but otherwise continuous. All other joints are of the kind illustrated in Figure 3.13(g), so that they have a limited ability to develop tensions as well as compressions.[22] There is no vertical loading on the frame other than that of the roof bearing

directly on the columns. The intention was, therefore, merely to brace the columns against sidesway. It seems probable, though, that the 'truss' could easily, if called upon, have carried some vertical load as well, and that its capacity to do so was increased by its fairly rigid connection to the columns.

Space frames

All the forms described so far have been two-dimensional arrangements of struts and ties or columns and beams, and have been designed primarily to carry loads – whether their own weights or additional imposed loads – acting in their own planes. Each has, nevertheless, had to exist in a three-dimensional world, and has had to be sufficiently braced in other directions to remove the risks both of buckling out of its own plane and of being overturned by a transverse load such as the wind. Sometimes, particularly in the case of most early roof trusses, this bracing has been provided largely by substantial end gables, interconnecting purlins, and the like. But in all the more mature forms it has been provided more directly by special bracing for just this purpose in another plane or planes. Such bracing – often called wind or sway bracing – may be seen, for instance, in Figures 9.6 and 9.9. To this extent, all these trusses might, in their complete three-dimensional form, be described as space frames. So also might the long line of skeletal domical or vaulted roofs stretching from the kind of hut shown in Figure 1.1 to the much more complex iron-ribbed dome shown in Figure 9.12 and the later similar one of the Capitol in Washington, DC.[23] Consisting, as these do for the most part, of intersecting sets of

Figure 9.12 Dome of St Isaac's Cathedral, St Petersburg, near the completion of construction, from Ricard de Montferrand, *Église cathédrale de Saint Isaac*, Paris, 1845. (Source: author)

curved ribs, often themselves trussed in their own planes, these are even more three-dimensional. In all these forms, however, the individual trusses or ribs and any additional bracing systems meet only at right angles, and can readily be conceived as independent elements of the total form.

Today the term 'space frame' is usually reserved for types of truss or rigid-jointed frame that cannot be so conceived as a number of largely independent elements. Three types can be distinguished. First, there are the forms most closely resembling the trussed forms considered hitherto and typified by the sketches (c), (f), and (i) in Figure 5.11. Second, there are the tri-angulated counterparts of the type of framed analogue of a domical or other shell that is illustrated in Figure 9.12. Third, there are the double-layer spatial grids typified by the sketches (j) and (l) and the analogous beam grid shown at (k). All these forms are comparatively recent developments, since none of them could be efficiently conceived without a fairly precise understanding of the relevant three dimensional conditions of equilibrium, and some more recent steel forms have been dependent also on new types of cast joint.

This understanding was first arrived at in the latter part of the nineteenth century for some of the simpler statically determinate examples of the first and second types of frame.[24] Just as a statically determinate plane truss can be built up to carry loads acting in its own plane if we start with a pinned triangular set of bars and then add two more bars in the same plane to locate each fresh pinned node as in Figure 5.11(g), so can its statically determinate spatial counterpart (i)

Figure 9.13(a) Gilded polyhedron above the lantern of the dome of the Medici Chapel, San. Lorenzo, Florence. (Source: author)

Figure 9.13(b) Three structures exhibited at the Museum of Modern Art, New York, in 1959 (Fuller). (Source: Museum of Modern Art/ Alexandre Georges)

built to any height or scale without the usual limitations on overall slenderness or flexibility conveniently overlooked the fact that buckling is equally a phenomenon associated with overall stiffness in bending and that they are inherently no stiffer in this sense than more conventionally trussed alternatives. Though Motro and others have experimented with 'tensegrity' vaults and domes,[26] they have therefore remained outside the main stream of development. This has kept more closely to the path opened up by Bell as in the example of the first type of frame that is illustrated in Figure 9.14.

Long before Bell's experiments, Schwedler, in 1863, had already constructed in Berlin the first fully triangulated framed dome. With the introduction of other patterns of triangulation on the surface, such domes had become fairly common by the beginning of the present century, and triangulated framed analogues of barrel vaults were also being built.

The principal objective in devising fresh patterns of triangulation was to arrive at one of maximum regularity or uniformity – one composed of members as nearly as possible identical in length and uniformly distributed over the surface rather than, in a dome, converging on one another towards the crown as in Figure 9.12. Though the latter arrangement does follow precisely the directions of the principal stresses under self-weight loading in a continuous shell of the same form [2.14(a)], it has two drawbacks. It calls, in the first place, for circumferential members of different lengths in each ring. Second, if material is not to be wasted, it calls also for repeated diminutions in the cross-section of the radial members as they converge, because there is no increase in the radial compression per unit length of circumference to correspond to the convergence. A near-uniform spacing of members calls, on the contrary, for much more uniform cross-sections to match the loads to be carried. It promises, therefore, both maximum economy of material and maximum ease of fabrication and erection. At least, it does so with little qualification in the case of a shallow dome or a continuously supported barrel vault. In a full hemispherical or similar dome, the circumferential loads to be carried are, of course, much less uniform in the lower part where they change from compression to tension. But, even here, much of the advantage of a uniform spacing of members can be retained if the triangulation is devised in such a way that, throughout this region of non-uniform loads, one set of members runs as nearly as possible horizontally to carry the horizontal circumferential tensions while the others share the radial compressions and most of any additional asymmetric loading due to wind or snow.

In a barrel vault a uniform distribution of members of identical length was easily achieved. Figure 9.15 shows one fine large-scale example (now destroyed) constructed in reinforced concrete.[27] Because of the uniformity of

be built up to carry loads acting in any plane if, using joints allowing free rotation in any direction, we start again with three bars and then add three more in different planes to locate each fresh node. So, again, can the statically determinate framed analogue of the domical or other shell be built up by locating each fresh node by only two more bars whose lengths are chosen so that, as assembly proceeds, their nodes all lie on the desired continuously-curved surface and not on a single plane. Three-dimensional stiffness will come in the first type of frame from the inherent stiffness of the skeletal tetrahedral (triangular pyramid) units of assembly, and in the second type from the curvature of the surface and the planar stiffness of the basic triangular units.

Alexander Graham Bell was among the first to experiment with the actual construction of beam- and column-like forms from skeletal tetrahedral units. Among his experiments were kites and a 24 m-high tower erected in 1907 which had three trussed legs arranged as a tripod.[25] More recently, in a similar pioneering experimental spirit, Buckminster Fuller went a step further by designing so-called 'tensegrity' structures like the mast seen in the left foreground of Figure 9.13(b). These had the novel characteristic that the individual struts and ties making up the component tetrahedra were so disposed that only the ties were continuous through the structure: there was no path for the continuous transmission of compression. Prestressing of the ties against the struts stiffened the structure and made it self-contained. Because the struts were all short, there was little risk of them buckling individually. But the claim that such structures could, for this reason, be

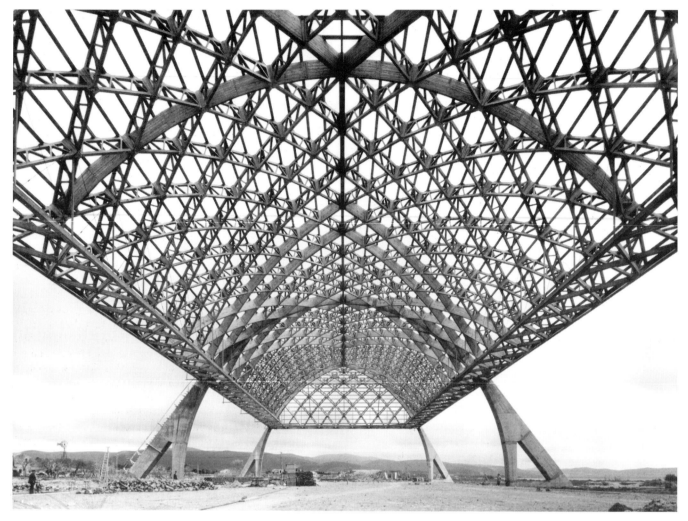

most of the members, they lent themselves readily to precasting. This made it economical to obtain the desirable depth of construction by giving them a relatively light trussed form. The overall pattern of triangulation may be compared with that in Figure 7.20.

In a hemispherical or other part-spherical dome a similar uniformity of the pattern of triangulation is virtually unattainable. As was known already to Plato, there are only five ways in which the surface of a sphere can be divided with complete regularity. Of these only the two that divide it into the largest number of identical equal-sided faces are of much relevance here. If the arcs bounding these faces are straightened out to give plane faces lying just inside the sphere rather than part-spherical ones, the resulting three-dimensional forms are the regular dodecahedron with twelve equal-sided pentagons as its faces [1.5] and the regular icosahedron with twenty equal-sided triangles as its faces. Further subdivision cannot be completely regular. If we start from one or other of these basic regular forms, or from a hybrid quasi-regular form bounded by twelve smaller equal-sided pentagons separated by twenty triangles (which is derived from the regular dodecahedron by joining the mid-points of the sides of all its constituent pentagons to give the smaller pentagons and then taking away the vertices of

the original form to leave instead the triangles), it is possible to proceed with the subdivision in various ways.[28] But each resulting multifaceted form, when interpreted as a skeletal linear framework, will then inevitably call for members of more than one length.

A small Renaissance example of such further subdivision – of no structural significance in itself, but designed by Michelangelo – is shown in Figure 9.13(a). Here each pentagon of the regular dodecahedron is divided into five identical, but not quite equal-sided, triangles. These, if viewed in groups of six centred on each vertex of the original dodecahedron, may also be seen as subdivisions of a set of overlapping hexagons. In the still relatively small structure illustrated in Figure 5.12 the starting-point was one half of the hybrid quasi-regular form referred to in the previous paragraph. Its pentagons were similarly divided into five triangles each, giving again two different lengths of side and therefore calling for two different lengths of bar.

Fuller, in his 'geodesic' domes, chose as his starting-point the regular icosahedral division of the spherical surface. He then subdivided its constituent basic spherical triangles by networks of members running roughly parallel to their sides (right-hand side of Figure 9.13(b)) or to the great-circle bisectors of their angles, spacing the members as uniformly as the surface

Figure 9.15 Aircraft Hangar, Orbetello (Nervi). (Source: Vasari)

geometry allowed.[29] Within each basic triangle the smaller triangles of this network may be grouped into hexagons, but this hexagonal 'honeycomb' is inevitably disrupted by an intrusive pentagon centred on each vertex of the basic icosahedral division, since only five of the basic triangles meet there. In a dome with a large radius of curvature calling for multiple subdivision of the basic triangles to give members of reasonable length, the hexagonal pattern nevertheless predominates. Its individual hexagons have then been found to be convenient units to prefabricate as single components. When the radius of curvature has been large enough to call also (to guard against buckling) for a greater depth or overall thickness of construction than can readily and economically be provided by a single-layer triangulation, the hexagons have either been 'dished' at their centres or a full double-layer system has been substituted. The latter alternative was adopted for the St Louis Climatron [9.16]. Here the hexagons of the dome proper are double compression-rings strutted apart about 0.75 m and having all their corners individually tied to a single central point to complete the triangulation and give three-dimensional rigidity. Within them, a lighter and independent secondary triangular mesh was suspended solely to carry a Plexiglas weather shield.

Experience showed that the restraints to relative rotation of the members at the joints were always such that not even the simplest of these forms behaved in a truly statically determinate manner. Nevertheless, before more accurate analysis by computer became possible, a sufficiently accurate picture of the likely distribution of forces to serve as a basis for preliminary design could often be obtained by analysing a statically determinate analogue. The distribution of forces in a triangulated framed dome was, for instance, frequently estimated from that in the analogous continuous shell. Sometimes, this shell analogy was considered close enough even where it was necessary to depart from a simple single-layer triangulation in the actual structure.[30]

No comparably simple design procedure was ever possible for frames of the third type, whether the beam grid of Figure 5.11(k) or its trussed equivalents, (j) and (l). These are all statically indeterminate in the highest degree. In fact it is precisely in this indeterminacy – or in the high degree of mutual support and interaction between the members that gives rise to it – that the structural merit of these forms lies. In the beam grid, it arises through the rigid interconnections between the beams. No one beam is able to deflect in carrying a load without, to some extent, causing most of the others to deflect with it and thereby share some of the load. In the trussed equivalents, it arises, even without rigid joints, through the fact that many of the individual struts and ties are simultaneously members of triangulated sub-frames in two or more different spanning directions. In the sketches, two-way arrangements are shown; but three-way systems (as seen in the background of

Figure 9.16 Climatron, St Louis, (Fuller). (Source: author)

Figure 9.13(b)) are now also common. The result, again, is a wide diffusion of any load throughout the system.

Because this behaviour must be understood and analysed in its full three-dimensional complexity if the potential economy of material is to be realized sufficiently to justify the somewhat higher-than-usual unit costs of fabrication and jointing, the practical exploitation of these forms has had to wait longer. Most of it has taken place since about 1960, with increasing momentum as high-speed electronic computers have become available to explore possible new framing patterns and cope with the very lengthy calculations required for adequate analyses of the designs.

Early examples are seen in Figures 9.17 and 9.18.[31] The earlier of the two, fairly small in scale and constructed in 1961 for only temporary use, is unusual in that the major compressions arose in the lower-layer grid because the roof was very light and the worst loading to be catered for was the maximum upward wind suction. These compressions were resisted by the edges of prefabricated aluminium pyramids open to the bottom. This, as well as being visually exciting, was economical because the sheet sides of the pyramids both provided the roof covering and stiffened the edges against buckling. The other, constructed almost ten years later and subject to more normal loading, was also a two-way trussed grid, but now with individual bays about 9 m square and 4.5 m deep. The equivalent beam grid, usually part of a ribbed floor system, has already been referred to in the previous chapter and a further example is illustrated in Figure 13.15.

By the late 1980s it became possible to build steel double-layer space-frame counterparts of virtually all the forms that had previously been built as continuous shells, and to do so over increased spans because of the reduction in weight. The reduction in weight, coupled with developments in jointing, has also made feasible the construction of large retractable and deployable roofs.[32]

Figure 9.17 IUA Congress Headquarters, London (Crosby. Newby, Makowski). (Source: author)

Figure 9.18 The new McCormick Place Exhibition Center, Chicago (C. F. Murphy Associates). (Source: author)

Chapter 10: Supports, walls, and foundations

The previous four chapters have all been concerned with spanning elements. These all require support; usually above ground level and always ultimately from the ground itself. Above ground level the support usually takes the form of columns, piers, or walls, conforming more or less to the norms of Figure 5.5 (d) and (e). To assist in resisting side loads (from the wind if nothing else), the shear wall (c) is also found. The tensile hanger (a) was relatively uncommon in the past, but is now being used to an increasing extent. Below ground level the requirement is essentially to spread the load and carry it down far enough for the natural strata to be able to bear it safely and without undue deformation. The foundations which meet this requirement may, in principle, take a fairly wide variety of forms.

This last chapter of the present part of the book will be concerned primarily with the forms serving as above-ground support, with the wall in its further role as a screen or encloser of space, and, rather more briefly, with foundations. Some further comment will also be made on the strut and tie considered as individual elements of the trusses and space frames already reviewed, since they are indistinguishable, in the simple terms of Figure 5.5, from the column and the tensile hanger.

To the extent that their action does correspond closely to that shown at (a) and (c) to (e) in this Figure, these forms are, structurally as well as geometrically, the simplest of all elemental forms. For the column and wall, problems of design and construction have arisen chiefly in three ways: first, from a frequent requirement to resist some transverse load in addition to the predominant axial load; second, from the ever-present risk of buckling under compression if the form is slender; third, in brick and stone masonry, from the need to achieve and maintain a reasonable homogeneity of construction. In either type of masonry, joints between the numerous blocks are only one source of difficulty. Another is the secondary transverse tension that always accompanies a primary axial compression. For simplicity, this was ignored in Chapter 5; but it cannot be wholly ignored in practice. It arises because much of the deformation associated with the carrying of any load is a change in shape rather than a change in volume. Thus a shortening under axial compression tends to be accompanied by an increase in the transverse dimensions; and this tensile strain induces a tensile stress which may burst the element open.

Allied to the basic simplicity of action of the column is (as with the arch and dome) a natural expressiveness that has been exploited architecturally not only in the Classical orders but also, for instance, in the soaring piers of the Gothic cathedral and in some of the best modern work in steel and reinforced concrete. The wall, being less naturally expressive, has been treated in a more varied manner. Sometimes its structural role has been emphasized by the application of columns or pilasters to the otherwise unbroken surface, or by giving an exaggerated emphasis to the massiveness of the masonry. At others it has been played down further by covering the surface with an insubstantial sheath of thin marble slabs or making it serve as a field for painted, mosaic, or relief decorations. But at all times the structural development has been directed chiefly to ensuring the necessary strength and stability economically with the means available. Foundations, being always out of sight, have of course always been designed solely to play their purely utilitarian role as efficiently and economically as possible.

Early forms

The earliest columns were, almost certainly, saplings, the branches of larger trees, or bundles of reeds, simply thrust into the ground like the stakes of a modern fence to give them some resistance to overturning. The similar use of large slabs of stone, either undressed or with just the rough dressing found at Stonehenge [3.5], may have been an independent development. However, the surviving evidence strongly suggests that, as in the case of the stone lintel, it grew out of a prior use of timber and did so in situations where the greater durability of the stronger and harder material justified the extra labour involved. Timber, used in this way, would have been far more prone to decay than when used above ground as a beam, because its vulnerable end grain would be constantly exposed to the varying dampness of the ground. Such decay, leading to overturning through inadequate fixity at the base, was probably the chief cause of failure. Columns of stone would not be immune from a similar risk of overturning, but would be more likely to remain stable if adequately restrained in the first place. They would also, as a general rule, be much stiffer and stronger and thus able to bear greater weights without assistance.

For walls, a wider choice of materials was available, at least in some localities. Logs, stone, turf, or mud-brick could all be piled up, to serve both as enclosure and to support some kind of roof. Or earth could be rammed into a solid mass to serve likewise. Reeds and some other materials could also be used if only a screen or enclosure, stiffened by something more substantial, was required. All are still so used in some contemporary huts in ways that can reasonably be assumed to reflect much earlier use.[1] Nevertheless it is only the early uses of stone and mud-brick that were sufficiently relevant to later developments to be worth looking at in any detail here. They alone had to face from the outset the third of the problems referred to above – that of achieving a reasonable homogeneity.

Today the obvious answer to that problem is to use uniformly sized units dressed, cut, or formed, with well-squared faces so that, with relatively thin joints of good mortar, they fit closely together; and to assemble them so that, as far as possible, the vertical joints are not in line with one another. In all probability a similar ideal, but without the use of mortar to assist in achieving the final fit, might have been formulated at quite an early date if builders had then been inclined to reflect in this way. But the requisite skills for accurate shaping or forming before assembly took time to acquire, and the aim, in practice, seems to have been to approach the ideal no closer than experience showed was necessary.

The very early stone masonry at Saqqara had closely-fitting straight joints on the exposed face. Out of sight, it was only very roughly dressed with wide irregular joints packed with gypsum mortar.[2] A century later, as we have seen, the excellent granite masonry of the Chephren Valley Temple at Giza [3.4] was composed of much larger blocks very closely fitted together. But most later Egyptian masonry of pre-Roman date reverted to the use of smaller blocks and weaker forms. The part-ruined wall seen in Figure 10.1 owes its survival only to reasonably well-fitted horizontal joints and to proportions that are still massive in relation to its height and the superimposed loading. If, as was usual in this kind of construction, the internal void was originally filled with rubble, this can never have contributed significantly to the total strength.

Figures 10.2 and 10.3 show two other typical early stone forms – both of a massiveness that was partly accounted for by their role as walls of defence. In the first, roughly contemporary with the wall in Figure 10.1, there was no pretence of close jointing on the faces of the

blocks; the dressing was all minimal. But there was a more useful attempt at breaking the line of the vertical joints to bond the blocks together. It was not wholly successful, because the very uneven bearing of the blocks on one another clearly imposed greater bending stresses on both the longer blocks in the foreground than they were able to bear. Both have broken as a result. In the second a very different manner of fitting the blocks together with the least amount of dressing was adopted about a thousand years later. As in some of the earlier examples, it was largely confined to the surface, but the irregularity of the rear faces of the blocks and their varying backward projections helped to bond them to the fill behind.

These walls all employed large blocks of stone; and the success that was achieved with the earlier examples led fairly naturally and directly (via some tentative experiments with attached pilasters at Saqqara) to the construction of free-standing columns not from single blocks of stone but built up from superimposed drums

Figure 10.1 Detail of the wall between the sanctuary and hypostyle hall, Ramesseum, Thebes. (Source: author)

Figure. 10.2 (top right) Internal gallery of the east wall of the citadel, Tiryns. (Source: author)

Figure 10.3 (centre right) Defence wall and gateway, Segni. (Source: author)

Figure 10.4 (bottom right) Boundary wall, Karnak. (Source: author)

as seen to the left and right of Figure 10.1. The earliest monolithic columns [1.6] had had neither capital nor base. These features now also appeared, at first presumably imitating the same features in the more developed forms of column made from timber or bundles of reeds – where they served respectively to distribute the load coming on the column from above and to raise its foot above the dampness of the ground. A feature of a more developed timber form that was not similarly carried over into stone construction was the inversion of the trunk, setting its broadest end uppermost, as in the Minoan Palace at Knossos, so as to increase further the bearing area over which the load was distributed at the top.

Where much smaller blocks of stone, mud bricks, or even fired bricks were used, rather different techniques had to be adopted to ensure the integrity and stability of the masonry in a wall of any size. Chief among these was the introduction, at fairly frequent intervals, of bonding or framing timbers. Sometimes these ran only longitudinally in the bed joints; sometimes both longitudinally and transversely forming fairly rigid horizontal grillages; and sometimes they constituted virtually complete vertical frames in relation to which the masonry was little more than an infilling. While the timber remained sound, it gave both cohesion and rigidity to the masonry; but the voids that it left became a source of weakness if it rotted. An Egyptian variant on the use of horizontal bonding timbers (though sometimes combined with it) was the placing of matted reeds in, say, every fourth or fifth horizontal mortar-joint in masonry of mud bricks.

Another technique, typical of Egyptian mud-brick masonry from the New Kingdom onwards, was the laying of the bricks not in horizontal courses but in courses that were concave up-wards or, in long walls such as that illustrated in Figure 10.4, that were alternately concave and convex in successive sections along the length of the wall. In the shorter walls the concavity of the courses would have contributed to the stability during construction, as it does to that of a modern builder's brick clamp. It would also have tended to compensate for any subsequent settlements due to inadequate foundations on the alluvial soil. In the longer walls the alternating concavity and convexity may similarly have helped during construction if the concave sections were built slightly ahead of the convex ones. In addition it probably assisted in taking up longitudinal contractions due to the drying out of the mass, and subsequent movements due to temperature changes, without extensive cracking. The illustration also shows the very massive proportions typical of defence walls of unframed brickwork often with a width greater than the height and with battered faces. Other unframed walls (as may be seen from Figure 4.16) were less massive, but still very thick by modern standards.

177

Later masonry and concrete and reinforced-concrete columns, piers, and walls

The later development of the masonry column was almost entirely a matter of using it in new ways and of refining the earlier forms by making them more expressive of their roles as supports and actually or seemingly less massive. The first steps in this direction were taken already with the construction of relatively slim columns of polygonal or even fluted form during the Middle Kingdom in Egypt. The Greek orders, almost invariably constructed of superimposed drums [10.5], and the monolithic or near-monolithic shafts of Achaemenid Persia [11.8] and the Roman Empire [10.6 and 12.6], then took the form itself almost as far as it was possible to go. The great columns of the Temple of Jupiter at Baalbek shown in Figure 10.6 were almost 20 m in height with a diameter of just over 2 m.

Walls of the finest ashlar masonry, with all the blocks fitting as perfectly as in the ideal referred to above but without even the assistance of thin beds of mortar, were first built when Classical Greek builders turned to the use of Pentelic marble – a stone which allowed the joints to be ground together as finely as desired. Since the highly uneven stressing of the blocks characteristic of earlier large-block masonry was thereby almost eliminated together with the major likely cause of instability, walls so constructed could be given a much-reduced thickness without loss of strength [10.7].

The Romans, who seem to have valued firmness in construction more than most people but to have sought to achieve it with the maximum economy of effort, profited from the Greek example in their own *opus quadratum*. But they also introduced new forms of construction and,

Figure 10.5 Ionic columns of the north porch, Erechtheum, Acropolis, Athens. (Source: author)

Figure 10.6 Giant columns of the Temple of Jupiter, Baalbek seen from the cornice of the Temple of Bacchus. Note the figures at the feet of the columns. (Source: author)

faced with having to use different materials in different places, learnt to use each where it was most appropriate according to the precise structural function and the demands of the construction process.[3]

Chief among their innovations were walls of well-fired large flat bricks set in good mortar, and walls of pozzolanic concrete merely faced with a thin skin of broken bricks, small stones or something similar [3.10]. The brick-faced concrete wall, in particular, allowed a greatly increased freedom in the placing of openings and in the planning of multi-storey buildings. It had the further advantage, since concrete was also eminently suitable for the construction of vaults, that these could be built integrally with it. Large vaults, however, were sometimes supported at a few separated points rather than by walls continuous around their periphery. They then exerted greater weights and thrusts than the typical column could bear, and thus called for another form – the more massive pier. Figure 10.8 shows an example from the second century A.D. which helped to carry the concrete annular vault of the substructure of a rotunda. It is constructed largely of fine ashlar or *opus quadratum* with a partial core of concrete or mortared rubble, whereas the continuous walls which carried the inner and outer springings of the vault were constructed of much smaller blocks of stone less accurately cut and fitted. Other examples may be seen in Figure 4.7 and, with attached half columns, in Figure 6 9.

Little evidence remains of the failures which may have marked some of the more pioneering phases of these developments. A thousand years or so later there is, on the other hand, ample evidence of the collapses and partial failures which attended the attempts of Romanesque and Gothic builders to emulate, with other materials,

Figure 10.7 Defence tower of the Attic frontier fortress, Aegosthena. (Source: author)

Figure 10.8 Detail of piers in the Sanctuary of Asklepios, Pergamon. (Source: author)

some of these Roman forms and methods of construction.

These later builders lacked chiefly a good pozzolanic mortar capable of binding together the rubble core of a wall or pier into a mass comparable in strength with that of an ashlar facing. A straight lime-mortar took a very long time to harden in a large mass under the best of circumstances. If it was poorly mixed in the first place, if rain was allowed to penetrate and leach out the lime as sometimes happened, and if the fill as a whole was just loosely tipped in and poorly compacted, the core could be worse than useless. It acted then as little more than an additional load on the ashlar skin, tending to push it outwards and burst it open. When the primary loading was not vertical and axial, any lack of homogeneity of the cross-section also had the effect of throwing most of the compression still farther towards one outer face of the wall or pier, with an attendant increased liability to gross deformation, buckling, or overturning. At the same time, the alternative of going back to earlier forms of large block masonry was virtually excluded by the nature of the building stones available at many sites and by the impracticability, at the time, of transporting large blocks for any distance and of lifting them on-site.

Confronted with the actual failures that resulted, Romanesque builders usually resorted again to very thick walls, or to somewhat thinner walls stiffened at frequent intervals by thicker projections like flat piers or buttresses.[4] Openings in these walls were again restricted. Striving in particular to admit more light, Gothic designers progressively improved the quality of both the facing masonry and the filling and evolved forms in which the major supporting function of the continuous wall was concentrated on a new kind of pier resembling externally a cluster of columns and often appearing, therefore, less massive than it really was. In Figure 10.9 the cluster of attached columns remains in place only on the pier at the extreme right but has fallen away from the other piers. Without their two tiers of attached shafts, the two free-standing central piers here have, unusually, a slenderness exceeding even that of the slimmest earlier monolithic shafts. This could be achieved only by ensuring that the loading was truly axial.

No advance was made on the best Roman and Gothic forms until, with the invention of Portland cement, improvements in brickmaking, and the development of reinforced concrete, it became possible to construct walls and columns in the latter material and to make brick-work of more consistent and much higher strength than before. To a great extent these have been twentieth-century developments.

In brickwork, the higher strength was achieved well before full advantage was taken of it, for want of adequate experimental data. Quite recently it was still the almost universal

practice to give a wall a thickness that was not less than, say, 1/16th of its height – a proportion not very different from that of the Greek wall in Figure 10.7. Today walls only 225 mm (or, with an internal cavity, 275 mm) thick at the base have been built up to heights of 34 m [15.20] and more. By using stronger bricks and mortar at the lower levels, the necessary crushing strength is easily obtained. Apart from this it is largely a matter of providing intermediate restraints against buckling, and disposing the walls in relation to one another so that the structure as a whole has the necessary lateral strength and stiffness. A limited amount of reinforcement is introduced where there is a risk of vertical tensions arising under wind loads.

In reinforced concrete, the development of the column preceded that of the wall, just as the beam was developed before the beamless slab. Again it was possible – and has become increasingly possible – to achieve the required crushing strength of the column by appropriate choices of materials while maintaining quite a small cross-sectional area in relation to the overall height and the magnitude of the load. Additional variables at the designer's command were the proportion and disposition of the reinforcement. Longitudinal reinforcement could not only assist in carrying the vertical load by directly sharing

Figure 10.9 Eastern transept (chapel of the nine altars) of Fountains Abbey Church seen from the south. (Source: author)

it. It could also confer an increased bending stiffness which reduced the liability to buckling and the restriction on the vertical load that stems from this in any column. Circumferential reinforcement was found to permit a further increase in load by preventing the concrete from bursting outwards under the secondary transverse tension and thereby increasing its own effective crushing strength. Not all of this was realized at the outset. But development proceeded roughly in step with that of beams and floor slabs, outlined in Chapter 8.[5] If anything, it was rather slower.

Such columns were first used in buildings in the latter part of the nineteenth century. They have now been used, in association with reinforced-concrete shear walls, to carry up to ninety storeys. They are also widely used as supports for bridges and viaducts. Here a cylindrical form, sometimes slightly tapered, is commonly adopted [10.10], both for its high uniform bending stiffness and for its visual unobtrusiveness and unchanging silhouette when seen from different directions. Other forms, sometimes more expressive, have been adopted where the required stiffnesses are different in different directions and vary more from top to bottom [3.19 and 8.13].

A recently developed alternative to the normal method of reinforcing the cylindrical form with longitudinal bars wrapped around by lighter circumferential ones has been to envelop the concrete in a steel tube.[6] In this form all the steel is available to assist in carrying the vertical load, and all the concrete has its effective crushing strength raised by the containment provided by the tube.

The load-bearing concrete wall as now used in buildings takes one of three forms. It is a direct counterpart of the modern load-bearing brick wall. It is a less direct counterpart of this which serves primarily as a shear wall to give lateral stiffness to a column-and-beam frame. Or it is an element of a more self-contained box-like core structure, usually containing lift shafts and stairwells, such as those seen in Figures 4.12 and 15.26. As first developed, it was more of an up-ended slab or laterally extended column used as the side of a silo or other container, or as a bridge pier where this was required to be stiff transversely to the axis of the bridge but flexible in the direction of this axis. Examples (though not very early ones) may be seen in Figures 1.5, 4.20, and 14.8.

In earlier applications, loadings were indeed closely comparable with those of columns or slabs, or intermediate between them. But in more recent ones, particularly in the first of the forms referred to above, the load-bearing brick wall provides a closer parallel. With, again, appropriate restraints and dispositions of the walls, the risk of buckling can be largely eliminated and the loading largely restricted to direct compression plus some horizontal shear in the plane of the wall under wind loads. It can then be matched over the height of the building by suitably varying the strength of the concrete and introducing a limited amount of reinforcement

Figure 10.10 Road bridge near Bregenz. (Source: author)

to resist the net tensions that may be produced by the wind shear. No other reinforcement should be called for by these primary loadings. The chief reasons for introducing more are to control cracking due to shrinkage of the concrete, to provide continuity throughout the structure and thereby reduce the risk of the kind of collapse shown in Figure 5.14, and to make it easier to handle the wall units when these are precast.

Figure 10.11 shows a typical precast wall unit being lifted free from the tilting steel mould on which it was cast. The first experiments with similar counterparts of the modern load-bearing brick wall were all based on casting the concrete *in situ*. But the large-scale application of the form to low-cost multi-storey housing after the Second World War, particularly in Eastern Europe, called for the maximum rationalization of the construction process and favoured precasting.[7] Partly mechanized *in-situ* construction, using, for instance, sliding formwork as in Figure 4.5, is the usual alternative. For the other two recent forms, the maximum continuity is required between storeys and between adjacent panels of the wall, or between the panels and the adjacent frame, so that *in-situ* construction, usually more heavily reinforced, is normal.

Later timber and iron and steel columns and struts

The early development of the timber column – from a post whose foot was thrust into the ground to hold it upright to one that was set on a raised base or plinth to protect its foot from the dampness of the ground – was repeated as the arts of construction were learnt again in north-west Europe after the collapse of the Roman Empire. Up to about the twelfth century, the earlier form seems to have been usual. When the other took its place, the practice, in England at

Figure 10.11 Lifting a precast wall in the site factory at Thamesmead (Balency system). (Source: author)

Figure 10.12 Priory Barn Little Wymondley. (Source: author)

Figure 10.13 Branched cast-iron columns of the upper floor, Tobacco Dock Warehouse, London (Rennie).
(Source: author)

slight misplacement of the core on casting would reduce strength significantly by leading to variations in thickness. Examples of the first are seen in Figure 10.13 and of the second in Figure 15.8.

Such columns shared two of the characteristics of timber and masonry columns. They could easily be given similar forms and they had to be assembled in much the same way. The ease with which different shapes could be cast allowed similar decorative treatments such as fluting of the surface. Superimposed straight lengths simply stood on one another, usually with machined bearing-surfaces and a jointing compound or interposed sheet of lead to give good contact, and spigots to ensure proper location. Beams bore directly on prepared seatings and were sometimes continuous over a column with the spigots passing through. Occasionally more complex forms were adopted as seen in Figure 10.13. Each column here had two integral branching arms and, at the point where these diverged, also carried two Y-shaped branches seated on pads of lead to give, in all, a six-point support to the pairs of beams that carried the roofing trusses.[10] With this effectively articulated assembly, stability was dependent on the mutual support of the whole structure.

Cast iron remained the usual material until the mid-nineteenth century, partly on account of its greater compressive strength as compared with wrought iron. While it did so, variants on the cylindrical form also remained the norm [8.5 and 10.14(a)], though it became more usual to devise bearing details that permitted a more positive connection between column lengths and between them and the beams. The Crystal Palace joint illustrated in Figure 4.11 is one of the best examples. The H-section cast columns with integral seatings for the beams used in the Sheerness Boat Store [15.12] were a further improvement but were highly untypical.[11]

Riveting, which provided a more versatile means of jointing, became feasible only with wrought iron. Since wrought iron could be rolled, it also facilitated the fabrication of efficient H- or box-sections, either rolled directly or built up from plates and angles or Tees. Such sections were introduced from mid-century onwards, though the cylindical column had an intermittent after-life in the built-up wrought-iron Phoenix column consisting of four, six, or eight curved sections riveted together [10.14(b)] and the much larger tubular sections of the principal compression members of some large bridges [10.15]. For medium-sized compression members, angles were latticed together to form open boxes, while in roof trusses cruciform sections or Tees were the usual choice.

The next major change came with the introduction of mild steel. The section almost universally adopted then for buildings was a rolled I, similar to, but usually with relatively wider flanges than, that adopted for beams

least, was initially to set up the timber as it had grown with its broader base helping to steady it at the foot [9.11]. Only later was it inverted, as had been done earlier at Knossos, to provide a larger bearing area for the members that converged on it at the top. Usually it was then trimmed to a uniform section for most of its height, leaving just a projecting jowl at the top as seen in Figure 10.12. As with the timber beam, the lengths and girths of the available timber limited the heights of column that could be cut from it and the loads that they could bear without buckling. But, rather than attempting to overcome the limitation by joining several pieces of timber together, builders seem to have preferred to increase the heights of the masonry plinths or to substitute columns constructed of masonry throughout, as in some of the larger medieval timber barns.[8]

Cast-iron columns first took the place of timber ones in English textile mills towards the end of the eighteenth century.[9] They did so just before the timber beams were also replaced by ones of cast iron, and for the same reasons – chiefly in the hope of making the structure fireproof. A cruciform section was usually adopted. Before long a hollow cylindrical section became commoner. Being stiffer for the same cross-sectional area, it could carry greater load without buckling, although there was a risk that

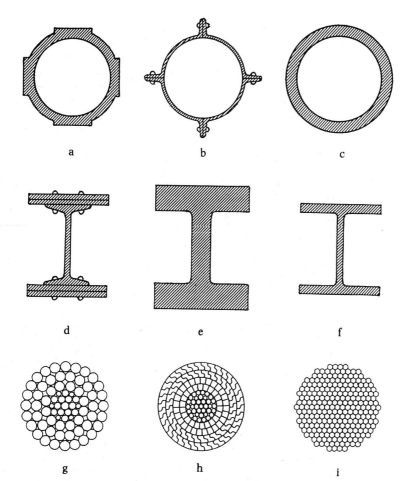

a   b   c

d   e   f

g   h   i

Figure 10.14 Iron and steel columns and ties: (a cast iron, Crystal Palace, 1850; (b) wrought iron, Phoenix, 1862; (c) cast steel, Centre Pompidou, 1975 (many early cast-ir columns were similar to this in section but less precisely formed); (d) mild steel, c.1885; (e) an (f) mild steel, widely available in an increasing range of sizes from 1930 onwards; (g) steel stranded wire cable, late 19th century onwards; (h high-tensile steel locked-coil cable, c.1950; (i) high-tensile steel parallel wire cable, c.1970 (show without protective jacket (Source: author)

[8.4(i)]. Increased strength was obtained by riveting extra plates to the flanges [10.14(d)]. More recently a broader flanged version of the still more efficient section shown at (j) in that figure and described in Chapter 8 largely took its place. By varying the thicknesses of flanges and web, a wide range of strengths could be achieved with little variation in overall depth and breadth [10.14 (e) and (f)]. Hollow sections, both square and circular, were reintroduced as widely available standard forms for relatively light structures. But larger cross-sections still had to be built up from plates and angles and similar sections. Figure 10.16 shows one of the slender towers of the modern suspension bridge along-side the older structure see in the previous figure. Here it is seen end-on where it is steadied by the pull of the cables acting to both sides. Transversely to the axis of the bridge it had to be given a greater stiffness able to resist a side wind without help from the cables. In this direction, as will be seen from Figure 4.21, it was therefore made much wider, with two columns braced together some distance apart. Each column or leg is cellular and was built up from three prefabricated rectangular-section boxes placed side by side and inter-connected by plates to make five cells in all.[12]

Recent developments in casting have permitted two more developments. One is that of large-diameter weldable cylindrical steel sections centrifugally cast with a reliably uniform thickness. Their first major use was in the Centre Pompidou in Paris [3.18, 10.14(c) and 13.18], where they had an external diameter of 850 mm throughout with different thicknesses to match the different loads. The other is use of cast joints to permit the construction again of multi-branched columns [10.17]. Since these joints can be rigid, the columns as a whole can, if desired, be far more rigid than their prototype illustrated earlier. Tying of the ends of the branches, either through the roof structure or independently of it, is nevertheless desirable to minimise bending, though ties would have been unnecessary in the finished masonry structure of Gaudi's church of the Sagrada Familia [5.4] because the inclined arms of the columns would then have been aligned with the inclined thrusts coming from above. Similar branched columns have more recently been constructed of laminated timber.[13]

Hangers and ties

The straight hanger or tie differs from the catenary only in that it is subject to little or no transverse load along its length. There is no need, therefore, for it to be able to flex freely in order to equilibrate such loads without inducing significant bending actions. On the contrary, it may even be advantageous (where, for instance, there is a possibility of the direct tension changing sometimes to compression) for it to have some bending stiffness. As with the catenary, though, the chief problems of design and construction relate to the choice of material, to fabrication, and to the attachments of the ends to develop the tension.

Up to the early nineteenth century, as we saw in Chapter 9, timber was used almost universally for the tie members as well as for the struts of roof and bridge trusses, its chief drawback being the impossibility of developing its full tensile strength with the connections then normal. To some extent this drawback was overcome by using forged iron straps bolted to the ends of the members to make the actual connections. Forged iron bars, usually a few centimetres square in cross-section, were however used almost as widely as timber from at least the sixth century A.D. as ties across the

Figure 10.15 Detail of a main span of the Forth Railway Bridge, near Edinburgh (Sir Benjamin Baker). (Source: author)

feet of arches and vaults.[14] The nineteenth-century substitution of wrought iron ties for timber ones in trusses might, indeed, be regarded as a simple extension of this earlier use.

The change in later steel trusses from eyebar ties to ties connected to the other members by riveted or bolted gusset-plates or by welding has also been noted in Chapter 9. Once the change had been made, it was usually convenient to adopt similar types of cross-section for both ties and struts, merely making the ties more slender as a rule in view of their immunity from the liability to buckle. Angles, I-, T-, U-sections and the hollow sections referred to above have all been used, the last chiefly since the introduction of welding and of special multi-directional connections for triangulated space-frames.

Vertical suspension as a means of support always means taking a load up in order to bring it down again, and this was hardly worth attempting until the hangers themselves could be fabricated from materials with the tensile strength of wrought iron or steel. Almost the same thing was true of straight inclined stays, although important additional factors have called for attention here in recent applications to roofs and bridges. These are the great import-ance of stiffness in addition to strength and, in bridges (as will be explained in Chapter 14), the much greater repeated changes in load to which the stays are subjected by heavy traffic as com-pared with those that arise in catenary suspen-sion cables.

For use in buildings, the greater stiffness of solid flats or rods has usually made them the preferred choice. But tubes, H-sections, and prestressed cables encased in concrete have also been used.

For the vertical hangers of some early suspension bridges, solid rods were used, but stranded wire cables of wrought iron or steel then became the simplest and usual choice. After earlier uses of eyebar chains or wrought-iron flats riveted together, similar cables [10.14(g)] were also used for most nineteenth-century and later bridge stays until they were largely superseded by the locked-coil cables [10.14(h)] already referred to in Chapter 6. Sometimes these were protected from corrosion by being encased in concrete that was pre-stressed by them. This encasement allowed full use to be made of a high tensile strength of the cable and, by increasing its effective cross-section, reduced its changes in length with changing load.[15] The more direct recent approach to obtaining maximum stiffness, and (by reducing voids and relative movements of the wires under changing loads) indirectly reducing corrosion, has been to use parallel-wire or parallel-strand cables [10.14(i)], or stranded cables with only a very small angle of lay or twist, or even to spin parallel wire cables in situ in much the same way as spinning a catenary. It is also now usual to encase such cables in protective sheaths or jackets.[16]

Figure 10.16 North tower, Forth Road Bridge, near Edinburgh (Freeman Fox). (Source: author)

Foundations

Foundations do not really comprise a further set of elemental forms, distinct from those con-sidered so far. Rather than this, they employ a very similar, though more restricted, set of elements in an almost inverse role; just as, from a structural point of view, the roots of a tree are almost an inverse reflection of its trunk and branches. Invariably also, the use of a particular form as part of a foundation has developed only from its prior use above ground. These are two reasons for dealing only briefly with foundations as a whole. Others are that to deal with them more adequately would call for a prior discuss-ion of the very varied nature of the ground at least as full as that of the materials used in construction given in Chapter 3, and that this whole book is primarily concerned with the structural forms of the superstructure. As a basis for the discussion in the next six chapters of the

Figure 10.17 Branched steel column giving 48-point support to the sloping continuous space-frame roof of the main terminal building, Stuttgart Airport (von Gerkan Marg, Weidleplan). (Source: author)

still standing, from the pyramids onwards.

Most other natural strata, particularly peat, clay, silt, and loose sand, are much less suited to bearing directly large concentrated loads. When they are found, it is necessary to spread the loads more widely, to carry them down to rock or some other firmer stratum such as firm sand or gravel, or to compensate for them by first excavating a like weight of soil and leaving the excavation as a basement. In theory, when this is done, the loads of the superstructure merely take the place of the similar loads previously imposed by the excavated material, and the ground below should not know the difference. In practice it is not quite as simple, because changes inevitably take place in the ground during the period when neither set of loads is acting, so that the compensation is only partial. Foundations of this kind are usually known as buoyant. Those which achieve their purpose primarily by spreading the loads or carrying them down to firmer strata are usually known by names referring more to the structural form – spread footings, grillages, rafts, piles, cylinders, and caissons being the commonest.

The objective is not merely to keep the superstructure above ground, but to support it without undue movement. Like any other structural form, every foundation will deform to some extent as it takes up the load put upon it. Most of this deformation will usually be a slight vertical sinking, and it is often more important that this should be uniform (rather than varying from place to place) than that its average magnitude should be below a particular limit. Sinking of the structure as a whole can be inconvenient if levels in relation to the surrounding ground are important. Differential sinking can be much more serious. If it occurs over the width of a single footing it will lead to the tilting (as in the notorious case of the Tower of Pisa) or the collapse of whatever is built on it. Once it starts it will tend, moreover, to worsen, because the tilting will throw more of the weight towards the edge that has sunk farthest. If it occurs also, or primarily, as a greater sinking of some footings than of others, it will lead to deformation and probably disruption of the interconnections between the elements built on the different footings. The commonest instance of this has been the sinking of a heavy tower in relation to a lower and lighter part of the same structure. Figure 10.18 shows an instance of this where the supporting piers were also sinking to different extents. The later addition of the scissor arches would have provided extra support to the pier or piers that were sinking most from those that were standing firm. Another common instance is the falling away of a heavy tower or pier from a wall which it was intended to buttress. Where horizontal thrusts or pulls are exerted on the ground, horizontal deformations are also to be expected. They will nearly always be important chiefly in relation to one another, as when they involve the spreading of the base of an arch or

development of these forms as complete structural systems, it is necessary to know something of the limitations imposed at different times and places by the need to secure adequate support from the ground. But it seems unnecessary, for this purpose, to go further.[17]

The ideal foundation is usually the virtually ready-made one of sound rock at or near the surface, calling, at the most, for some trimming to provide a level base on which to build or to key in whatever is to be built upon it. This was the foundation chosen for his house by the wise man in the gospel parable:

> And the rain descended, and the floods came, and the winds blew, and beat upon that house; and it fell not: for it was founded upon a rock.[18]

It is also the foundation that has ensured the survival of most of the earlier structures that are

dome or the pulling towards one another of the anchorages of a catenary. The latter caused the early collapse of Navier's Pont des Invalides in Paris in 1826.[19]

Up to the nineteenth century the only foundations, other than natural rock, were spread footings, rafts, and timber piles. Excavation to rock rarely went down more than a metre or two, and footings and rafts were also rarely much deeper than this, though there were exceptions such as the broad footings of Amiens Cathedral which Viollet le Duc found to extend more than 7 m below ground level.[20]

Figure 10.19 shows a foundation that might probably be taken as representing the best pre-Roman practice – though it was built up from fairly steeply-sloping natural rock with little or no prior excavation. It is, in effect, a cellular stone raft constructed from uniformly sized and accurately cut blocks not only bonded together but also interlocked at their intersections to give the maximum resistance to any relative lateral displacement. An outstanding later counterpart was Smeaton's foundation for the third Eddystone Lighthouse. Here each of the lower courses consisted of large blocks of stone cut to an interlocking dovetail form with joints breaking in successive courses and with inset marble joggles to link course to course. Each of the six bottom courses was also keyed back to the rock itself, which had previously been precisely cut back in level steps with dovetail recesses to receive it.[21]

The Romans seem to have taken similar care with the foundations of their earlier major structures, though the elaboration of some of those shown by Piranesi in his *Antichita Romane* can only be a figment of his fertile imagination.[22] Their important innovation was

Figure 10.18 Strainer arches between the western crossing piers, Wells Cathedral. (Source: author)

Figure 10.19 Substructure of the Temple of Apollo, Delphi. (Source: author)

188

the massive concrete footing or raft. This could be constructed even under water, thanks to the ability of their pozzolanic mortar to gain strength away from contact with the air. Good later examples of such footings are the massive substructures – continuous solid walls several metres thick – now laid bare in the naves of the so-called Red Basilica at Pergamon and the Church of St Polyeuktos at Istanbul.[23] Later still are the equally massive continuous footings of the eastern part of the eleventh-century York Minster. These were constructed of rougher masonry interlaced, like many walls, with stout oak timbers.[24]

The use of concrete was revived in the late eighteenth century. It was used then for bridge foundations in France, and, not much later, for

thrusts from the arch as well as vertical loads to be carried, and since the continued lateral support of the existing buildings at the two ends of the bridge had also to be ensured, the piles at each end were driven in a stepped formation and then capped by several stepped but interconnected layers of stout timbers [10.20].

Piles were also used in two ways to facilitate the construction of spread footings under water rather than to take the loads of the completed structure directly.[28] The better procedure was to set the piles close together, either in a single line or in two lines spaced a little apart and packed in between with clay, so as to form a continuous wall or 'cofferdam' around the intended footing. The space within was then pumped dry to allow excavation and construction of the footing to

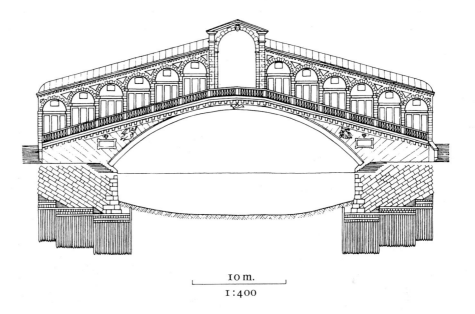

Figure 10.20 Elevation of the piled foundations of the Rialto Bridge, Venice, as described in contemporary documents. (Source: author)

10 m.

1 : 400

the first modern continuous raft foundation. This was constructed, in a notable rescue operation, by Sir Robert Smirke at the Millbank Penitentiary in London, previous foundations having failed to hold fast in the poor peaty soil.[25]

In such poor soil and worse, piles have the advantage that they can reach down much farther without calling for any excavation. Whereas spread footings and rafts distribute the loads coming on them by acting, to some extent at least, as cantilevered or continuous beams or slabs, piles act as extended columns or struts, either transmitting the whole load to a lower firmer stratum or distributing it vertically over some or all of their depth.

Timber piles seem to have been used from a very early date in marshy ground, and were certainly in common use when Vitruvius wrote of them late in the first century B.C.[26] Without them it would have been quite impossible to build on the marshy sites of cities like Venice and Amsterdam. An outstanding and well-documented example of their use was that made in the late sixteenth century to carry the present Rialto Bridge.[27] Since there were outward

proceed as if on dry land. The other, not involving pumping the water out, was simply to dredge a fairly level bed within the cofferdam and then fill the space by tipping in stone rubble or concrete, the piles serving to retain these in place.

An alternative underwater procedure, employed chiefly in the eighteenth century, was first to pile the whole area, then saw the heads of the piles level with one another close to the level of the river bed and build upon them. Since they were under water, something tantamount to a cofferdam was required for this purpose and it was provided by an open-topped box floated out and sunk in place.[29]

At best, these foundations only partly overcame the deficient bearing capacity of the poorer types of ground. No timber piles or spread footings that were not carried down to rock would have continued to carry the great weight of the central piers and arches of the Pont du Gard [4.7] and withstand the scouring action of the river in flood. And this is but one of the many structures that has owed its survival to being founded directly on surface outcrops of good rock. Similarly it is unlikely, for instance, that any of the more structurally daring Gothic cathedrals of the Île de France would have stood for long on the sort of piled foundations used for St Mark's in Venice. Their undoing, in this case, would have been an excessive differential sinking of the bases of the principal piers and columns, both in relation to one another and, more seriously, of individual bases under pressures concentrated towards one side as a result of the vault thrusts. There were limits, therefore, to the structures that could be built on some sites even with the best-known types of foundation. The complete absence of any reliable technique for quantitatively assessing in advance the ground conditions and the precise requirements for safely spreading the loads meant, moreover, that these limits were never fully known in advance. Doubtless, this often encouraged a tendency to unwise economies and even a deliberate skimping of work that would always remain out of sight.

Nevertheless it was not until the mid-nineteenth century that a pressing need was felt to overcome some of the limitations in the case of underwater foundations, and not until towards the end of the century that a similar need made itself felt in foundations for tall buildings.

The need in the first case stemmed from a desire to build bridges in places where the adequate founding of some of the piers called for excavation to rock at depths well below those that could be reached with the usual open cofferdam. The answer was a new form of caisson. This, like the cofferdam, was a continuous wall within which the necessary excavation could take place. But it was constructed away from the site of the pier and was provided with a cutting edge at the foot. After being towed to the site, it was allowed to sink to the river bed. As the ground was excavated within it, it gradually sank to the firm rock. The pier was then constructed within it, but it remained as a permanent outer wall to the pier. Usually, therefore, it was constructed of steel or, later, of reinforced concrete. The excavation could be carried out by dredging under water, but it was more satisfactory to do it in the dry. The major innovation was to close the top of the caisson, leaving air locks for access and for removal of the excavated material, and then to expel the water by means of compressed air. Such compressed-air caissons were first used for major bridges in England, notably by Brunel, and then by Eads to found the piers of his bridge over the Mississippi at St Louis, Missouri [10.21].[30]

The need in the case of tall buildings was a consequence partly of the relaxation, with the development of the mechanical lift or elevator, of the restriction on height previously imposed by the staircase as the only means of vertical access, and partly of the increasing pressure for more intensive development of the confined central areas of cities like Chicago. Where, as at

Chicago, natural rock was not easily reached, the point came at which the traditional spread footing occupied practically the whole area of the site. It therefore took up nearly all the space that might otherwise have been used as a basement, and there was no possibility of extending it to permit a further increase in height. To escape from this impasse, new forms had to be devised. Chief among these were the steel grillage, the reinforced-concrete slab or continuous raft, the deep piled foundation, and the more substantial reinforced-concrete cylinder or pier extending even farther down to rock.[31]

In the present century, there have been further developments in a number of directions. In terms of actual structural forms, most, but not all, of them have been improvements on the forms already in existence at the end of the

Figure 10.21 Sectional view of the compressed-air caisson for constructing the east pier of the Eads Bridge, St Louis, from Woodward, C. M., *A history of the St Louis Bridge*, St Louis, 1881. (Source: author)

nineteenth century, replacing timber piles, for instance, by piles of steel, reinforced concrete, and prestressed concrete [10.22].[32] One important new form, though not entirely unprecedented, has been the buoyant foundation referred to earlier.[33] This calls for the design of the whole basement as a stiff raft, but, more than anything else, has raised the height ceiling for very tall buildings on poor ground. Another, useful chiefly for very tall towers of unusual slenderness, has been an inverted conical or similar shell-like form which is more closely analogous than any other form so far devised to the spreading roots of a tree. In terms of construction methods, the development already under way in the nineteenth century of power-driven plant for such operations as digging, pile driving, and boring holes into which piles are then cast has continued with increased momentum. But most far-reaching, probably, was the development of a completely new science of foundation behaviour, at the outset largely by one man, Karl Terzaghi. Thus, for the first time, it has become increasingly possible since about 1950 to design foundations with a fair knowledge of what is required. Limitations on the bearing capacities that can be achieved remain, but in most places they are now economic restrictions on the choice of foundation type, rather than physical ones.[34]

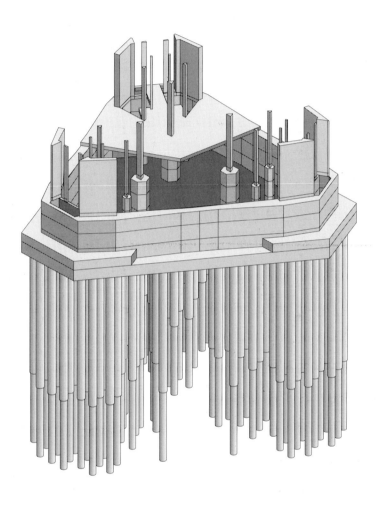

Figure 10.22 Piled foundation of the Commerzbank headquarters, Frankfurt (Arup). The weight of the tower (whose base is seen at the top) is distributed by a deep cellular substructure to 111 large-diameter piles cast in-situ in pre-bored holes and the whole foundation was designed to minimise settlements and interference with nearby structures. (Source Ove Arup and Partners from the *Arup Journal*)

Part 3: Complete structures

Chapter 11: Early forms

Having looked in turn at all the principal types of structural element, we can now widen the focus and look at the larger wholes – the complete structures – of which these elements were or are only parts. In tracing the developments of these complete structures we shall, as has already been pointed out, be concerned with the contexts within which the individual elements were, at least until very recently, themselves developed. But in again approaching the developments from the standpoint of our present understanding of structural actions and interactions, it will be more useful to concentrate on the ways in which different elements were associated with one another to meet the primary requirements for shelter, protection, passage, or even display. In structural terms this means

concentrating on the ways in which loads were carried down from spanning elements to a safe foundation, horizontal thrusts or pulls resisted without endangering the stability of the supports, and an adequate margin of overall stability ensured against the actions of wind and earthquake.

The much greater diversity of form characteristic of complete structures, as compared with structural elements, means that it will be necessary to be much more selective in the choice of a relatively small number of developments as illustrations of much broader trends. Also (as previously noted in Chapter 5) it will not be possible to fit these into as neat and comprehensive a classification as that in Figure 5.5. One basic distinction was suggested in this

Figure 11.1 A modern Egyptian mud-brick house south of Abydos. (Source: author)

Figure 11.2 Wall painting
showing an internal cross-
section of an earlier
Egyptian house, Thebes.
(From a photograph by an
Egyptian Expedition of
the Metropolitan Museum
of Art retouched by the
author to subdue the major
areas of damage)

earlier chapter. This was the distinction between
forms that are broadly analogous to one of the
simple elemental forms plus the necessary
support and those that, in important respects, are
too complex to be considered only in this way.
Some of the structures that will be discussed in
Chapters 12 and 13 will be found to be of the
first kind; while nearly all those to be discussed
in Chapter 15 are more complex. But these
classes are not clearly defined, and there is
much diversity within them.

Moreover (as was noted in Chapter 1), the
actual pattern of development was not a series
of fairly straightforward linear progressions as
the development of individual elements largely
was. The repeated cross-currents, borrowings,
influences, and interactions referred to there can
hardly be ignored. In tracing the pattern, more
emphasis must therefore be placed on the chronological sequence.

The approach that will actually be adopted
in the next five chapters is a compromise, based
partly on broad chronological subdivisions and
partly on equally broad groupings of structural
forms. In the chronological sense it takes
account chiefly of two far-reaching changes in
direction – those brought about by the Roman
introduction of concrete and of wide-spanning
arches and vaults; and by the introduction of
iron and of a new quantitative analytical and
experimental approach to design around the
beginning of the nineteenth century. Chapter 12
will discuss the development of the two
structurally most important types of building
between these two changes in direction. Chapter
13 will discuss the development of their more
recent counterparts after the second change.
Chapters 14 and 15 will discuss that of two
other types of structure – the bridge and the tall
building – from the first change to the present
day, with the main emphasis on the more recent
developments. In the last three of these chapters,
attention will also be paid to the further
widening of choice that has stemmed from the
availability of other new materials and new
possibilities of achieving an active structural

response (as distinct from the usual passive one), and to the new freedoms gained by the use of increasingly powerful computers in quantitative analysis and form-finding.

This leaves for discussion in the present chapter a much less coherent set of earlier developments. They range from the most primitive forms of shelter – very impermanent forms – to more developed but still impermanent house forms and the much more lasting forms devised for such uses as tombs and temples. Structurally they were mostly quite simple, or their complexity (as in the case of the pyramids) was of a kind that has little relevance to later developments. It will be sufficient, therefore, to look at them quite briefly.

Primitive shelters and later house forms

We need not linger over the primitive shelters of early nomadic hunters, fishers, and collectors of wild fruits, berries, and nuts. All the evidence, sparse though it is, suggests that, over many millennia, they barely advanced beyond the simple types of hut illustrated in Figures 1.1 and 7.1.[1] No doubt very long experience went into the perfecting of these forms, but they remained structurally simple and elemental as well as being the simplest forms to conceive spatially and the simplest to construct. As prototypes for the dome, they have been considered already in Chapter 7.

Around the eighth millennium B.C., after the major climatic change of the passing of the last Ice Age, a movement from a nomadic to a more settled way of life increasingly dependent on agriculture seems to have begun in the so-called 'fertile crescent' of the great river valleys in the Near East. As one aspect of it, the more permanent house began to replace the temporary

shelter. Initially all that this meant was probably the substitution, in many cases, of rammed earth or mud-brick for less permanent reed or reed plastered over with mud. But soon a different form did appear (or perhaps reappear). In this, the wall and the roof were clearly distinguished from each other. There were four walls on a rectangular plan, and they were presumably first roofed with straight horizontal timbers set side by side and given a covering of mud to provide added protection from the sun. The walls were thick enough both to carry the roof and to give lateral stability. Figure 11.1 shows a modern example, virtually identical with the earliest form save that a second storey has been built above the first at the left. To the right is a mere wind-break of reeds probably reproducing a much earlier form of shelter.

The chief limitation here was that imposed by the lengths and girths of timber available for spanning between the walls. It was overcome in two ways: by introducing columns as intermediate supports, and by using only a few long heavy spanning timbers to carry a large number of lighter shorter ones. The heavy timbers spanned directly between the walls or between wall and column or column and column, and the lighter ones, more closely spaced, spanned only from one heavy timber to another or to one of the other walls. The walls again had to provide all lateral stability, including, through the heavy principal timbers, that of the columns.

The final pre-Roman development of this form may be represented by the Egyptian house shown in cross-section in the wall painting from a New Kingdom tomb reproduced in Figure 11.2. Some allowances must, of course, be made for artistic distortions of the proportions and simplification of the details of, for instance, the floor construction. Excavations of roughly contemporary houses have shown that the mud-brick walls were nearly 1 m thick; that the columns, standing on stone bases, were substantial timber ones; and that the floors were probably constructed with palm trunks as the principal beams. A common form of planning, at least in two-storey buildings, was for most of the rooms to be arranged around either a tall central hall or an open central court, perhaps with a balcony carried by columns. Two storeys were probably much commoner than three, this being the limit.

Where, in Asia Minor, Crete, Greece, and neighbouring countries, climatic conditions were sometimes different, and stone (often with framing or bonding timbers) was used in place of mud-brick for the walls, there were variations. In general there was probably less use of columns to increase the width of rooms and a tendency to plan more by the simple agglomeration of units. But the basic structural form of beam-type floors supported by columns and walls, and given all lateral stability by the latter, was still the same. Barrel vaults (such as those seen in Figures 4.16 and 4.17) were an occasional alternative to timber floors, but were

Figure 11.3 Ground-floor detail of a house, Um-al-Jimal, Jordan. (Source: author)

not built with spans large enough, in relation to the thicknesses of the supporting walls, to have led to serious problems of thrust containment. The non-thrusting alternative illustrated in Figure 11.3, in which the span to be roofed was reduced by short cantilevered slabs set on top of the walls, was probably found only in regions where suitable flat slabs of stone were readily available.

Column-and-beam temples and palace and public halls

The more precarious and at the mercy of the will and whim of the gods ordinary life on earth seemed to be, the more important often, in relation to ordinary human habitations, were the temples of the gods. Similarly important were the provisions made for the after-life, particularly that of the priest- or god-king. Thus it was in the building of temples and royal tombs that more permanent forms of construction than those based on the use of materials like mud-brick and timber were first developed on a large scale.

Permanence, in this context, meant the use of stone; and, of all the countries in which earlier developments in construction and in social and economic organization had proceeded far enough to make it possible to take another step, the one most plentifully provided with easily accessible supplies of a variety of suitable stones was Egypt.

The actual change from mud-brick, reed, and timber apparently took place with little or no prior experiment in the building of Zoser's great funerary complex at Saqqara in the first half of the twenty-seventh century B.C.[2] In the centre of this complex were the burial chambers of the king and other members of the royal family, excavated deeply in the rock and surmounted by the step pyramid to be discussed later. Surrounding it, though, were not only a mortuary temple but numerous other structures within a high-walled enclosure almost 300 × 600 m in extent. Though most of these were only dummy replicas of contemporary less permanent structures – mere facades with dummy doors behind which was only a solid filling – they included a long roofed-in entrance hall leading from the only gateway in the perimeter wall to a large internal courtyard. The roofing consisted of slabs of the local limestone set close against one another and dressed on their underside to resemble the rounded logs of a timber roof. To reduce the distance that the slabs had to span from the overall width of the hall of about 6 m to about 1.5 m, walls terminated by fluted three-quarter columns projected towards one another at close intervals from each long side-wall, creating the effect of a long sequence of alcoves. Like all the other masonry except the larger roofing slabs, they were constructed of quite small, easily handled, blocks of stone. Similar attached columns occurred also in many other

Figure 11.4 Detail of the west portico of the second court, Temple of Ramesses III, Medinet Habu. Thebes. (Source: author)

parts of the complex ; and it seems most likely that their attachment to a wall in every case – and even the very existence of the projecting walls in the entrance hall – stemmed from a lack of confidence in the strength and stability of a similarly constructed free-standing column. Be this as it may, the crucial step in the change to a form constructed throughout of stone had been taken.

In the next dynasty, the difficulties of reliably jointing small blocks were avoided by going over to the use of very large blocks from which entire columns as well as beams and roofing slabs could be formed. The result, as seen in Figure 1.6, was a very different counterpart of earlier mud-brick and timber structures. Both columns and beams were uncompromisingly and unmistakably cut from stone with no hint of counterfeiting any other material, and their proportions were such that there was really no need, for the sake of lateral stability, for the massive surrounding wall.

Figure 11.5 Colonnades
of Amenophis III, Temple
of Amun-Mut-Khons,
Luxor. (Source: author)

A thousand years later, in the earlier dynasties of the New Kingdom, this form was still used, probably for the last time in the Osireion of Sety I at Abydos.[3] This had squatter proportions than the Chephren Valley Temple, but otherwise closely resembled it structurally. By then, though, the usual column was one of superimposed drums, large enough individually to give no cause for anxiety about the stability of the column as a whole. With it emerged the definitive Egyptian temple form which remained virtually unchanged into the Ptolemaic period.[4] This was essentially inward-looking. A tall and continuous outer wall enclosed a series of courts leading to roofed-in pillared halls and finally to the small completely enclosed sanctuary lit only through its doorway.

This later form was also constructed throughout of stone: thick walls, columns that were still very bulky by later standards, and heavy stone beams and roofing slabs. It is seen in the partial views of the Ramesseum and the Temple of Ramesses III, both at Thebes, in Figures 10.1 and 11.4. The walls, as previously noted, were less substantial than they appeared, but again there was little need of them to provide lateral stability where the column foundations were adequate. This is particularly clear from the perfect preservation over more than 3300 years of the two colonnades, over 15 m high, seen in Figure 11.5. Though they must have been intended as the central supports of the roof of a wider pillared hall, perhaps like the even taller hypostyle hall constructed about a hundred years later at Karnak, this hall was never completed and they have long been free-standing. The roofs, as will be seen from all three figures, were of the simplest possible form. Main beams, often coupled slabs of stone to make up the width, spanned between the columns or between them and the walls. The roofing slabs then spanned between these beams.

The spanning capacities of the stone dictated the spacing of the columns, which was usually

Figure 11.6 Unfinished temple, Segesta. Note that the final dressing of the columns and stylobate has not been carried out and that the blocks of the stylobate retain the projecting bosses used for lifting. The metal reinforcements of the architrave are modern. (Source: author)

Figure 11.7 East front of the Parthenon, Athens. (Source: author)

not much greater than their diameter. Inside any of the larger pillared halls the extremities were, therefore, visible only if one looked parallel to a line of columns. Their bulk soon closed the view in any other direction. But there was probably little concern for the resulting darkness of the interiors in a land where the light outside was usually excessive. Clerestory lighting was provided without any difficulty in the hypostyle hall at Karnak by dropping the roof level above the first row of lower columns that flanked, on the outside, each of the tall central rows. Elsewhere, doorways and small slits in the roofing slabs seem to have admitted all the light that was considered necessary.

Outside Egypt, similarly plentiful supplies of good building stone were first exploited in a comparable manner in Greece and Asia Minor. At about the same time as the building of the temples just considered, the fortified palaces of Mycenae and Tiryns, for instance, were built on the Greek mainland [8.2 and 10.2]. But the only structural achievement of this time that really challenged the Egyptian one was the great beehive or tholos tomb to be considered in the final section of this chapter.

According to the Homeric poems, the Mycenean Greeks had remarkably little need for temples as distinct structures. Their gods were much more human than the Egyptian ones and quite at home in the palace itself. The building of temples as distinct structures seems to have begun only during the long Dark Age following the collapse of Mycenean power, and to have begun on a very modest scale using mud-brick

and timber set on a stone base for the main structure, and thatch for the roof, and modelling the form on an oblong hall with a columned porch that had been typical of the earlier palaces. Only from the sixth century B.C. onwards, and quite independently of the Egyptian development, were larger-scale versions of it – the Doric temple on mainland Greece and the Ionic on the seaboard of Asia Minor – transposed into stone.[5]

Two of the better-preserved Doric temples, the first not in Greece itself but in Sicily, are illustrated in Figures 11.6 and 11.7. Figure 8.1 shows a detail of the second one and Figure 10.5 shows a detail of the more slender and graceful Ionic 'order' – that is of the basic unit of column and beam as elaborated architecturally in the Ionic form.

Whereas the Egyptian temple could be described as inward-looking, these later Greek temples were much more outward-looking. Indeed they were mostly set in elevated positions so that they could be seen from afar. Architecturally each of them remained essentially a large oblong hall (or cella), open at one end and containing a statue of the god at the other. The roof, however, projected on all sides well beyond the walls enclosing this hall, being carried partly by outer rows of columns. In smaller simpler temples like the so-called Temple of Concord at Agrigento, there are no other columns. At Segesta likewise [11.6], no others may have been intended. But it was never completed and possesses only the outer colonnade and the entablature and pediments that this

always carried. In many larger temples like the Parthenon [11.7], there were also other rows inside the hall, sometimes double-tiered, to give the roof further support. In the largest Ionic temples all the outer colonnades were doubled. Unlike the Egyptian temples again – whose roofs were always flat apart from the very slight falls that sufficed for drainage in a dry climate – both the Doric and Ionic temples had gently pitched roofs as indicated by the slopes of the pediments.

Despite their independent development and the resulting differences of architectural form, they nevertheless were very similar to the Egyptian temples in their basic structural concept and action. The roofs, though pitched, do not seem to have been envisaged as thrusting or to have exerted much outward thrust in practice. They seem to have been regarded simply as vertical dead weights. The columns were, perhaps, regarded more as the primary supports of the roof than in the Egyptian case if the design of some of the later timber-and-mud-brick prototypes (in which columns set close inside the walls seem to have been intended to take most of the load) may be taken as a guide. Certainly, as shown by the survival intact of the unfinished structure at Segesta, there was no more need at that stage of construction than there was at Luxor [11.5] for the further lateral stability that was provided in completed temples by the walls. Indeed, in both Egyptian and Greek temples, it was probably an essential requirement of the construction sequence that each column should initially be stable when

free-standing. But in all completed temples the walls were also present and, when favourably aligned, they could have contributed to the stability of the whole in the event (more likely than in Egypt) of an earthquake.[6]

The major structural differences that did exist were to be found in the character of the masonry and the type of roof. The masonry showed a marked improvement on almost any earlier masonry in the precision with which the blocks of stone – whether the squared blocks of the walls and entablatures or the drums of the columns – were fitted together.[7] Perhaps as an added safeguard in the event of earthquakes, the individual blocks were further extensively tied together by iron cramps as described in Chapter 3. Though this cramping could be paralleled by the Egyptian use of wooden dovetail cramps to hold the blocks together during construction, it was certainly intended to be more permanently effective.

Whereas even the roofs of the Egyptian temples had been of stone, the use of stone in the Greek ones usually stopped short at the entablatures and pediments – that is, at the point reached at Segesta – and the ceilings of the perisytle. The Greek roofs were tiled and, with few exceptions, were supported by gently sloping timber beams. Since the timber was much better than any native Egyptian timber, it permitted greater spans and thus wider spacings of the supports in the direction of the principal span and a more open and spacious interior. The central span between the internal colonnades in the Parthenon was 11 m.

The presence of these internal colonnades makes it virtually certain that the whole roof was a beam-type structure. It probably resembled the roof of a fourth-century naval arsenal at Piraeus, not far away, which is clearly described in the earliest surviving record of a Greek roof. Here simple horizontal beams were the principal spanning elements, and the ridge beam and heads of the inclined rafters were propped up from them at the centres of the spans. The clearly visible level of the original ceiling in the Temple of Concord and an opening to provide access to space above seem also to be consistent only with a similar beam-supported roof.

The wide cellas of the largest temples at Selinunte, Agrigento and Didyma were never roofed. At Didyma, roofing was clearly not intended. At Selinunte and Agrigento the intention is less clear because the temples were never completed. It could be argued, from the absence of internal colonnades, that trussed roofs were intended there to shorten the spans. But to do so hazardous in the absence of more positive evidence would be hazardous.[8]

Contemporary with these temples were a variety of public halls, usually square in plan, with external stone walls, and with rows of internal columns parallel to the walls and spaced closely enough to carry timber roof beams.

The outstanding examples of this form of pillared hall are the two great audience halls of the Achaemenid kings – the rulers of the Persian empire that was overthrown by Alexander the Great – at Persepolis in southern Iran.[9] Though not temples, their function was religious rather than secular in that the king appeared in them more in the capacity of high priest than in that of earthly ruler. They had very thick external walls constructed largely of mud-brick but strengthened by door jambs and inset panels of cut stone. All the mud-brick has now perished and little more than the partly-restored base of the walls of the older hall – the Apadana – can be seen in the view from the mountain behind that is reproduced in Figure 11.8. Several stone jambs and inset panels belonging to the wall of the slightly larger and later Hall of a Hundred Columns can, however, be seen in the foreground. Both within these enclosing walls and outside them on some sides to create open porticoes, columns were set on regular square grids. The thirty-six internal columns of the Apadana were spaced almost 9 m apart on their axes and, including the capitals, were over 20 m high. Those of the Hall of a Hundred Columns were spaced about 6 m apart.

The basic form was, therefore, almost identical with that of the hypostyle hall at Karnak referred to above. But the columns, besides being much more slender, were, in the Apadana at least, considerably more widely spaced. As in the Greek temples, the wider spacing, was made

Figure 11.8 Looking down on the Apadana of Xerxes, Persepolis. Note the figures at the feet of the columns. (Source: author)

possible by the use of timber beams. Nothing remains of these, the whole palace having been pillaged and then set alight by Alexander in 330 B.C. But there is ample evidence that they were of cedar imported from the Lebanon, and detailed representations of the complete roofing system are conveniently preserved in the facades of the rock-cut royal tombs at nearby Naqsh-i Rustam [11.9]. Continuous beams running parallel to one pair of walls were set in prepared recesses in the column capitals. Similar beams were then set above these running parallel to the other walls. Finally, closely-spaced lighter beams were set in the first direction, spanning simply from one main beam to the next to carry the weathering surface. It was, therefore, much more of a two-way system than the roofs considered previously and, by being built into the thick external walls, would have served to stabilize the columns in both directions.

Tombs, pyramids, and some other structures

The tomb, even more than the temple, usually reproduced in essence and in a more durable

Figure 11.9 Rock-cut tomb, Naqsh-i Rustam. (Source: author)

manner the contemporary house form or some earlier house form hallowed by association. Indeed examples cut into natural rock and closely modelled on contemporary timber house or palace forms often provide today the best evidence of these forms as just seen in Figure 11.9. Other notable ones are to be found in Lycia on the southern coast of Asia Minor and in some Etruscan tombs in Italy. Here, though, we are concerned with tombs that did more than just reproduce an existing form. These fall into two classes: the Egyptian pyramid and the Mycenean tholos or beehive tomb.

Though the Mycenean tholos was much the later of the two, dating only from the fifteenth to thirteenth centuries B.C., its origin can be traced back to the presumed earliest hut form of all, to the earliest transpositions of this form into the false dome of small blocks of undressed stone [7.2], and to a fairly widespread subsequent use of this form for burial purposes.[10] The reason for selecting the Mycenean version is that – particularly in the finest example, the so-called 'Treasury of Atreus' or 'Tomb of Agamemnon' [11.10] – it achieved a scale and a perfection that went far beyond that achieved elsewhere either in a tomb or in any other use of the form. Today at least, this tomb marks the culmination of vaulted construction in pre-Roman times, in addition to which its diameter at the base of 14.5 m exceeds by an appreciable margin the largest roofing-span of any of the structures so far considered, even of those roofed in timber.

When such domes were built above ground, they had to be very thick over much of their height, like the arches in Figure 6.4. As already explained in Chapter 7, this was necessary to resist the outward thrusts that, in spite of the horizontal coursing of the blocks, would otherwise have caused the lower courses to slip on one another and thereby allowed the upper ones to tip inwards and fall. In the 'Treasury', as can be seen from the present illustration, the blocks were themselves very deep in the direction of the bed joints. Further buttressing was provided by heaping up large masses of earth against the outside, presumably as construction proceeded. Externally, this buttressing mass created a flat conical mound, thick enough at the base to accommodate a smaller side-chamber, and terminated by a peripheral stone retaining-wall. Some idea of the internal scale can be gained from the doorway seen at the left, which gives access to the side chamber. It is a little over the height of a man.

The Egyptian pyramid, as seen at Giza and Abu Sir in the fully developed form of the fourth and fifth dynasties (twenty-seventh to twenty-fifth centuries B.C.), has certain structural affinities to the large tholos tomb. At its heart was a burial chamber roofed in some cases at least – notably in the pyramid of Sahure at Abu Sir – by slabs inclined against one another as at the entrance to the principal access gallery of the Pyramid of Cheops at Giza [6.2]. The

201

weight of masonry above these slabs would, to some extent, bear down upon them and increase their outward thrusts. But these thrusts were resisted, in a similar manner to those of the tholos vaults, by the even greater masses of masonry to each side.

The proportions were, however, very different. The burial chamber was always quite small, certainly in relation to the size of the pyramid, and there was no need for the latter to rise high above it for the purpose of resisting outward thrusts. The height was clearly sought for its own sake, and it introduced its own structural problems rather than solving others. The resulting huge mass both imposed extra weight on the roof of the burial chamber and could itself become unstable. Possible problems with the burial chamber could be, and sometimes were, partly evaded by excavating this deeply in the natural rock on which the pyramid was built. The main problem was to ensure the lasting stability of the huge mass. This problem would not have arisen if it had all been carved from the solid, in the same way as some surprisingly pyramid-like rock masses have been carved by natural erosion in the Nubian Desert farther south. Instead, it was always put together from innumerable separate blocks which were individually quite small in relation to the whole, and it was impracticable to dress more than a small proportion of them to fit closely together. Such heterogeneous masses have natural 'angles of repose' which are related to the shapes of the individual blocks. But these angles are usually fairly flat. The mass will be unstable if an attempt is made to give it a greater slope unless some additional provisions are made to hold it together.

Figure 11.10 The so-called 'Treasury of Atreus', Mycenae. (Source: Hirmer)

Fortunately, from a historical point of view, evidence of successive modifications of the designs of the earlier pyramids of Zoser and Seneferu at Saqqara, Meidum, and Dahshur (accompanied by a partial collapse at Meidum) allows the sequence of experiments through which the final form was evolved to be traced. We must, nevertheless, be content here with a quick look at Zoser's step pyramid at Saqqara [11.11] and the relation to this of the final form.[11]

The step pyramid clearly began as an unusually large square mastaba built over the burial chambers. The mastaba was a flat-topped bench-like structure of which there were already numerous examples nearby, each constructed of solid mud-brick over earlier burial chambers. Apart from its size and square plan form (the usual form having been oblong), this particular mastaba differed from the others in being constructed of stone, like the other structures in the same funerary complex discussed above. Again, only the facing was constructed of dressed blocks: the masonry behind was much more loosely fitted together, To help stabilize it, the side walls were given a slight inward batter so that they leant inwards on the filling. Twice this mastaba was increased slightly in area, but not in height, by additions to its sides, perhaps partly to stabilize it further. Its final extent

Figure 11.11 Step Pyramid, Saqqara. (Source: author)

Figure 11.12 Temple 1, Great Plaza, Tikal, now lacking some of the outermost stepped facings and most of the final flight of stairs leading to the 'roof comb'. (Source: author)

corresponded very closely to the area on the south face seen in shadow in the figure. Then, as the first major modification and simultaneously with a slight further lateral extension, it became the proto-pyramid by the superimposing of three other similar mastaba-like structures, each with its facing walls set back somewhat behind those of the one below. When this experiment succeeded, it was then given its final form by a considerable extension of the base on two sides (the left and rear when viewed as in the figure), corresponding extensions of the stage already built, and the addition of a further two stages at the top. It retained the stepped profile.

Subsequent experiments seem to have been primarily directed towards transforming this stepped form into the true pyramid form with flat triangular faces. Each time, as far as can be ascertained without partly dismantling the better preserved examples, the process of building up the main mass that came about largely by accident at Saqqara was retained. But it was followed more systematically, giving rise to a basic stepped form consisting of six or more inward-sloping layers on each side, each stopped short at a lower level than the one built previously just inside it. Finally the steps were filled in by adding another layer, varying of course in thickness, and sometimes bedded horizontally but elsewhere inclined inwards towards the stepped core. At Meidum, in the first of these subsequent experiments, this last layer collapsed almost to the foot, taking with it some layers of the core. But, partly thanks to growing skill in and attention to the dressing and fitting of the blocks, and the use of much larger blocks at least for the facings of the layers, the later experiments were more successful. The almost perfect preservation of some of these later pyramids, despite the continuing invitation that they have offered to later builders to use them as a convenient quarry, makes them by far the most enduring major constructions up to now.

Of related forms, the one that most closely resembled the step pyramid at Saqqara was the Mesopotamian ziggurat. Indeed this was even more the product of an accidental or ad-hoc process of accretion – the successive enlargement and heightening over a long period of a temple platform as one temple was superseded by another built over its filled-in shell. But even in the form given to it by Nebuchadnezzar in the biblical Tower of Babel, it did not approach in height the largest Egyptian pyramids built some 2000 years earlier. That no ziggurat has been as well preserved as these pyramids must be attributed, at least partly, to the use of brick – mostly unfired mud-brick – rather than stone.

The earlier Mayan pyramids were also the product of successive accretion over considerable periods, but, like their Egyptian forerunners, they were again built of stone. Although they came no nearer than the ziggurats to equalling these forerunners in size, the forms

they finally assumed were more structurally daring and accomplished.[12] Their steeply battered faces and crowning 'roof combs' gave some of them an almost tower-like appearance. This is best exemplified by Temple 1 at Tikal [11.12], which rises to a height of 44 m and seems to have been built in one sustained operation and, in other ways also, in much the same way as the later Egyptian pyramids. Subsequent pyramids and tower-like temples elsewhere marked no further structural advance.

More tower-like forms ranged from defence towers and the pylons that served, from the New Kingdom onwards, as monumental entrances to Egyptian temples, to one of the seven wonders of the Greek Classical world – the lighthouse of Alexandria. Many of them, including the lighthouse, have perished. But numerous temple pylons and some defence towers are well preserved; something is known of others from, for

Figure 11.13 Central tower of the gateway in the boundary wall, Temple of Ramesses III., Medinet Habu, Thebes. Note that over the entrance and over the upper window, the upper one of the pair of superimposed lintels bears on the lower one only near to each end. A small gap was deliberately left between the two along the central parts of the spans to relieve these of load. (Source: author)

instance, Assyrian palace reliefs; and the general form of the lighthouse was recorded in medieval Arabic descriptions after it had suffered some damage and reconstruction but before it finally disappeared in the fourteenth century A.D.

With the exception of the temple pylons, these forms resembled most closely the forms of multi-storey houses of the kind illustrated in Figure 11.2, but with more substantial walls and smaller rooms that eliminated the need for internal columns. The 20 m-high tower built in

The temple pylons, being also largely solid masonry, were really enormous walls enclosing one end of a courtyard and containing the entrance to it. Like many other walls and like the pyramids before them, they were only faced with carefully fitted squared blocks of stone. Inside was a much looser fill. Figure 11.14 shows, at the upper right, the lower part of one end of the inner face of the largest of several pylons at Karnak – one that was never quite completed.

Figure 11.14 Detail of the inner face, First Pylon, Temple of Amun, Karnak, showing the remains of a construction ramp. (Source: author)

the early twelfth century B.C. that is shown in Figure 11.13 was probably used partly for residential purposes and may have been more a monument to the Syrian campaigns of Ramesses III than a genuine defence-tower. It nevertheless closely reproduced the form used in contemporary fortifications. A later example, from the fourth or third century B.C., has already been illustrated in Figure 10.7.[13]

The lighthouse, which in medieval times was more than 130 m high, must have been constructed on similar principles, but on a broader base and with even more substantial walls. It was constructed in the mid-third century B.C. The first stage was apparently 30 m square at the foot, tapering slightly inwards as it rose. Above this rose two further stages, the first octagonal and the other circular. Probably it was largely solid masonry apart from the necessary internal ramps and staircases.[14]

The greatest interest in this photograph is not, however, the pylon itself but the rather shapeless mound piled up against it. This is what remains of the only surviving example of a mud-brick ramp built up against the pylon as its construction advanced. Up it, all the blocks of stone would have had to be hauled manually in the absence of any mechanical handling equipment. On completion it would then have had to be laboriously dismantled and taken away. Comparable temporary works would have been required even to set upright a large monolithic column, and similar works on an even greater scale would have been needed for the construction of the pyramids. This was the penalty, for most of the period reviewed in this chapter, of being dependent on little more than the power of human hands assisted by simple devices like the lever and the sled for all the mechanical operations entailed in construction.

Of those builders using large-stone masonry rather than bricks, the first to have turned to the use of hoisting devices based on ropes and pulleys seem to have been the Greeks of the sixth or fifth century B.C. U-shaped cuttings or projecting bosses on the faces of the blocks for the attachments of lifting ropes furnish some of the evidence for this. Such bosses may be seen on the stylobate of the temple at Segesta [11.6]. To the wider exploitation of such devices must be attributed at least some of the progress in construction made by the Romans.

Chapter 12: Pre-nineteenth-century wide-span buildings

If multi-storey buildings are excluded, the two structurally most important types of building from the Roman period to the end of the eighteenth century are those that might be considered as the successors of the earlier column-and-beam temples and of the domed tholos tomb. Examples of the first type range from Roman basilicas and early basilican churches to the Gothic cathedrals of thirteenth-century France. They also include the once very numerous large timber-framed halls and barns. The second type is best exemplified by the Roman Pantheon, the sixth-century Church of Hagia Sophia in Istanbul, and a number of large centralized churches, mosques, and mausoleums built from the early fifteenth century onwards.

All these structures were primarily space-enclosing. In that sense they were inward-looking, usually with less, and sometimes with very much less, emphasis on the exterior. But though they shared this emphasis on the interior with, for instance, the Egyptian temple, the interior itself was now very different. The aim seems always to have been to make it as open and unobstructed by intrusive columns or piers as possible, whether for simple functional reasons like the provision of unobstructed sight-lines for large numbers of people, for the sake of monumental grandeur, or in the pursuit of less mundane and wholly non-structural objectives such as those referred to in Chapter 1. To achieve it called for the use either of wide-spanning roof trusses or of equally wide spanning vaults or domes. Even if the truss itself was probably not, and the vault and dome were certainly not, Roman inventions, it was only in

Figure 12.1 Principal support systems for concrete and masonry domes and vaults prior to the nineteenth century. Heavy full lines show the directions of the principal thrusts; heavy broken lines indicate ties. In the more skeletal forms the principal thrusts necessarily follow the lines of the ribs. (Source: author)

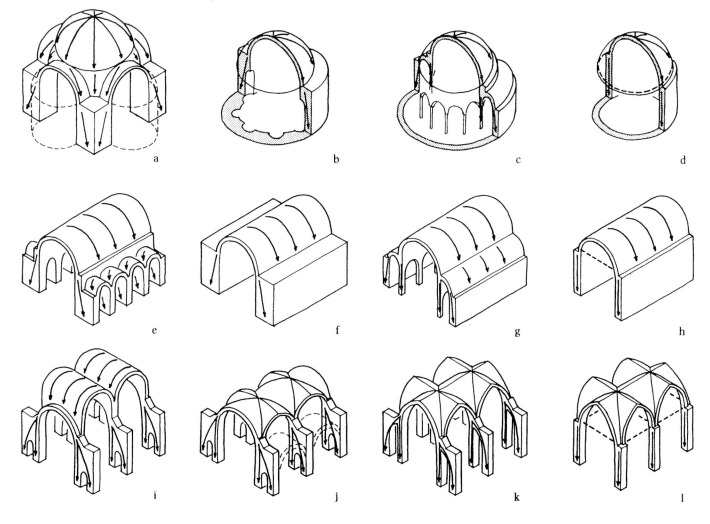

late Republican and early Imperial Rome that both were fully developed and exploited. In addition, in order to expand further the total spatial enclosure by linking other peripheral and subsidiary spaces with the main one, extensive use was made of the arch to permit openings to be made in the walls that directly carried the principal spanning elements or to reduce these walls to arcades or a few isolated columns or piers.

When trusses were used as the principal spanning elements, they themselves exerted no side loads on their supports. Provided that their horizontal tie members could be relied upon to remain effective, they were, in this respect, no different from the timber and stone beams that had been used earlier. Buttressing of the supporting walls or piers was required only to resist overturning by externally imposed side loads due to the wind or a possible earthquake.

When, on the other hand, vaults or domes were so used, and when the supporting walls were opened up by means of arches, the dead weights of these elements did generate outward horizontal thrusts of which full account had to be taken in the design of the supports or in the design of the whole structural system – for the most efficient system was usually one in which, as far as possible, the thrust of one spanning element was directly balanced by the opposing thrust of another similar element. Some of the possibilities are illustrated in a very diagrammatic way in Figure 12.1. Of them, only (b) and (f), with some variations on (b) referred to in Chapter 7,[1] were exploited on a relatively small scale in pre-Roman times. The rest were all characteristic of the period now under review, and it is with their detailed realization in a number of typical or outstanding buildings that we shall now chiefly be concerned. In one instance – the Gothic cathedral (k) – there was a close approach to a skeletal form, and the fully skeletal form of the large timber-framed hall and barn will be considered alongside it.

Roman timber-roofed basilicas and large vaulted and domed halls

The Roman basilica was a place of general assembly, rather like a covered forum. In this sense it resembled most closely, among earlier forms, some Greek public halls, though it was less restricted in its manner of use than these. Structurally it was more akin to the Greek temple stripped of its external colonnades. It usually differed chiefly from this in making full use of the timber truss to obtain a wider uninterrupted central span. At least we can infer as much from the description by Vitruvius of a basilica that he constructed at Fano,[2] and from the widths of the central naves of basilicas elsewhere of which just enough has survived to indicate their ground floor-plans.

The greatest and probably the most influential of these truss-roofed structures was Trajan's Basilica Ulpia constructed early in the second century AD.[3] In plan it was a large rectangle with one of its longer sides closing one end of his Forum in Rome and a projecting semi-circular apse opening off each shorter side, and with porches and the other main entrances along the side facing the forum. Within the enclosing rectangle were double colonnades forming two aisles on each side of a central nave, 25 m wide. Though only a few re-erected single columns now stand, the colonnades around the nave were originally double-tiered and were spanned at each level by horizontal entablatures in the same manner as the two-storey internal colonnades of some of the larger Greek temples. The lower entablatures carried vaulted terraces over the aisles, and the upper ones carried a trussed timber roof over the nave. Clerestory lighting entered through the open upper colonnades.

Since so little of it can now be seen above ground, the form can be best appreciated today by looking at early churches that seem to have been closely modelled on it. One is illustrated in Figure 12.2 and another in Figure 12.18. If horizontal entablatures were substituted for the arcades carried by their columns, and open vaulted terraces for the galleries of the first and above the aisles of the second, there would be close resemblances to the earlier form, apart from the much reduced scale of St. Demetrius.

The parallel development of large vaulted and domed halls was linked more, until the construction of the Basilica Nova, with palace halls, garden pavilions, large public baths, and mausoleums. The forms adopted for these owed much less to any earlier precedents than to the exploitation of the new space-enclosing possibilities offered by Roman concrete. The development of the vaults and domes themselves was, in fact, so central to it that most that needs to be said about it has been said already in Chapter 7. Here we need look only at those aspects of the design of the supports that were touched on only very briefly there.

The simplest form of vault to be widely used was the semicircular barrel. Mostly it was used over fairly modest spans, but it is possible that some at least of the large state halls of Domitian's Domus Flavia were so vaulted over spans approaching or exceeding that of the Basilica Ulpia. Almost always the outward thrusts were resisted by giving the vaults continuous support from thick walls as in Figure 12.1(f), or from somewhat thinner walls buttressed at intervals by projecting piers. Sometimes though, the outward thrust of one vault does seem to have been deliberately countered, to some extent, by constructing another vault alongside it – a procedure of which full advantage could have been taken only if both vaults were constructed together or, where a wide vault was flanked by narrower ones, if the narrower ones were constructed first. A good example of the use of narrower flanking vaults as on the left-hand side of Figure 12.1(g) is the so-called 'Temple of Diana' at Nîmes. This still stands with much of its main vault intact, spanning about 10 m and constructed of cut stone stiffened by inwardly projecting ribs.

In general, the type of support illustrated in Figure 12(f) may be regarded as virtually a downward continuation of the arch or vault, within which the thrust line becomes progressively more vertical as it nears the ground. Equilibrium is theoretically possible, as for the voussoir arch itself, as long as the thrust line remains within the thickness throughout.[4] But, since this ignores both the deformations that will inevitably take place in the support and the slight tilting of the base that is likely if most of the pressure is exerted on the foundation near one of its extremities, sufficient extra thickness is usually necessary to ensure that the theoretical thrust-line will remain somewhat closer to the centre.

When there are flanking vaults as on the left-hand side of Figure 12.1(g), most of the outward thrusts of the main vault are, in effect, carried over to the outer supports, which may then similarly be regarded as downward continuations of the composite arched form. In this case, the intermediate supports need be proportioned to carry little more than a vertical load. Clearly the alternative form sketched at the right-hand side of this figure would have been more efficient, since it conforms better to the thrust line within the vaults. The usual Roman practice of placing a thick concrete filling over the haunches of any vault meant, however, that there would have been little real gain from the adoption of that alternative form.

In many ways a more efficient alternative to the simple barrel vault for covering a large rectangular hall was the groined vault, in which the main barrel vault running along the length of the hall was intersected by a series of shorter ones crossing it at right angles [12.3]. This was the favourite type of vault for the largest halls of Imperial baths, and well-preserved examples remain in the Baths of Diocletian (where the frigidarium was converted in the Renaissance into the Church of S. Maria degli Angeli). It was also used for the main hall of Trajan's market adjoining his Forum not far from the Basilica Ulpia, and, finally, for the last of the large basilicas – the Basilica Nova started by Maxentius and completed by Constantine. It had the advantage of concentrating the need for support at only a few widely separated points along the longer sides of the rectangle. This allowed both for clerestory lighting below the transverse barrels and for the opening out of the main room at floor level into smaller side-rooms below the windows. Of the Basilica Nova, only the supports of the main vault on the north side and the side rooms contrived between them now stand to their full height [12.4]. But some impression of its original scale can be gained from these.

Figure 12.2 St Demetrius, Salonika (extensively rebuilt to the original design after a fire early in the present century). (Source: author)

Figure 12.3 Nineteenth-
century reconstruction
drawing by Professor
Cockerell of the
frigidarium of the Baths of
Caracalla, Rome.
(From *Transactions of the
Royal Institute of British
Architects*, 1889)

Structurally, the result of concentrating the need for support at the feet of the groins was, of course, that the supports had to be proportionately more substantial than the corresponding lengths of continuous side-walls. But they still acted in essentially the same manner in carrying the vault thrusts down to the foundations, except that the end supports had to be given sufficient thickness to resist also the unbalanced longitudinal thrusts of the outer transverse sections of the main vault plus those of any lower transverse barrel vaults over the side rooms [12.1(j)]. More than enough depth of support in the transverse direction was automatically provided by adding the side rooms, so it became the usual practice to cut arched openings through the supports, both at floor level to interconnect the side rooms and above the vaults of these side rooms to provide there a continuous passageway over the roof. The effect of this was to split the resistance to the thrust of each section of the main vault into three parts in a manner more or less analogous to the splitting of the resistance to the thrust of a simple barrel-vault by constructing a lower and smaller vault alongside it [12.1(g) and (j)].

Though the concrete dome, given a sufficiently high quality of concrete, need not have thrust outwards in the same way, we have seen that it always seems, in practice, to have done so. That this was appreciated at the time is

indicated by the adoption of much the same kinds of support. The commonest was that illustrated in Figure 12.1(b), or some variant of this. The third-century mausoleum known as 'Tor de' Schiavi' [12.5] illustrates it very well, since it is just sufficiently ruined to show the complete form in cross-section. The support is the direct equivalent, adapted to a circular plan, of that illustrated in Figure 12.1(f). It acts in essentially the same way, even with some of its thickness hollowed out internally into eight alternately square and semi-circular niches at the level where the thrusts pass over towards the exterior. Characteristically also, it extends well above the internal springing-level of the dome as a counterpart to the usual filling over the haunch of an arch.

The greatest of all Roman domed structures – the Pantheon – might be described as merely a larger and more elaborate earlier version of the form typified by the 'Tor de' Schiavi', were it not that this would hardly do justice to the internal complexity of its construction. But that complexity has been described already in relation to the dome in Chapter 7, and it will be sufficient to concentrate here on the supporting drum. Its complexity resulted likewise from the much greater and quite unprecedented scale of the structure [12.6]; from consequent attempts to lighten the total mass by incorporating within it a considerable number of hidden voids in

Figure 12.4 Basilica
Nova, Rome.
(Source: author)

Figure 12.5 'Tor de' Schiavi', Via Praenestina, Rome. (Source: author)

Figure 12.6 Pantheon, Rome. (Source: author)

addition to niches open to the interior like those of the 'Tor de' Schiavi'; and from the further incorporation of ranges of large brick arches to bridge safely over both the hidden voids and the larger visible niches. The principal arches run continuously in two tiers around the whole circumference, the lower tier corresponding to the heads of the internal niches and the upper one to the hidden voids above these niches [7.7]. Each arch abuts its neighbour on each side and absorbs its circumferential thrust, just as did the arches of the circumferential arcades of the typical theatre or amphitheatre [6.9]. At least the thrusts are almost absorbed in this way. Because no two adjacent arches are quite in line in plan, there is a small unbalanced outward thrust from each, which adds to the outward thrust of the dome itself. But if we ignore this and some lesser complications of the precise manner of transmission of loads to the foundation, there was no major departure from the basic type of support illustrated in Figure 12.1(b). The foundations were appropriately massive, consisting of a solid ring of concrete 4.5 m deep, though probably not quite massive enough to prevent all outward tilting of the drum after it had cracked radially (in the manner seen in

Figures 2.14(c) and 2.15) under the combined thrusts of the dome and the circumferential arches.

The chief variants on the circular drum-like support of Figure 12.1(b) prior to, or contemporary with, the Pantheon were ones in which, usually on a much smaller scale, a more sinuous plan form was adopted to correspond to, or to be matched by, a scalloped and groined modelling of the inner surface of the dome itself in the manner already referred to in Chapter 7. If the dome had then been just a thin shell, this modelling would have largely precluded the development within it of hoop tensions to absorb its outward radial thrusts, and would thus have called for a greater thickness of the support than was ideally required for a circular dome. But the much greater thickness typically given to circular drums like those of the Pantheon and the 'Tor de' Schiavi' was already ample.

In some later structures, notably the so-called 'Temple of Minerva Medica' [7.8], there was a more significant change in the manner of support. It was largely reduced to a number of separate piers, between which were arched window openings and externally projecting niches.[5] But this was never carried far enough to call for a mode of action very different from that already introduced by the voids and niches and bridging arches in the more continuous support of the Pantheon dome. It merely called for a proportionately greater radial thickness of the piers to absorb both a more concentrated thrust from the dome itself and the additional unbalanced outward components of the thrusts of the bridging arches.

At a fairly late date though – in the Constantinian mausoleum now known as the Church of S. Constanza – a method of support akin again to that of Figure 12.1(g) was adopted. This is shown in (c) in the same figure. Directly beneath the dome springing-line the only supports were twelve radially-aligned pairs of monolithic columns which could hardly have withstood by themselves its outward thrust. Moreover, they carried the dome not on a horizontal architrave such as was used to span between the columns in the earlier timber-roofed basilicas and may be seen in the background of Figure 12.7 but on a ring of arches forming a light arcade like the slightly earlier one seen in the right foreground of that figure. Like those of the bridging arches considered above, the thrusts of these arches would, to a large extent, have cancelled one another out. Indeed it was this that permitted the substitution of relatively slender columns for the type of pier seen in Figure 6.9. But they would also, as before, have had residual unbalanced outward components which would have added to the total outward thrust to be resisted. To resist this total thrust, an outer barrel vault, now annular in plan, was constructed below the springing-level of the dome over an outer ambulatory. This vault effectively carried nearly all the thrust over to a

Figure 12.7 Diocletian's Palace, Split. The arcaded open colonnade of the peristyle and, behind it, Diocletian's Mausoleum, now the Cathedral. (Source: author)

substantial outer wall, leaving the columns to take only the vertical weight of the dome. (It is of course possible that, in practice, the idea of the ambulatory came first and that advantage was then taken of the potential buttressing action of its vault to reduce the immediate supports of the dome to the double ring of slender columns.)

Early domed churches and related later churches and mosques

It was from these Roman structural forms and their counterparts in a wide empire that most of the later developments to be considered in this chapter largely stemmed, either directly or indirectly. But the needs that these later developments served were no longer the predominantly secular ones of Imperial Roman times. They soon became much more those of the Christian Church and, later, those of Islam also. And the resulting pattern of development was far less consistent and coherent.

Initially its diversity did little more than reflect that within the later empire itself – a diversity that stemmed from the use, away from Rome and the Campagna, of brick or cut stone in place of brick-faced concrete for walls and piers, and of brick, cut stone, or hollow tubes in place of concrete for vaults and domes [12.7]. The usual architectural preference at the time,

both in Rome itself and in Constantine's new Rome (Byzantium as it was, Constantinople and now Istanbul as it became), was for the basilican form to serve as the church. The domed form was chiefly reserved for use on a smaller scale for mausoleums and martyria. From about the sixth century onwards, though, the centralized domed form was developed much further in Constantinople and the East to serve as the church itself and later as a mosque, while it was, with minor exceptions, dropped in the West. There, an almost independent development of the basilican form led, instead, to the Gothic cathedral. Only in seventeenth-century Italy and eighteenth-century France were these two different lines of development effectively brought together again to give rise to other new forms.

This is, of course, a gross simplification. The real pattern was much more complex and the mutual influences and stimuli passing between East and West were by no means negligible even when these two seemed to be farthest apart. But, in terms of the limited objectives of this book and the judgements of structural significance on which it is based, it seems the best framework to adopt. In spite of the facts that most attention was paid initially to the basilican form and that the particular development of the centralized domed form that was associated above all with sixth-century

Figure 12.8 SS. Sergius and Bacchus, Istanbul (now a mosque) looking north east. (Source: author)

Constantinople culminated only with the Ottoman Imperial mosques of a thousand years later, it was this latter development that showed the greatest continuity with what had gone before. It might almost be regarded, in fact, as the logical fusion of the two principal Roman developments – those of the domed hall and of the groin-vaulted rectangular hall. It will therefore be appropriate to consider this domed form first.

Something has already been said in Chapter 1 about the underlying non-structural factors that led to it being chosen for a church. It was not an obvious choice on simple functional grounds since, as built hitherto, it offered far less planning freedom than any of the other possibilities. Given a certain limit on the size of dome, all that could be done was to adopt an essentially circular plan with an overall diameter somewhere between that of the dome itself and, say, twice this. To gain a greater freedom it was necessary to do one of two things. Either a number of domes had to be set alongside one another in the same manner as a series of groined vaults. Or a single focal dome had to be incorporated into some other roofing system, with the space below it expanded to a greater extent than hitherto by externally projecting niches or an ambulatory or in some other way.

The second alternative, as first realized in practice, entailed no more than adding an outer ambulatory to a space already expanded by large niches or (as they are usually called) exedrae. The first experiment on these lines may well have been the Church of S. Lorenzo in Milan built in the second half of the fourth century.[6] This was modernized in the sixteenth century and its present central dome dates only from then. But much of the original structure was retained, and shows that there was, in plan, a central square expanded first by large semicircular exedrae and then, beyond these, by an ambulatory running around the whole. Since the exedrae were not bounded by continuous walls but only by open colonnades (as also were those of the 'Temple of Minerva Medica' originally), they communicated directly with the ambulatory. Twelve isolated piers in four groups of three supported the vertical load of the central vault, but left its outward thrusts to be carried over to four corner towers on the far side of the ambulatory. This manner of support and buttressing was so much better suited to resist the thrusts of a groined vault than those of a dome that it strongly suggests that the original vault was a groined one.

The best preserved later examples are the early sixth-century churches of S. Vitale in Ravenna and SS. Sergius and Bacchus in Istanbul. In both of these, the central dome is directly supported on eight piers, with part of its outward thrust carried over, again, to the lower outer walls. In S. Vitale there is an open exedra (extended at the east into a projecting apse) between each pair of piers, and the enclosing outer wall is octagonal. In SS. Sergius and Bacchus [12.8] there are four exedrae only, the alternate spaces between the piers being either filled with straight colonnades recessed slightly behind them or left open, so that the whole is fitted into a square enclosing outer wall. In both buildings the colonnades are two-tier, carrying a gallery as well as an upper vault spanning over to the outer wall.

The first alternative – setting a number of domes alongside one another – called for nothing less than the reduction of the number of direct supports for each dome from eight or more to the four that automatically sufficed for a groined vault. These four supports could be similarly bridged by arches. But, if the arches were straight in plan, as unreinforced masonry arches almost have to be, the result was still only a square rather than, say, an octagon as in S. Vitale and SS. Sergius and Bacchus or a decagon as in 'Minerva Medica'. Whereas the octagon and decagon approximated sufficiently closely to a circle for the transition to be made very easily [7.6 and 7.8], the square did not.

Two pointers to possible transitional elements were the sort of corbelled construction seen in Figure 7.3, and a simple enlargement of the radius of the dome and flattening of its curvature to allow it to extend right into the corners as seen in Figure 7.9. But neither was suitable in itself for construction on a larger

Figure 12.9 Apse of the Church of Qalb Louzeh. (Source: George H. Forsyth Jr., University of Michigan)

scale. It was necessary to substitute secondary diagonal bridging-arches for the horizontal corner-cutting courses of the first example, or to abstract from the second the spherical triangles below the level of the crowns of the main arches, to give the transitional elements now known respectively as the squinch and the pendentive and already described in Chapter 7. The squinch was first definitely used in this way in the fifth century, and the pendentive not until the sixth; but it was the pendentive that was by far the more important in the practical realization of this alternative.

With the pendentive to serve as the transition from the square to the circle, the complete domed unit of construction became that which is illustrated in Figure 12.1(a). Below the springing-level of the four arches and the pendentives, the need for support was just the same as for a single-bay groined vault. There was a significant difference only above this level, in that the radial thrusts of the dome (assuming again that there was insufficient tensile strength in the lower part to contain them within its own masonry) would act equally in all directions around the circumference instead of being concentrated, by the geometry of the vault itself, in the directions of the piers. It was necessary, therefore, for the bridging arches to be broad enough to carry to the piers the outward horizontal components of the thrusts acting on them near mid span as well as deep enough to carry the vertical loads. They had, in other words, to be capable of acting as arches in a horizontal plane as well as in a vertical one. Given this ability on the part of the bridging arches, such complete domed units could either be set side by side in the same way as groined vaults, or they could be combined with other forms of vaulting (such as extended barrel vaults) on one or more sides.

In the sixth-century churches of the Holy Apostles in Istanbul (long since disappeared but reproduced fairly closely in the present St Mark's in Venice and, less closely, in the churches of St Front in Perigueux and St Anthony in Padua) and of St John at Ephesus, five or six units were set side by side over cross-shaped plans.[7] More commonly a single dome was used in combination with other kinds of roof or vault.

The outstanding Byzantine achievement Justinian's Church of Hagia Sophia in Istanbul built between 532 and 537 – exploited also another possibility that had long been implicit in the use of semicircular niches, exedrae, or apses crowned by semidomes. A very early example of this use, standing in isolation and virtually carved out of a very solid mass of concrete, was in the Serapeum of Hadrian's Villa at Tivoli. The somewhat later use of exedrae to expand the space covered by a larger dome has just been referred to above. A more widespread use of semicircular apses was as the eastern terminations of basilican churches, a use which merely followed earlier similar use in secular basilicas like the Basilica Nova.

When the crowning semidome was constructed of solid concrete, it was possible for it to act as a series of adjacent arches parallel to the open forward edge. But when it was constructed of brick or cut stone laid in the usual way with the joints at right angles to the surface – and especially if there were any window-openings in the lower part – the internal compressions had to act more nearly radially and circumferentially as in a full dome. This meant that there was inevitably a tendency for the free forward edge to move farther forward near the crown until a horizontal arching action was established in the neighbourhood of this forward edge to resist the otherwise unbalanced radial compressions there. Figure 12.9 shows the resultant movement in a large semidome constructed of cut stone in about AD 500 – at least it shows the displacements of some of the wedge-shaped blocks of stone and the dropping of the crown that accompanied the forward movement.

Clearly though, the tendency of the forward edge to move farther forward could alternatively be turned to advantage by opposing it to some thrust in the opposite direction. The ideal state of balance would be achieved simply by setting two semidomes together to form one complete dome. But almost the same thing could be achieved by setting the two semidomes on opposite sides of another complete dome as indicated by the light broken lines in Figure 12.1(a) (where four semidomes are thus placed). Though they are, for convenience, shown there as abutting the deep bridging-arches otherwise necessary to resist the outward thrusts of the dome, their real potential value was, of course, in largely doing away with the need for these arches.

In Justinian's Hagia Sophia, semidomes of the same diameter as the main dome were set against its pendentives only at east and west to maintain the east-west longitudinal axis of the previous basilican church [7.10 and 12.10], and they were themselves opened out into smaller semidomes above colonnaded exedrae, and into short barrel vaults of the same diameter as these exedrae on the axis [12.11]. The dome's outward thrusts to east and west were thereby largely carried over to the substantial secondary piers at each side of the short barrel vaults. Only at the north and south had these thrusts to be resisted solely by broader arches spanning between the main piers as in Figure 12.1(a). To stabilize these piers against the thrusts, they were linked, across the aisles and galleries, to other piers. It was these buttressing piers that, rising above the gallery roofs, became the boldly projecting masses that are still one of the most conspicuous features of the exterior (A,A in Figure 12.11). To north and south, in other words, a support system essentially similar to that of the Basilica Nova was retained, with similar arched openings

Figure 12.10 Hagia Sophia, Istanbul, looking east. (Source: author)

216

introduced to provide continuous passageways along the aisles and galleries and over the gallery roofs.[8]

The result, architecturally, was a breathtakingly vast central expanse of completely uninterrupted space 30 m wide, more than twice as long and open, for most of its length, to broad aisles and galleries that were once filled with light much more than they are today. Only the relatively insubstantial screens of marble colonnades defined the boundaries of these aisles and galleries. Such details as the failure of their columns at gallery level to answer directly

great Imperial liturgy, to exchange the kiss of peace and share the chalice.

The choice and use of materials also showed considerable qualitative understanding of what was needed to meet the structural requirements. Up to the level of the springings of the main arches and semidomes, for instance, the piers' which directly carried them were constructed throughout of large blocks of stone very closely fitted together. Only above this level, and for the vaults and arches of the aisles and galleries and part of the outer walls, was brickwork used instead – and this brickwork was of uniformly

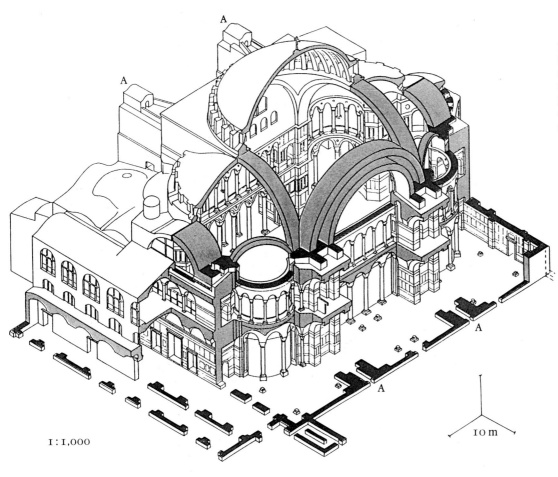

Figure 12.11 Hagia Sophia. Istanbul, part-cutaway isometric of the structure omitting later peripheral buttresses and other additions. For A,A see text. (Source: author)

1 : 1,000

10 m

to those below, coupled with the casing of the piers in thin slabs of veined marbles (an earlier Roman practice) and of all curved surfaces including the dome and main semidomes in gold mosaic, had the effect, moreover, of apparently dematerializing such structural mass as there was and creating a marked sense of ambiguity about the nature of the true system of support. They even made it seem, in the well-known words of Procopius, as if the dome did not rest on solid masonry but covered the space with its golden orb 'suspended from heaven'.[9] With all our greater resources it would indeed be hard to conceive today a more fitting and sublime setting than under this dome for the meeting of Emperor and Patriarch, at the climaxes of the

high quality where it would be subject to most stress. There was, moreover, a considerable use of both timber and iron ties. The latter – which mostly took the form of cramps between adjacent blocks of stone in the string courses provided at the springing-levels of the main arches and semidomes and of the dome itself – must have been intended to contain some of their outward thrusts.

But a complementary quantitative understanding of the requirements was lacking. The tensile strengths actually provided by the cramped string-courses were far below the required minimum strengths of about 1.5 MN at the base of the dome (for its presumed original design) and much more than this across the feet of the deep arches at north and south. This allowed the dome and semidomes to crack, just as the Roman ones had done, and to transmit their full outward thrusts to the supports as shown in Figure 12.12. Secondly, the buttressing outward extensions of the main piers to north and south (A,A in Figure 12.11 again) were too much weakened, as initially constructed, by the arched openings left in them, and a similar weakness was introduced at the east and west by insufficient inter-connections between the main piers and the secondary piers on the far sides of the exedrae. These short-

little from the ultimate achievement as they did from the lasting impact of its example on later patrons and designers.

In the Byzantine Empire itself, that impact was primarily an architectural one in that the basic centralized domed form was henceforth adopted almost universally for all kinds of church, but only on a much smaller scale and with only the simpler system of support of Figure 12.1(a) without the semidomes [1.8, 1.9]. Likewise, in the Islamic east until the conquest of Constantinople in 1453 A.D., the achievement acted more as the spur to the building of a

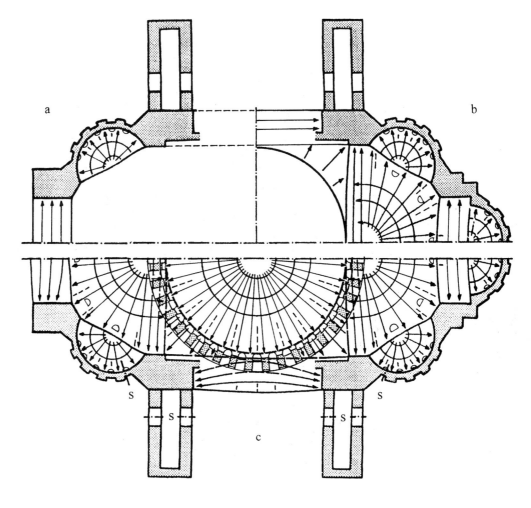

Figure 12.12 Hagia Sophia, Istanbul, part plans of the main vaulting system showing typical thrust lines and the general pattern of associated deformations and cracking at successive stages of construction: at (a) after completion of the small semidomes (assuming that they were completed before the large semidomes); at (b) after completion of the large semidomes and pendentives; at (c) after completion of the dome. Thrust lines are arrowed, cracking is shown by broken lines, and 's' denotes the approximate location of a shear failure. (Source: author)

comings were accentuated by an excessive speed of construction which left too little time for the brickwork to develop its strength before major loads came to bear on it.

The results of these weaknesses were shearing failures at the points marked 's' in the Figure. These allowed excessive spreading of the feet of the east and west main arches at the forward edges of the main semidomes. Within twenty years of the completion of construction this spreading, accentuated by earthquakes, led in turn to the most serious mishap – a partial collapse of the eastern arch and that part of the shallower original dome that it carried. A complete rebuilding of the dome to its present height followed. Yet such mishaps detract as

considerable number of large domed prayer-halls and mausoleums of a simpler design than as something to be directly emulated [7.14].[10] In these again, the same simpler system of support was adopted for the dome, with the two differences that pointed were preferred to semicircular arches and squinches to pendentives [7.11]. In an interesting Russian variant a tall spire, easier to construct, took the place of a dome [12.13].[11]

After the conquest of Constantinople, the implied challenge of Hagia Sophia was inevitably more keenly felt by the Ottoman victors, and it led to the building of a series of Imperial mosques that might, in a double sense, be regarded as constructive criticisms of the design of Justinian's church.[12] They cannot quite be compared architecturally, since Muslim prayer called for a single unified space, with any aisles and galleries either brought fully into this space or reduced to a very minor role indeed. The greatest spatial comprehensibility and structural clarity seem to have been sought, rather than the subtleties of the spatial interpenetrations and the structural dissimulations of Hagia Sophia. These differences apart, they all adopted the basic centralized domed form of Hagia Sophia and played almost all the possible variations on the two systems of support – one at east and west and one at north and south – adopted for its dome.

A few, notably the Suleymaniye Mosque in Istanbul, adopted the same combination as at Hagia Sophia but took advantage of the absence of any requirement for continuous side galleries to link the outer buttressing more effectively to the main piers [12.14].[13] Or, as at the Selimye Mosque in Edirne they adapted it to a more structurally-efficient octagonal arrangement of eight main piers [12.15]. Others, again with more structural logic, adopted only one of the two systems and employed this on all four sides or adopted a single system throughout on a hexagonal or octagonal plan. Notable among examples of this approach with only four piers is the Mosque of Sultan Ahmet, better known today as the Blue Mosque [12.16]. Here there are buttressing semidomes on all sides with further clusters of smaller exedrae semidomes around each of these.

Inside all these mosques, in contrast to the interior of Hagia Sophia, there is no doubt whatever about the lines of support. Closer examination shows two further aspects of 'constructive criticism' – changes in the manner of construction to reduce the extent of the structural deformations that were so apparent at the upper levels of Hagia Sophia. Typically there was greater use of well-fitted blocks of cut stone in place of more deformable brickwork and a greater use of iron tie rods, the latter particularly across the feet of the exedrae semidomes.

In the west, meanwhile, after the early experiments at Milan and Ravenna referred to above, the domed form had, until the fifteenth century, been much less used, and never in ways that marked any further advance on Roman prototypes. Here, most of the larger medieval domes, except those already mentioned as having been closely modelled on Byzantine ones, were really domical vaults polygonal in plan. They were carried either on polygonal bases, as in a number of north Italian baptisteries, or on substantial piers at all the corners.

The octagonal dome of Florence Cathedral [7.15] could be regarded as the last example of such a vault carried in the latter way, though it remained only a project until the second quarter of the fifteenth century and was, in some

respects, more the first major achievement of the Renaissance. The octagonal drum from which it rises is carried by eight substantial piers joined together in pairs above the aisles to provide virtually continuous supports for the four diagonal sides of the octagon.[14] The piers were buttressed at the west by the nave vaults and arcades, and at the north, east, and south by large projecting apses. But, in relation to the springings of the dome, this buttressing was provided at a much lower level than that provided by the main semidomes of Hagia Sophia and the later Ottoman mosques. Most of the real buttressing action was provided by the very substantial piers themselves. From the present point of view, the attempts made to contain the outward radial thrusts within the dome itself by the provision of circumferential stone chains', as discussed in Chapter 7, are therefore of greater interest.

The truly Renaissance sequel, first in Italy and then beyond, was more significant architecturally and even ideologically than it was structurally.[15] The ideal church had then not only to have a dome; it had also to be as fully centred on this dome as possible. A good

Figure 12.13 Church of the Ascension, Kolomenskoe. (Source: author)

220

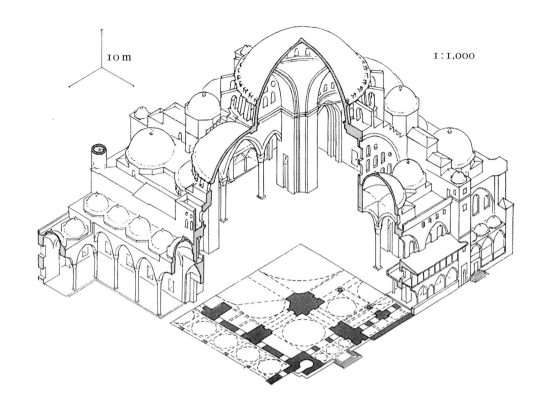

Figure 12.14
Suleymaniye Mosque,
Istanbul, part-cut-away
isometric.
(Source: author)

10 m

1 : 1,000

Figure 12.15 Selimiye
Mosque, Edirne.
(Source: author)

illustration of the ideal actually realized is the Church of S. Maria della Consolazione at Todi [12.17]. There can be few buildings that are quite as harmoniously composed. But, as a structure, the support of the dome is simply that of Figure 12.1(a), complete with all four semidomes and with a circular drum interposed between their crowns and the springing of the dome.

Elsewhere this simple system of support was varied and elaborated in a number of ways, much as that of Hagia Sophia was by Ottoman designers. But even in the new St Peter's no new principle was introduced unless the chaining-in of the dome, already noted in Florence, should be counted as such in this context. The chief difference, as compared with Florence, was that, in place of the eight paired piers that there directly carried an octagonal drum, there were four more massive single piers that carried bridging arches and pendentives and a circular drum [12.31]. These piers were backed by other massive piers set back across the main aisles, but the support was again essentially that of Figure 12.1(a), now without the buttressing semidomes.[16]

Pre-Gothic timber-roofed and stone-vaulted churches

It is now time to return to the parallel development – initially from the other Roman prototype, the timber-roofed basilica – of the basilican church and its successors.

One fifth-century church bearing a fairly close resemblance to the prototype has already been illustrated [12.2]. Piranesi's engraving of the late fourth-century church of St Paul Outside the Walls [12.18] shows the largest of the earlier churches that survived almost unchanged until fairly recently. Like the original basilica of St Peter,[17] it had double aisles with partly-trussed timber lean-to roofs on each side of a nave that was about 24 m wide. In the seventeenth century the nave roof was fully trussed as shown, and it was probably similar to this from the start. There would, therefore, have been little side thrust acting on the inner walls, and these walls were accordingly made not much more than 1 m thick for a height of about 30 m. With such slender proportions, the margin of stability against lateral bowing and buckling (of the kind illustrated in Figure 2.16) along the considerable length of the nave would have been slight. The substitution of light arcaded colonnades for the lower parts of the walls would have reduced it further. But for the lateral support provided through the aisle roofs by the thicker outer walls, it is, in fact, unlikely that these inner walls would have stood for anything like the length of time they did. Elsewhere, as in the similar but smaller late fifth-century single-aisled church of S. Apollinare Nuovo in Ravenna, a pronounced horizontal bowing can still be seen today at the clerestory level [12.19]; while at St Peter's the distortions had become so bad by the fifteenth century that Alberti expressed the opinion that some slight shock or movement would soon cause the south wall to fall.[18]

Since, in the West, the basic architectural form of a rectangular nave flanked by narrower aisles along its longer sides, and terminated usually by a vaulted apse at the east end and some kind of entrance porch at the west end, proved to be eminently suitable both for the congregational worship of the church and the rather different requirements of its monastic communities, it continued, in the main, to be preferred there to more centralized forms. While the trussed timber roof was retained, there were two possible ways of improving the stability of the nave walls without eliminating completely the spatial interconnection of nave and aisles. The first was to leave the structural form and proportions of the early churches unchanged but to increase the effectiveness of the structural interconnection of the walls, and thereby the lateral bracing that they could provide for one another. This was the usual Italian approach, as seen in the Cathedral of Torcello [12.20]. Here, timber ties run across both nave and aisles just above the capitals of alternate columns of the two arcades, and there are also arches above these ties bracing the nave walls against the outer walls. Something which is, in principle, almost identical may be seen in large numbers of Islamic mosques whose prayer halls are of the multi-aisled type rather than the domed type considered above. As an alternative to the combined use of ties and arches as at Torcello, the Church of S. Miniato in Florence had bracing-arches at intervals across both nave and aisles. The other way of improving the stability was to increase the thickness of the walls and to make the slender columns of the arcades into substantial piers. This was more typical of the

Figure 12.16 Mosque of Sultan Ahmet, Istanbul. (Source: author)

Figure 12.17 S. Maria della Consolazione, Todi. (Source: author)

Romanesque churches of France, Germany, and England and probably owed something both to the need, there, for the structure to be able sometimes to withstand attack and to a lesser familairity with the true-truss type of roof.

It was by fire, though, that St Paul's Outside the Walls was eventually destroyed. As long as an open timber roof was used, there remained an ever-present risk of destruction in this way, particularly to a church under attack,. The use of transverse masonry arches separating one part of the roof from another would have limited the risk; but that was all. It could be removed only by substituting an incombustible vaulted roof, though some risk would still remain if this was really only a ceiling above which a timber roof was still needed to keep out the rain. Some of the Romanesque walls were already strong enough to resist the thrusts of masonry vaults, at least over the aisles. It is hardly surprising, therefore, that large numbers of experiments in substituting masonry vaults were made in the period of growing material prosperity in the eleventh and early twelfth centuries.[19] Among these experiments must be included the churches in southern France that derived ultimately from the sixth-century Byzantine churches with multiple domes. St Front in Perigueux, which was mentioned above, was unfortunately largely rebuilt in the nineteenth century. The Church of St Ours at Loches is an interesting, well-preserved, and unique variant, in which each square bay is covered by an octagonal pyramidal vault that is both ceiling and roof. Each corner

Figure 12.18 St Paul Outside the Walls. Rome. Etching of 1748 from the *Vedute di Roma*, G. B. Piranesi, Rome. (Source: Victoria and Albert Museum/ John Webb)

Figure 12.19 S. Apollinare Nuovo, Ravenna, looking up to the clerestory on the south side. (Source: author)

Figure 12.20 (top right) Torcello Cathedral. (Source: author)

Figure 12.21 (bottom right) Aisle vault of the upper narthex, St Philibert, Tournus. (Source: author)

pier here terminates in what is virtually a downward continuation of the pyramid.

Of much greater significance, in relation to the Gothic sequel, were numerous variations on the use of barrel and groined vaults. Since the simplest vault to construct over the nave was the semicircular barrel, it was the form most widely used initially. Sometimes timber ties were placed across the springings as in Figure 12.1(h), presumably to assist in resisting the outward thrust. But numerous mishaps with and without such ties clearly showed the need for effective buttressing.

One common way of providing this buttressing was by means of quadrant vaults over the aisles as shown in Figure 12.21 and at the right of Figure 12.1(g). Another was to transform the aisles into short transverse bays vaulted at right angles to the nave as in Figure 12.1(e). This was essentially the same buttressing-system as used, for instance, in the Basilica Nova, but with the vaults over the aisle bays playing a role that was not required of them in that earlier structure – that of directing the continuously distributed thrusts of the nave vault to the transverse walls. It was used in the Abbey Church of Fontenay [12.22] and elsewhere, usually with pointed rather than semicircular vaults, though as a structural system it can be traced back a good deal earlier in the Middle East.

Yet another ingenious alternative, also adopted on a smaller scale much earlier in the Middle East, was to vault the nave itself transversely to its axis in a number of bays as in

Figure 12.1(i). This had the advantage, not possessed by any of the other systems, of making it easy to provide clerestory lighting. Figure 12.23 shows the best example, where the vault replaced an earlier timber roof. But it was still structurally far from ideal. The opposed thrusts of the pair of vaults meeting above each transverse arch would balance only when both vaults were complete. Each arch must therefore have been made strong enough to resist an unbalanced thrust from one side only as each fresh vault was constructed. Since, moreover, nearly all the weight of the vaults was eventually brought down to the piers through the transverse arches, the final outward thrust just above the springing level of these arches was not diminished as much as might have been hoped.

Until the late eleventh century the greater practical difficulties of constructing groined vaults were faced only over lesser spans than those of the central naves of large churches. They were used, for instance, in porches and narthexes, and over those aisles where one of the buttressing systems described above was employed at a higher level or, as in the church seen in Figure 12.23, where no great buttressing action was called for by the original roof. But by the time that both simple longitudinal barrel vaults and a series of barrel vaults placed transversely had been tried over the nave, if not before, the potential advantages there of the groined vault – which was, of course, half one and half the other – must have been apparent.

The groined vault seems to have been first used in this position towards the end of the century in Speyer Cathedral and then, not much later, in Durham Cathedral and in several places in France and Italy.[20] (The earlier date that used to be claimed for S. Ambrogio in Milan is very doubtful.) None of these vaults was a true inter-penetration of horizontal barrels. The vaults at Speyer are markedly domical like the one seen in Figure 4.18, so that they called for substantial transverse arches. Those at Durham are less precise and consistent in their geometry. But a significant step had been taken. Equally significantly, the projecting diagonal rib was introduced in some of the vaults.

With these churches we now are, in one sense, back to the alternative structural system of the Roman basilica – to the vaulted system of the Basilica Nova and Figure 12.1(j). But it was a return with important differences, and it was these that made it also a springing-point for the Gothic system.

For one thing, the new forms lacked the benefit of the excellent quality of Roman concrete, and used instead masonry with little tensile strength. The vaults were made considerably thinner and depended, for their stability, almost as much on their slight double-curvature as on the relative immobility of their supports. The basic support, moreover, was now the isolated pier rather than the deep transverse wall. From these points of view, there was a

closer resemblance to the structural system of the Byzantine church with multiple domes, each supported as in Figure 12.1(a). This resemblance was closest in a church like Speyer Cathedral, whose nave consisted of six such bays in a row, flanked on each side by twice the number of similarly vaulted narrower and lower aisle bays. (Alternate piers of the nave – which also originally had only a timber roof – therefore carried only the aisle vaults and the clerestory wall above these.) But it was a good deal less close in Durham Cathedral, on account chiefly of the ribbing of the vaults there and the subdivision of the nave vault into narrower oblong bays of the same width as the aisle bays. The introduction of the rib, allied with the substitution of the isolated pier for the transverse wall, unmistakably marked a trend towards a more linear concentration of the working structural mass.

Figure 12.22 Nave, Fontenay Abbey. (Source: author)

226

Timber barns and Gothic stone-vaulted churches

Where the materials could resist only compression, the structurally logical goal of this trend would have been forms like those designed by Gaudí for the Güell Chapel and the Church of the Holy Family [5.3 and 5.4]. But these were nineteenth-century conceptions, arrived at in a way that would never have occurred to those responsible for design in the twelfth and thirteenth centuries.[21] Their only models for a truly skeletal form would have been large barns, barn-like halls, and even barn-like churches, constructed throughout of timber, or of timber standing on masonry footings and low masonry walls.

Today, these timber forms can be studied best by looking at slightly later examples that have been preserved well enough to indicate all details of their construction.[22] One of the earliest of these is the barn illustrated in Figure 9.11, dating from about AD 1200. The examples shown in Figures 10.12 and 12.24 are both later; but they do not differ greatly in their basic structural system from earlier examples, and the greater height of the second makes its proportions more comparable with those of the larger churches. The main frames, consisting of the tall columns and the horizontal members running between them both longitudinally and transversely, act, by virtue of their bracing, in much the same way as tall longitudinal arcades joined by high transverse arches. At about mid-height, each column is braced over the adjacent aisle to one of the outer walls, and further transverse bracing against wind-loading and possible instability under vertical load is provided by the buttressing action of the lower principal rafters running in straight lines from the heads of the columns to the tops of the side walls.

Whether models of this kind played any direct role in stimulating or guiding the actual further development of the Gothic system from the beginnings seen at places like Speyer and Durham is open to doubt. But it is certainly possible, in retrospect, to see in them not only suggestive prototypes for a more skeletal form but also hints, at least, of ways in which this might be braced to give it the necessary overall stability. The necessarily diagonal alignments of the principal ribs of the masonry vault, in contrast to the transverse 'arches' of the timber counterparts, should not have invalidated these hints. Architecturally, the chief spur was probably the desire, when rebuilding the choir of the Abbey of St Denis in the mid twelfth century, to unify chancel, ambulatory and radiating chapels in a way that the Romanesque structural system did not allow.[23]

Three things characterized the development. The first was the linear concentration of load and resistance on a skeleton of ribs, arches, columnar shafts, and piers, that was discussed in rather broader terms in Chapter 1. The second, a necessary corollary of the first, was the disposition of these elements, in both plan and elevation or vertical section, in such a way as to permit everywhere a purely compressive balance of thrust and counter-thrust. The third was the use, almost everywhere, of pointed profiles for arches and ribs to facilitate three-dimensional planning and setting out, to reduce centering requirements, to assist in achieving visual coherence, and perhaps in the hope of improving stability.

The sequence of successive experiments – spread over much of the twelfth century and continuing into the thirteenth – that made possible the cathedrals of Chartres, Paris, Reims, Amiens, Bourges, and Cologne, to name but a few, can be followed in considerably more detail than most of the developments considered hitherto in this chapter. But it has already been

Figure 12.23 Nave, St Philibert, Tournus. (Source: author)

227

so fully documented that it should be sufficient here to make a few comments on the second of these three characteristics.[24]

In plan, a balance of the rib thrusts was fairly easily achieved, even at the centre of an apse vault where most of the ribs inevitably converged from one side [12.25]. At least it was easily achieved in the completed structure. During construction, it could sometimes have been less easy to achieve along the line that marked, for the time being, the limit of progress.

The chief difficulty would have arisen in vertical sections, where thrusts exerted by the high vaults and wind loads acting on the tall timber roofs (which were almost always retained above the vaults) had to be safely brought down to a lower level. At Speyer and Durham there was little but the thickness of the piers available to do this. As the springing-levels of the vaults were increased and the piers were made more slender, something else was needed. The answer, as is well known, was the flying buttress – an inclined prop arched on its underside. Notre Dame in Paris was previously thought to have been the first church to have had such props provided from the start, bridging over the aisles and galleries of its nave. But subsequent changes both here and elsewhere make it difficult to establish precise priorities with certainty. At Reims Cathedral [12.26] they are seen fully developed in the usual two tiers to abut both the thrusts of the vaults and wind loads on the roof.[25]

In principle, this answer should not have been difficult to conceive. It was implicit both in the ramping quadrant vaults noticed earlier in some Romanesque churches [12.21] and in the lower principal rafters over the aisles of timber

Figure 12.24 Tithe Barn, Great Coxwell. (Source: author)

barns. But many existing flying buttresses are clearly additions made after construction of the vaults they help to support, and it is possible – even probable – that the first ones were built only as hasty expedients to prevent a threatened vault collapse or added during reconstructions after such collapses.[26] Much as at Hagia Sophia, the quantitative understanding needed to assess in advance whether extra buttressing was needed was lacking. Indeed, even when the whole structural system including the flying buttress (as illustrated in a simplified manner in Figure 12.1(k)) was fully developed, troubles still arose through ignorance of the safe limits of height or slenderness in particular circumstances.

There were three possible causes of serious failure: local shearing of the masonry; local crushing of the masonry; and such gross deformation of the slender supports that equilibrium of the vaulting ribs or the supports themselves could no longer be maintained by purely compressive actions within the masonry.

Local shearing was most likely to occur where two members met at an angle without a third to stabilize the joint. This happened particularly at the ends of a flying buttress. Where, for instance, the lower end met the outer

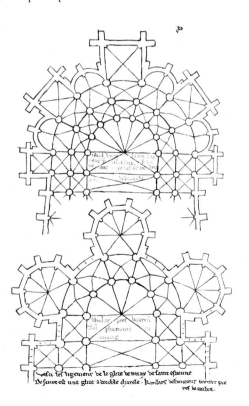

Figure 12.25 Apse vaulting plans drawn by Villard de Honnecourt. (From MS. Français 19093, Bibliothèque Nationale, Paris)

Figure 12.27
Marienkirche, Lübeck.
(Source: author)

buttress pier, there was a risk of it simply pushing the top off the pier. But this risk was easily removed by extending the pier above the end of the flyer or capping it with a pinnacle.

Local crushing of the masonry and gross deformation of supports both stemmed, to a large extent, from significant non-axiality of the lines of action of the resultant compressions coupled with non-homogeneity of the masonry.[27] In the corresponding members of large timber barns, such non-axiality could occur with impunity because the timber was strong and stiff enough to resist the bending moments to which it gave rise. Masonry was unable to do this. Local crushing was liable to occur if the compression was concentrated near to one face. Or the bending deformations could lead to buckling or to some other failure.

The most obvious sign of distress today in those structures that are still standing is a pronounced outward bowing of the upper parts of the main piers supporting the central vaults. In their lower parts the piers are usually bowed inwards to a lesser extent as a result of the eccentric compressions due to the unbalanced inward thrusts of the aisle vaults [1.12]. But under the action of the larger outward thrusts of the main vaults, the resultant compressions

above the level of the aisle vaults have all been to the outside and have led to the more pronounced outward bows seen there. Figure 12.27 shows one of the large Hanseatic brick churches in which, partly because of the use of brickwork that is more readily deformed than finely-jointed cut stone, the outward bow is particularly noticeable in the top third or so of the piers. Though flying buttresses over the aisle roofs have limited it they could do no more, because the external buttress piers are subject to the same sort of eccentric compression and consequent bowing.

Similar outward bowing was almost certainly the main cause of most of the serious vault collapses that occurred. A good many such collapses are recorded, the most significant being probably that of part of the main vault of the choir of Beauvais Cathedral in 1284, twelve years after it had been completed. Here the main piers rose almost 40 m to the springings of the vault and were unusually slender. Measurements made in 1903 showed that they were then bowed to such an extent that the original clear span of the vault of about 13 m had increased to almost 14 m.[28] It is reasonable to assume that the bowing was not much less immediately prior to the collapse. If so, it could have brought the slenderest parts of the piers close to the point of buckling and would have made the stability of the ribs of even a highly deformable Gothic vault precarious. The precise manner of collapse is not known. Since the rebuilding began with the construction of a second set of main piers in the straight bays, halving the spacing between the original ones, and a partial reconstruction of the buttress piers in the middle bay, it was clearly attributed then, at least partly, to some weakness of these piers. Detailed examination

Figure 12.26 Reims Cathedral. Interior of the nave with the glass of the clerestory windows removed. (Source: Professor Kenneth J. Conant)

of the present fabric and the history of its construction has thrown further light on it by identifying a particularly weak outer pier on the north side and finding that, only after construction had advanced to triforium level, the height of the whole structure was increased well beyond what had originally been planned.[29] The painting of the interior reproduced in Figure 12.28 shows the intercalated main piers that now take about a third of the weight of the reconstructed vault. The external buttressing was not doubled-up to correspond. It seems to have been left more or less as it was, apart from the addition (perhaps later) of iron tie rods, some of which may be seen in Figure 12.29.

The collapse at Beauvais was significant because it marked the end of the road for the Gothic structural system as hitherto considered. It was never again pushed so close to the limits of practicability, although construction of Cologne Cathedral continued until much later with proportions only a little less daring. Later structures did develop the system a little further in some respects, but never in such a way as to make it possible to build something that could not have been built in the way that, say, the cathedrals of Reims and Amiens were built.

The development, or modification, of greatest structural interest was the use of iron tie rods, no longer simply to cramp the masonry together, but to tie together separate elements of the structure. The tying together, in this way, of the buttress piers at Beauvais has just been noted. Less obtrusively, the piers of the Sainte-Chapelle in Paris had been similarly tied together rather earlier. More frequently, and particularly in Italy where flying buttresses seem not to have been liked [12.30], the ties ran between the springings of the vaults to obviate or reduce the need for external buttressing of their thrusts as shown in Figure 12.1(l). A similar use at a slightly lower level may be seen in Figure 12.27. Sometimes such ties were added during construction and were even taken away later. Often they seem to have been added as remedial measures only after the vaults had been completed and were seen to be cracking.[30]

Most other developments were of much more importance architecturally than structurally, or involved little more than the vaults themselves and do not, therefore, call for further comment here.

Seventeenth- and eighteenth-century vaulted and domed churches

If, now, the Gothic achievement is compared with that represented by the domed east ends of Florence Cathedral and St Peter's in Rome, the two will be seen to be very different from each other. The difference is highlighted by their very varied divergences from Roman prototypes like the Basilica Nova and the Pantheon. The replacement, in the Gothic cathedrals, of the massive walls of the Basilica Nova by what are

Figure 12.30 Frari Church, Venice. (Source: author)

virtually point-supports stabilized by independent skeletal buttresses had no real counterpart in the replacement, in the Italian domed churches, of the massive wall that carried the Pantheon dome by groups of equally massive piers.

Seventeenth- and eighteenth-century designers were never confronted by the structural challenges faced by the designers of the major Gothic cathedrals or by Brunelleschi and those successively in charge at St Peter's. They mostly built on a smaller scale and concentrated on the purely formal and spatial innovations that are so characteristic of the Baroque, though Neumann went further than most in developing and exploiting interpenetrating vaults supported by warped arches [1.15].[31] A few others – notably Longhena at the Salute in Venice and Bähr at the Frauenkirche in Dresden[32] – took further the late Roman form of a dome carried only by a ring of free-standing columns with its thrusts carried over an ambulatory to the outer walls [12.1(c)]. Longhena kept fairly closely to the prototype as represented, for instance, by the church of S. Constanza. The dome proper was given a slightly pointed profile, raised on a

previous pages:

Figure 12.28 Interior of Beauvais Cathedral, from a painting by an unidentified artist.

Figure 12.29 Beauvais Cathedral from the east. (Source: author)

drum, and crowned with a lantern in typically Baroque manner. It also carried a timber-framed outer dome. Bähr's design was more innovative in several ways. The tall triple dome, carried like that of the Salute on a ring of eight piers, was of masonry throughout. It also had an unusual cross-section with the low inner shell kept well below the drum, the two upper shells carried by the drum rising to almost twice the height, and the outer one of the two much thicker than the inner one. The buttressing was also different. Though it was again provided by the outer walls, now square in plan, it was concentrated, more as in San Lorenzo in Milan, on four corner towers to which the piers were connected in pairs. Finally the whole structure was unified externally in such a way that the dome appeared to rise directly from the walls.

confine ourselves largely to the structures themselves, concentrating on Wren's St Paul's in London, Guarini's S. Lorenzo in Turin, and Soufflot's Panthéon in Paris (originally the Church of St Geneviève), noting at the outset that the designers of all three had, in different ways, made close prior studies of the Gothic structural system.[33]

The skeletal quality of the Gothic structural system of support is least apparent in late seventeenth-century St Paul's, probably partly because of the way in which the design evolved as construction proceeded – the foundations and the piers having been laid out when there was still a possibility of a much heavier dome than the one actually constructed.[34] But, if the composite quarter-plan of the dome and of its supporting piers at the top left of Figure 12.31 is

Figure 12.31 Composite quarter plans to a common scale of the domes of St Paul's, London; St Peter's Rome; the Panthéon, Paris; and the Invalides, Paris, from Rondelet, J., *Mémoire historique sur le dôme du Panthéon Français*, Paris, 1797. The heaviest tone indicates the plan of the dome at the level of its drum; medium hatching indicates the supports at floor level (in part only for St Peter's); and light hatching indicates bridging-arches and pendentives below the drum.. (Source: author)

Of greater interest structurally, however, are a number of churches in which the lessons of the Gothic achievement – and particularly the principles of its system of support – were applied to the basic centralized domed form as seen at Todi and, elongated by a nave, at Florence and St Peter's. Two of these churches are important also on account of the rational scientific approach (backed up in one at least by calculation and experiment and attested by public scientific controversy) that was, for the first time, brought to bear on detailed practical choices of form and proportions. But it will be more appropriate to look further at this latter aspect in the concluding chapter. Here we can

compared with the corresponding plan of St Peter's dome at the top right and it is remembered that the latter, because of the common scale, does not show the outer buttressing piers, a considerable reduction of the relative mass of the piers supporting the dome will be seen. Even more significant are the design of the dome itself and the way in which the tall columnar drum at its base (through which the plan of the dome is drawn) is related to the piers below and to the dome proper above. The twin shells of the last were structurally independent over most of their height and much thinner than the interconnected shells of the dome of St Peter's [12.32 and

16.6]. As at St Peter's, they were chained in at their common base in an attempt to contain their outward thrusts. Below them, however, the drum was designed not only to afford some buttressing action, but also, unlike the drum at St Peter's, to serve as the first stage in directing the uniformly distributed weights of the shells to the isolated piers below. Finally, the bridging-arches between the piers, which served as the last stage in this collection and concentration of the weights, were topped not by a solid fill but by a carefully contrived series of small hidden inverted arches and there were hidden buttresses above the aisle vaults.[35] Within a structure that, to a casual glance, was not very different from St Peter's, there was, therefore, at least something of the typically Gothic balanced concentration of load and resistance.

In Guarini's work, this Gothic characteristic was more openly expressed. In S. Lorenzo in Turin, begun in 1668 when Wren was just starting the lengthy process of designing St Paul's, the dome itself was reduced to a two-tiered skeleton of interlaced ribs, and was seemingly supported only by arches and pendentives curving in towards it from clusters

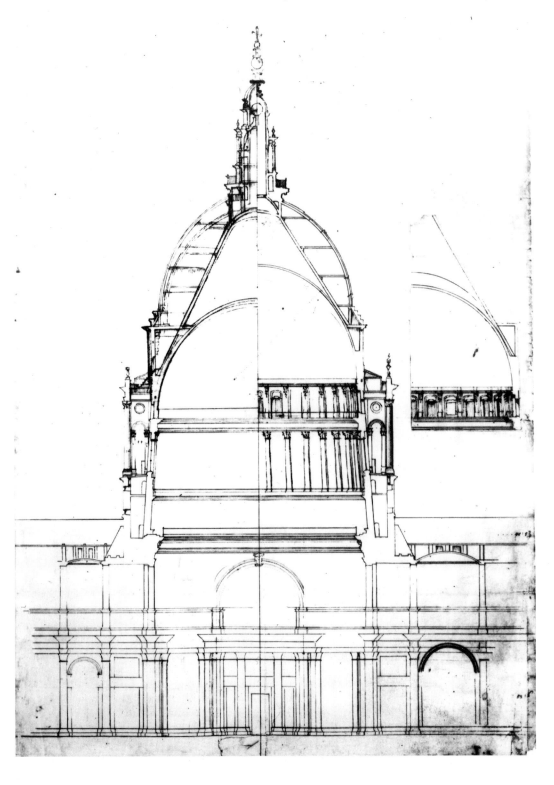

Figure 12.32 Late design studies of St Paul's, London, showing transverse sections through the dome. The dome as finally constructed corresponded closely to the left-hand section, but with an open eye at the top of the inner shell. (Source: St Paul's Library).

Figure 12.33 Composite plan of S. Lorenzo, Turin. The ribs of the lantern, dome and lower vaults are shown in broken line, and their supports are shown outlined and shaded and at the levels of the springings of the lantern and dome and at floor level. (From Guarini, G., *Architettura civile*, Turin, 1737)

Figure 12.34 Detail of the dome, S. Lorenzo, Turin. (Source: author)

VESTIGIVM S.LAVRENTII TAVRINI

of pilasters and free-standing columns disposed farther back around the periphery of the space below. Guarini's own plan [12.33] shows as clearly as is possible in any single drawing the interrelation of these levels, and emphasizes (especially in comparison with the super-imposed plans of domes and supports in Figure 12.31) the withdrawal of all ground supports from the dome itself. Ultimately much of the weight of the dome and all its outward thrusts were carried right over to the masonry of the outer walls by arches which must be envisaged as spanning within the empty spaces of the plan immediately above and below and to left and right of the springings of the main dome.[36] Secondary squinch-like arches embedded in the pendentives spanned diagonally between these arches in the remaining empty spaces in the plan.

Figure 12.34 is a vertical view of the dome and lantern, which may be compared with their representation in the plan. The striking re-semblance of the pattern of the ribs to that in the Moorish dome seen in Figure 1.11 has naturally suggested a fairly straight copying. But the resemblance is only superficial. The Moorish dome is little more than half the size of that in S. Lorenzo (which has a span of about 14 m) and its ribs carried nothing but continuous infills like those seen in Figure 7.18. The ribs of the main dome of S. Lorenzo were largely free of infills and, above the octagon formed by their intersections, they carried the lantern. Guarini's design was really much more closely related to the late fifteenth-century unexecuted projects by Leonardo for a dome over the crossing of Milan Cathedral and, perhaps through these, to some central European High-Gothic ribbed vaults, and skeletal towers.[37] One might even liken the entire structure of S. Lorenzo right down to ground level to the diaphanous spires at Freiburg im Breisgau, Vienna, Ulm [15.3], and Strasbourg – a comparison that could be used to demonstrate very tellingly that what is fundamentally a single structural system can serve very different ends, both functional and aesthetic.

Soufflot, under the influence of neo-Classical architectural ideals a hundred years later, employed the lessons he had learnt from Gothic structures in yet another way. The basic architectural form of the French Panthéon was, like that of St Paul's, an aisled cross with a high dome over the crossing.[38] The dome as finally built [12.35 and 12.36]] also closely resembled that of St Paul's except that it was given a third masonry shell in place of the strutted-timber one sheathed in lead that Wren used as an outer covering. It was, in particular, similarly chained-in around the base – details of the iron bars used having been illustrated in Figure 3.15. But the neo-Classical ideal rejected even the sort of solid piers placed under the dome of St Paul's or, nearer home, under that of the Invalides. Only free-standing columns carrying horizontal

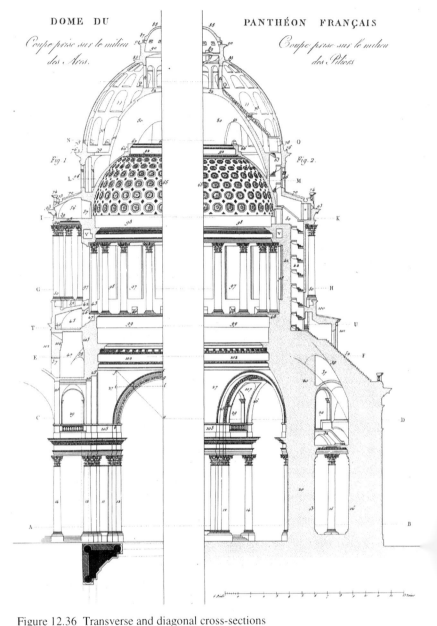

Figure 12.35 Dome of the Panthéon, Paris. The upper surface of the innermost shell is seen at the lower left, the intermediate shell rises above it, an outer circumferential iron tie is just visible above the arched opening in this, and the inside of the outer shell is seen in relative darkness at the upper right and brightly lit at the centre below. (Source: author)

entablatures (which could only be constructed as reinforced flat arches) conformed to the ideal.[39] Accordingly, Wren's eight piers were replaced, as immediate supports, by little more than four groups of three almost free-standing columns, each disposed around the mere vestige of a pier [12.31, bottom left]. Largely hidden buttresses carried most of the outward thrusts over the aisles to the outer walls where these met at the angles of the cross-shaped plan [right-hand section of 12.36].

Never before had such a dome been, to all appearance, carried so lightly – for it was appreciably larger than that of Guarini's S. Lorenzo. In fact the final design was, as will be seen later, considerably more ambitious than the earlier projects.[40] Also the supports did, before long, have to be strengthened. Though this seems to have been due more to poor construction than to any fundamental unsoundness in the design, the experiment was not repeated: Soufflot's structure marked the end of a road. In its extensive reliance on iron reinforcement in particular, it went, if anything, beyond the natural limits of masonry construction.

Figure 12.36 Transverse and diagonal cross-sections of the dome of the Panthéon, Paris, and its supports, from Rondelet, J., *Mémoire historique sur le dôme du Panthéon Français*, Paris, 1797. (Source: author)

Chapter 13: Nineteenth- and twentieth-century wide-span buildings

The nineteenth- and twentieth-century buildings that, structurally speaking, have widened further the range of choice for wide-span enclosures have mostly been built of materials like iron, steel, or reinforced concrete, rather than of masonry. Since the 1970s, new high-strength fabrics and glued laminated timber have also been used. Nevertheless, the types of roof have mostly paralleled fairly closely the earlier forms. Various multiple-arch forms have replaced the barrel vault; doubly curved shells, cable nets, membranes, and space frames have replaced the dome and the groined vault; and various beam-and-slab, trussed, and other space-frame forms have replaced the more limited earlier range of timber beams and roof trusses.

To this extent, the requirements for support have been similar. The change to materials capable of resisting tension as effectively as compression, and of being jointed in such a way as to transmit tensions and bending moments has, however, made it relatively easy to meet them in ways such as those illustrated in Figure 12.1 (d), (h), and (l), which previously presented considerable difficulty or were quite impracticable. The relative lightness of the newer types of roof has helped also, as have the possibilities of prestressing and a reduced concern for a long maintenance-free life. In addition, the functional requirements have usually left open the choice of the precise geometric form of the enclosure and have permitted the adoption, for the complete structure, of forms a good deal more closely analogous to one of the elemental spanning forms. Provision of the required support has then been relatively straightforward if the problems arising from increased scale are ignored. Interest has therefore lain more in the design of the spanning element than − as it did, for instance, in Hagia Sophia and the Gothic cathedrals − in that of the supporting structure.

Most of these buildings have also differed in use from earlier ones. Few have been churches or designed for other religious use. Nearly all have been for some secular use − as conservatory, train shed, exhibition hall, factory, hangar, airport terminal, assembly hall, sports hall or arena. In addition, novelty of form has often been valued more highly than the symbolisms acquired over centuries of use by some more traditional forms. These changes, coupled with new requirements for servicing and internal environmental control, have not naturally made for buildings of comparable architectural quality. Sometimes they have led to a certain reticence; at others to a *tour de force* that has not been inappropriate to the activity housed. Structurally, the closest counterparts of many of the later structures considered in the previous chapter are, paradoxically, some of those with tensile cable-net and membrane roofs or canopies.

Halls with framed and trussed counterparts of arched and domed roofs

The adoption of a simpler total form was first seen in large glasshouses or conservatories in which the glazing was entirely carried by timber or iron ribs, the most notable early examples being the Great Conservatory at Chatsworth built between 1837 and 1840 and the Palm House at Kew built almost ten years later. Prior to this, timber-ribbed domes and vaults had, of course, been used for a very long time in much the same way as masonry or concrete ones, with merely some reduction in the thickness of the supporting walls on account of the reduction in weight. From 1779 onwards, iron ribs had also been substituted for timber ones in the hope of avoiding disastrous fires. But this had remained little more than a straight substitution with little or no simplification of the manner of support.[1] In the Great Conservatory at Chatsworth, and again at Kew [13.1 and 13.5(c)], the entire cross-section was reduced to a half-circle standing upon the tips of two flanking quarter-circles. It was like that of those Romanesque barrel-vaulted churches in which the vault over the nave was abutted by a quadrant vault over each aisle [12.21], but with the aisle vaults directly abutting the nave vault rather than having a clerestory wall interposed, and resting directly on the foundations at the bottom.

At Chatsworth, the ribs were made of laminated timber, but those of the central span rested on tubular columns of cast iron that also carried a narrow gallery and served for rainwater disposal. The central span was 21 m and the overall width 37 m, the ribs being spaced about 4 m apart. Longitudinal stability was ensured largely by constructing the two ends in a similar manner, with the ribs curving in towards

Figure 13.1 Palm House, Kew. (Source: author)

one another at opposite ends and meeting the transverse ribs along others ribs, elliptical in profile, running in diagonally from the corners of the rectangular plan. At Kew, both the ribs and the columns were of iron, but the dimensions of the cross-section were somewhat less. Similar construction was widely adopted elsewhere, both on a smaller scale, as at the Kibble Palace in Glasgow, and, on a larger scale, for exhibition halls like the Grand Palais and Petit Palais in Paris.[2]

Not long after the completion of the Palm House in 1848, the ultimate simplification of the arched form came with the great train sheds of some major railway stations and termini, particularly with that at St Pancras. At St Pancras, built between 1868 and 1869, the entire 73 m width of the terminal was spanned at intervals of 9 m by great trussed arches rising directly from platform level [13.2] and supported below it on massive brick piers. These piers might have been designed to resist all the outward thrusts of the arches. But since all the platforms and tracks were supported by wrought-iron beams above a basement store, it was more convenient to tie the feet together, as in Figure 12.1(h), at the level of these beams. The arches plus their hidden ties were then self-contained units. Above the level of the platforms, they were interconnected by trussed purlins that carried lighter intermediate arched ribs as well as stiffening the main arches laterally and providing some overall longitudinal stability. Further longitudinal stability was provided by diagonal bracing of all intersections of the main arches and the trussed purlins.[3]

The continuous arch tied down at its feet, as at St Pancras, and without significant freedom of rotation there, was, however, a structure with a fairly high degree of statical indeterminacy. An arch with a central pin and pinned supports was statically determinate. This had the great advantage, in the latter part of the nineteenth century, that the stresses in it and the reactions at the supports could easily be calculated and were little affected by slight movements of the supports or by thermal expansions and contractions of the arch itself. It therefore soon came to be preferred, both for large train sheds and for exhibition halls – where the strictly functional justification for a single wide span was, likewise, possibly less important than its value in suggesting superiority over rivals.

The first large-scale, and almost certainly the best, example of the use of the three-pinned arch was in the Galerie des Machines built for the 1889 Paris Exhibition [13.3]. In the illustration, the pins at the crowns of the arches can be clearly seen. The others were set just below floor level beneath the triangular feet of the arches. Here the span was 111 m, and there were no tie bars between the feet to absorb the arch thrusts. These were taken by large concrete foundations. In other respects the structure generally resembled that at St Pancras, except that the use of the pins made it possible to prefabricate the arches in much larger sections on the ground and then rotate them in a vertical plane into their final positions. A comparison of other contemporary illustrations of the two structures under construction shows that this permitted a considerable saving in temporary staging.[4]

The St Pancras roof was constructed of wrought iron; those of the Galerie des Machines and a series of later large airship hangars of steel. Most more recent counterparts have had continuous trussed arches constructed from tubular steel.

In the New Leipzig Fair building, completed in 1996 with a span of 80 m, these arches are

Figure 13.2 St Pancras Station, London (Barlow and Ordish). (Source: Science Museum, London)

Figure 13.3 Galerie des Machines at the 1889 Paris exhibition (Cottancin). (From: 'La construction en fer au 19me siècle', *Bâtir*, 1958)

two-dimensional as in the earlier structures, but they stand some distance above the glazed grid of the roof proper.[5] They increase in depth towards their splayed feet with only the lower chords following the roof's circular-arc profile.

Elsewhere, it has been a more usual practice to adopt a triangular cross-section for the trusses (broadly as shown in Figure 5.11(c)) to give them greater inherent lateral stiffness.

At the Kansai International Airport Terminal building [13.17(a)], they were given an undulating profile with slightly upturned cantilevered projections of 20 m and 31 m beyond the arched main span of 83 m.[6] This profile was also much shallower than the examples considered hitherto. With the cantilevered projections it was really a hybrid form, somewhere between an arch and a beam, with much reduced thrusts as compared with those that an arch would have exerted. Each support had four legs which were splayed outwards not only (as shown) in the direction of the span, but also transversely to it, the foot of each support being midway between a pair of trusses. Thus stability was conferred along the length of the building, by several continuous series of Vs.

Since all these structures were rectangular in plan, the design of the arches or trusses simply repeated along the length.

There was a different situation at the new Waterloo International Rail Terminal [13.4 and 13.17(b)]. This had to be built on a cramped site alongside an existing terminal. Its axis was a continuous curve and its width reduced from 48.5 m at the terminal end to 32.7 m at the open departure end. The internal profile also had to be highly unsymmetrical – much steeper on the outside of the curve in order to give adequate clearance to the trains with no platform on this side. For this reason, the arches were again three-pinned as in the Galerie des Machines but were, in addition, given profiles and widths that diminished towards each pin in accordance with the varying bending moments they had to carry. Since, moreover, these bending moments were hogging over the short steeper sections and sagging over the longer flatter sections, the compression booms were respectively placed on the inside and outside. They are single tubes (connected above to pairs of ties made of solid rod) in the steeper sections and double tubes (connected below to single ties) on the flatter

Figure 13.4 Waterloo International Rail Terminal (Grimshaw, YRM Hunt). Note the pinned joints at the crown and bottom right. (Source: Jo Reid and John Peck)

sections. Longitudinal stability is provided by diagonally-braced tubular purlins. The glazing is carried by a secondary structure hung from the compression booms of the steeper sections and carried above those of the flatter sections. Much ingenuity had to go into the design of individual members and numerous joints, both to allow the complex curved and tapered three-dimensional structure and envelope to be assembled from the minimum number of differently sized components, and to allow repeated small relative movements of the envelope with changes in temperature while still keeping out the weather. The reward was a structure of unusual interest and elegance.[7]

The chief nineteenth-century development of the simple domed form – if we exclude domes like that at St Isaac's Cathedral in St Petersburg which were merely framed counterparts of recent multi-shell masonry domes – may be represented by the second dome of the Halle au Blé in Paris and that of the Albert Hall in London [13.5 (a) and (b)]. Like the dome of the Roman Pantheon, both of these stood directly on drum-like walls, circular in Paris with an internal diameter of 39 m and elliptical in London with a diameter varying from 57 m to 67 m. The dome of the Halle au Blé was a fairly straightforward and relatively lightweight reconstruction in iron of an earlier ribbed timber dome that had been destroyed by fire. That of the Albert Hall consists of a ring of considerably deeper trussed ribs which increased in depth towards the centre and were interconnected by diagonally-braced trussed purlins. Each rib terminates at the top in a tall central ring, also elliptical in plan, and at the foot in a cast-iron shoe housed in a continuous wrought-iron channel-shaped curb. Pairs of ribs in line with one another should, when connected via the central ring, have been capable of acting in much the same way as the trusses of Lime Street Station illustrated in Figure 9.2(j). – i.e. without exerting outward thrusts unless restrained from expanding slightly. The curb was, however, designed to resist expansion under the weight of the glazing and suspended ceiling. In absorbing the resulting thrusts, it would have compensated for their varying magnitudes and directions by it own varying curvature.[8]

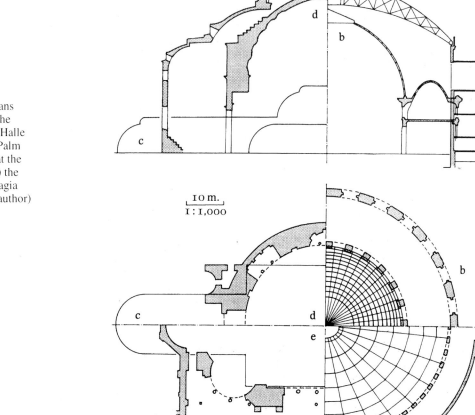

Figure 13.5 Part plans and sections of (a) the Albert Hall, (b) the Halle au Blé, and (c) the Palm House, compared (at the same scale) with (d) the Pantheon and (e) Hagia Sophia.   (Source: author)

10 m.
1 : 1,000

Most later framed domes have had their thrust similarly contained by tension rings at the base unless, like that of the St Louis Climatron described in Chapter 9, they have lacked a continuous ring at the base and have constituted virtually the whole space-enclosing structure. The thrusts of the Climatron dome was taken by five stub piers with inwardly inclined bearing faces [9.16].

Their complex geometry made a simple tension ring even less practicable for the timber-lattice shells of the Mannheim pavilion [7.29]. These had to be supported both around the whole boundary and along three other lines where the surface changed direction. The supports along the changes of direction were akin to ribs in a groined concrete shell. The boundary supports were concrete walls or timber beams around most of the periphery, with arches over entries. Over the remaining section there was a continuous boundary cable spanning between columns at five changes of direction.

Halls with reinforced-concrete arched and shell roofs

As a reinforced-concrete example of the 'ultimate simplification' of the arched total form seen at St Pancras, we might look again at the Orly Airship Hangars [6.19], cast *in-situ* from reinforced concrete as described in Chapter 6. Here the arches themselves were given wide trough-shaped cross-sections. When cast alongside one another, they thus formed a continuous corrugated barrel that called for no purlins, diagonal braces, or cladding. By means of a tall profile following the thrust-line for dead load, and a thickening of the troughs towards the feet, the horizontal thrusts there were kept down to about 15 per cent of the vertical loads. This permitted the adoption of extremely simple foundations without calling for horizontal ties. They consisted merely of continuous slabs of reinforced concrete 1 m thick resting directly on the ground 2 m below floor level. The entire hangars, when completed, therefore consisted of little more than sets of arches standing side by side directly on the ground. The lower sections, serving as abutments, were cast first, and the upper sections were then cast on them, one arch at a time, using the travelling centring shown in the illustration.[9]

While Freyssinet was building the Orly hangars in the early 1920s, Perret was building the church of Notre-Dame in Raincy, a northern suburb of Paris. Inside it, a shallow reinforced-concrete barrel vault extended the whole length of the nave, carried by widely spaced slender columns. At a slightly lower level, sets of barrel vaults of the same shallow profile were aligned at right angles to roof the aisles. All these vaults were thin shells only 3 to 4 cm. thick and there were no visible ties and no external buttresses. Thrusts were neutralized partly as in Figure 12.1(e) and stability was further ensured by thin

reinforced diaphragm walls set transversely above the main vaults and continued over the aisles to carry a third set of narrower barrels that constituted the outer roof.[10]

Good examples of the further development of the arched form and the corrugated barrel form used at Orly are the aircraft hangars built by Nervi between 1935 and 1941 and a large exhibition and two sports halls that he built between 1948 and 1960.[11] The diagonally ribbed vaults of the hangers at Orvieto, Orbetello [9.15], and Torre del Lago sprang from edge beams supported well above ground level. These edge beams assisted in transferring weights and thrusts to more substantial arched ribs that were continued downwards as inclined buttress columns at a few points around the periphery. These columns then transmitted the thrusts to the ground through hidden splayed-out feet.

The first of the other three structures was the exhibition hall in Turin whose vault was illustrated in Figure 4.8. Both this and the larger

Figure 13.6 Small Sports Palace for the 1960 Olympic Games, Rome (Nervi). (Source: author)

sports hall employed the type of precast 'voussoir' described earlier, the only major difference being that in the large sports hall these 'voussoirs' were tapered along their lengths and set radially to form a corrugated dome. The common characteristic feature of their manner of support was that every three corrugations converged, at the base of the vault proper, on a fan-shaped element which collected their thrusts and passed them on to an inclined buttress-column whose splayed feet then passed them to the ground as before. Since their inclinations again followed the directions of the thrusts (as modified to some extent in the sports hall by the additional downward loads of a separate projecting outer gallery roof), they

provided in 1957 and mobilized whenever desired.)

The CNIT Hall in Paris [7.28], built shortly before the sports halls, might also be regarded as a fairly direct development of the form used at Orly, though different again in its basic geometry. Its corrugated shells, made up of three fans of box-section arched ribs, rose from ground level without interposed piers of any kind. The vertical loads of the shells passed straight to the ground through foundation blocks, but it was not possible to design these, as in the Turin hall and the large sports hall, to absorb safely the outward components of the thrusts without assistance. Prestressed ties were therefore provided for this purpose, more as in

Figure 13.7 Assembly Hall, University of Illinois (Harrison and Abramovitz, Ammann and Whitney). (Source: author)

were, in effect, simply continuations of the vaults themselves.

The smaller sports hall was given a ribbed dome that differed somewhat from that of the larger one, but here also the thrusts were collected by similar fan-shaped elements to be passed on to inclined buttress-columns, now Y-shaped [1.2 and 13.6]. Here though, the outward components of the thrusts were absorbed by prestressed steel cables encircling the feet of the buttress-columns within a continuous concrete ring foundation. (In passing, it is interesting to compare the figure of about 1.5 MN quoted earlier as the minimum strength of a peripheral tie to absorb the outward thrusts of the presumed first dome of Hagia Sophia, with the corresponding minimum strength of about 2.5 MN required here for a flatter but relatively lighter dome of almost twice the diameter. Whereas it was not practicable in the sixth century to provide a tie of the former strength, and still less to prestress it, strengths of the latter magnitude, and even much greater ones, could readily be

the smaller sports hall.[12] (Now, though, their required minimum working strength was more than twenty times that which would have been required at Hagia Sophia.)

Where ties are used to absorb the outward thrusts, there is, of course, no need to carry the roof down to the ground either in the way that Freyssinet did at Orly and Esquillan did in his CNIT Hall or in the way that Nervi did at Turin and in his sports halls. Often it has been convenient to carry it down in the first of these ways where it could again form almost the whole enclosure, as many of Candela's shells did and as many of Isler's have done more recently. Or to do so more in the second way, as in Candela's Church of the Miraculous Virgin in Mexico City [7.27] where the grouped shells were merged elegantly into short piers that were slightly inclined against the resultant thrusts. But there is no reason why it should not be carried at a higher level by vertical supports as in Figure 12.1 (d), (h), or (l).

Torroja's market hall at Algeciras [7.24] was

Figure 13.8 and 13.9
Exterior and interior of the
TWA Terminal Building,
Kennedy Airport, New
York (Saarinen,
Ammann and Whitney).
(Source: author)

one of the first structures in which a flat concrete shell was actually carried in this way. Its vertical load was carried by the whitewashed corner columns, and all the outward thrusts were absorbed by ties encased, after they had been stressed, in the horizontal bands of concrete between the column heads.[13] The thin groined vaults seen in Figure 7.26 were also carried on vertical columns interconnected, just below the springing points of the vaults, by horizontal ties. Since there were three vaults placed side by side here, the structural system was exactly that of Figure 12.1(l).

The Assembly Hall of the University of Illinois [13.7] went even further in this direction. The entire weight of its giant folded-plate saucer dome – 120 m in diameter and cast *in situ* in twenty-four radial slices – was brought down to a ring footing less than 80 m in outside diameter. Between this footing and the ring at the base of the dome was interposed an outward spreading ring of 'buttresses in reverse', resembling, in their corrugated upper parts where they directly support some of the tiered seating, an inverted reflection of the dome rather than a downward continuation of its form. Almost a thousand kilometres of prestressed wire were wound round the upper ring to take the outward thrusts and stabilize the 'buttresses in reverse'.[14]

Most of these reinforced-concrete structures were conceived primarily, like the earlier iron and steel ones, by engineer-constructors, and they no doubt owed to this the clear structural logic of their designs. Underlying some other notable examples has been a different logic, more at odds with the structural requirements. Saarinen's Kresge Auditorium has been referred to already in Chapter 7. His TWA Terminal at Kennedy Airport and Utzon's initial vision of the Sydney Opera House were conceived as families of thin shells, not just placed side by side like those of the Air Terminal at St Louis or like the groined or ribbed vaults of earlier basilicas, churches, and cathedrals, but related to one another in an almost organic interdependence. But neither could be built in quite this way.

The TWA Terminal was conceived as a large-scale piece of sculpture, rather as some of Le Corbusier's buildings have been, except that even the final detailing of the forms was done by modelling them in three dimensions.[15] As one approaches the building, it is a great bird with wings outstretched [13.8]. Inside, one is presented with a continuous succession of smoothly-flowing curves and a spatial experience that remains unique [13.9]. To create these, the roof was conceived as consisting of four shells, each trapezoidal in plan, symmetrical about a central ridge, and cantilevered out from just two points symmetrically placed on each side of the ridge. The four ridge-lines were to meet over the centre of the building, and only there were the shells to be directly interconnected. Elsewhere, where their edges ran alongside one another, they were to be separated by strips of glazing, two of which may be seen in the interior photograph. The smallest shell was that over the entrance – the head and shoulders of the bird. The largest were the two identical wings, with their ridges both inclined slightly forwards in relation to the transverse axis so that they met at a wide angle at the rear. Here, in line with the entrance, was the fourth shell – the widely splayed tail seen from below in the interior photograph. As they approached ground level, the eight supports were paired, as may be seen alongside the entrance in the exterior photograph.

The structural problems presented by the design were many, though they all reduced to a statical impossibility of supporting the weights of the 'shells' and other loads on them solely by direct tensions and compressions without bending. There were three reasons for this: the presence of the ridges when there were no concentrated loads acting along them, the outward cantilevering of the shells from only two points, and the reliance on the central interconnection of the four shells to prevent the three larger ones from tipping outwards and the

smallest from tipping inwards. Considerable bending strengths and stiffnesses had, therefore, to be provided, particularly along the ridges and along the edges of the shells. Near the central meeting-point and near the supports, thicknesses had to be increased to around 1 m. But, thanks to skilful structural design and quite a feat of setting-out and formwork construction,[16] the sculptural integrity of the initial concept was preserved. Below ground, deep concrete piles were required to bear the considerable weights concentrated on the supports, and prestressed ties were provided between the supports to relieve these piles of most of the outward thrusts.

Utzon's initial vision of the Sydney Opera House – for it was little more than a vision when he won the architectural competition – was essentially of two main halls roofed, independently of each other, by groups of sharply pointed shells of no very precisely defined geometrical form [1.13 top]. The principal shells of each group were to be in line, were to be supported like those of the TWA Terminal at only one point on each side of the ridge, and were to stabilize one another. The problems set to Arup, the structural consultant, were therefore of the same kind, but they were more difficult to solve on account of the very sharply pointed ridges that were called for and the large scale of the whole structure. He has already admirably described his struggle with Utzon to realize the vision – his refusal to bring Utzon down to earth while there remained a chance of Utzon pulling him up to heaven.[17] Here it can be referred to only very briefly.

Three from a considerably larger number of tentative schemes together with the one finally adopted are also shown in Figure 1.13. The tentative schemes ranged from single- and double-skin reinforced-concrete shells with stiffening ribs to resist the large bending-moments induced by the sharply pointed ridges (second drawing) to steel space frames with reinforced-concrete skins (third drawing) and fans of triangular-section hollow arched ribs rather like those used by Nervi in the large Sports Palace (fourth drawing).

The final scheme (bottom drawing) was also of the last kind, but with the geometry of the principal 'shells' modified so that each half-shell was, geometrically, a triangle cut from the same sphere. This meant that the external profile of each rib was an arc of a circle of the same radius (75 m). One rib differed from another only in the length of the arc. Both dimensional control and the physical problems of providing temporary support to the ribs during construction were, thereby, greatly eased.

The ribs were built up from precast sections as in Nervi's dome, and, as the sections or voussoirs of each rib were assembled, they were supported on one side by the already completed adjacent rib and on the other by a steel erection arch swinging about the points from which all the ribs radiated. This arch had telescoping sides so that it could be accommodated to the changing lengths of the ribs. One such arch is still in position at the forward edge of each of the two smaller shells nearest to the tower crane at the left of Figure 13.10. A further box-section filled the gap between the opposite halves of each arched rib and, when construction was completed, constituted part of the ridge. Complete interaction of all the precast units was achieved by carefully phased prestressing across both the voussoir and rib interfaces.

In this way, pairs of principal shells were built up, set back to back and meeting only at the ridge. To unite and stabilize them more effectively, the intervening smaller shells of the earlier schemes were constructed from precast

acoustic requirements of an air terminal. In the Kresge Auditorium as in Sydney, his shell was seen from below only in the circulation space around the auditorium proper. And even in that structural system in which exterior and interior were, in other respects, most closely interrelated visually – that of the Gothic cathedral – the vaulting systems were almost invariably hidden externally by structurally independent roofs of a completely different character.

What is different about the Sydney Opera House and the Kresge Auditorium, and relatively unusual, is that the external roofs are the primary spanning elements from which ceilings of a different form are hung. The flat ceilings of many trussed timber roofs are really no parallel to this. The only close earlier parallels are those

box beams spanning more or less horizontally between them, either directly or via an intermediate arch. Considerably greater continuity of action was thereby obtained than that given by the single interconnection of the opposed pairs of shells in the TWA Terminal.

It is tempting to prefer the design of the latter on the grounds that the structural effort needed to turn the architect's idea into a practical reality had the double justification of creating simultaneously a distinctive and memorable exterior and a fine interior. For acoustic reasons, a ceiling of a completely different form had to be suspended below the Sydney roofs, which are now seen only as a magnificent exterior. But it should be remembered that Saarinen's achievement in the TWA Terminal was a rare one and was made possible partly by the less stringent

trussed timber roofs from which Baroque or Rococo ceilings (shaped as vaults but constructed of lath and plaster) were suspended, and some Islamic vaults, like that illustrated in Figure 3.8, which were left exposed externally but from which a stalactite vault (likewise constructed only of plaster on a timber framework) was similarly suspended. But in neither of these cases was the shape of the outer roof determined by anything but utilitarian considerations.

Halls with beam, slab, truss, and space-frame roofs

Unlike the arch and dome and related forms, the beam and slab and their related forms cannot alone enclose space. They must be supported above the ground and the enclosure must be

Figure 13.10 Sydney Opera House under construction (Utzon, Arup). (Source: *Architect and Building News*)

Figure 13.11 Crown Hall,
Illinois Institute of
Technology, Chicago
(Mies van der Rohe,
Kornacker).
(Source: author)

Figure 13.12 Central
Railway Station Naples
(Perillo).
(Source: author)

completed by other vertical or inclined elements. They do not necessarily exert horizontal thrusts on these supports merely by weighing down on them as the arch does. But nor are the resulting structures necessarily stable under horizontal disturbing forces such as the wind.

In the structures with wider-spanning trussed roofs considered in the previous chapter, stability against horizontal disturbances was provided by the massiveness of the supports, by the inherent stability of a box-like arrangements of walls, or by some kind of bracing such as the diagonal braces of timber barns, the transverse arches of churches like the cathedral of Torcello, and the flying buttresses (particularly the upper flyers) of Gothic cathedrals. Available jointing techniques did not allow it to be provided by rigid joints.

When, in the nineteenth century, iron trusses superseded timber ones as the spanning elements, stability usually continued to be provided by the enclosing walls or by bracing. The alternative of relying wholly on rigid joints was not yet feasible, though the time when it

would be was coming nearer. Indeed the joint used at the Crystal Palace [4.11] was semi-rigid and the riveted joints used in the Sheerness Boat Store [15.12] were more so. But the outer bays of both these buildings were multi-storey structures of modest span which it will be more appropriate to consider in Chapter 15. Their wider single-storey central bays had little or no lateral stability of their own and overall stability was conferred largely by the outer bays acting, from this point of view, in much the same way as the outer walls of earlier structures.

With the truly rigid joints made possible by welding mild steel, a simple column-and beam assembly does become a self-sufficiently stable portal frame [5.11(e)].

The first major single-storey building using the fully developed portal frame was the Crown Hall of the Illinois Institute of Technology [9.10 and 13.11]. Its span is only half that of the train shed at St Pancras and a third of that of the Galerie des Machines. But its proportions and Mies van der Rohe's consummate skill in detailing gave both its interior and its exterior an immaculate elegance to which they hardly aspired.[18]. All the side cladding is of glass, some clear and some opaque. All the lateral stability was therefore provided by the rigidity of the welded joints of the steelwork which enabled the beams to act with the columns virtually as arches, at the expense of exerting some side thrusts on the foundations. The construction photograph reproduced in the first figure shows this most clearly. The view of the completed structure shows how the flat roof is carried by the purlins which, in the construction view, are seen being welded to brackets on the underside of the girders of the main frames. Some diagonal bracing was provided in the plane of the roof.

In spite of the stimulus to its use resulting from the introduction of new 'plastic' methods for calculating its strength, the simple portal frame with members of solid cross section was not much used for wide-span structures after a more efficient alternative was offered by its trussed counterparts [5.11 (d) and (f) and 9.14]. Moreover, it is usually even more efficient to substitute a beam grid or horizontal space-frame deck spanning in two directions for the trussed beams of individual portals spanning one-way only.[19]

With such a grid or space frame as the roofing element an equivalent fixity of support to that given by rigid joints can be obtained by using branched supports. The branched supports of the Tobacco Dock Warehouse [10.13] did not serve this purpose because they were effectively articulated and because the roof trusses acted independently of one another. But in Naples Central Railway Station [13.12], built more than a century later, the inverted reinforced-concrete tripods cannot rotate independently of the three-way grid of the roof. Such a structure can therefore be regarded as the ultimate expression

of the portal-frame principle, having also the merit of two-way spanning in carrying vertical loads. For a particular pattern of loading – say just that of the dead weight of the roof – the supports can be arranged so that each leg is subject to a similar direct compression. When the load pattern changes, and particularly when there is a side load, they will act like the legs of simple portal frames except that they will able to resist overturning in any direction and not just one. And, by virtue of their branching, they should be able to do so simply by variations in the compressions in individual legs. The star heads of the columns carrying the stiff space-deck of the McCormick Place Exhibition Center [9.18] confer overall stability in much the same way as at Naples, but leave the columns to resist side loads purely by bending. Since the elegant trees that carry the roof of the Stuttgart Air Terminal [10.17] start to branch only well above floor level and (like the roof system) are less

stiff, they will provide a more elastic support, leaving stability to be ensured more by bracing of the roof and walls.

In their overall pattern of behaviour, all these structures were relatively simple ones. However, portal-frame action is not restricted to table-like structures with a single flat deck system carried on a number of legs. Reinforced concrete not only lends itself readily to rigid jointing: it also allows a more ambitious three-dimensional approach, even when the choice of elements is restricted to beams, slabs and walls. Clifton Cathedral [13.13], constructed a little after the McCormick Place structure, is one of the best examples of what can be achieved in this way and is almost a modern counterpart of Guarini's San Lorenzo. The deep beams, carried by columns set well back near the outer walls, span as a star over the main trapezoidal space and progressively close in towards an off-centre spire and source of indirect light. Their three-dimensional continuity thus creates an inherently stiff assembly, which perfectly met the architectural objective.[20]

The chief rigid-jointed alternative to a portal frame is a mushroom-headed cantilever with a securely fixed base and a capacity to resist bending moments at the head.

An outstanding early example was the Palace of Labour in Turin, seen under construction in Figure 13.14.[21] The huge reinforced concrete columns, each 20 m. high, were designed to resist not only a side load but also any eccentric distribution of the load on the section of the roof that each carried. Broad cruciform bases standing on piled foundations provided the necessary fixity at the foot. The finished roof, of which only half is shown, consisted of sixteen mushroom heads ribbed in steel, each supported independently of the others by a single column. As compared with the roofs of the Naples Station and the McCormick Place Centre, it lacks the benefit of equalization of moments that stems from continuity over a large number of supports, but it gains in that each section is statically determinate and, being independent of the others, could be constructed independently of them. This latter advantage, according to Nervi, was a principal reason for its choice, since only a short time was available for construction. It therefore provides an interesting parallel to the Crystal Palace, which it resembled also in having its walls entirely clad in glass. (Compare the photograph with the interior view of the latter reproduced in Figure 15.11, which embraces an area, including the aisles, only about twice as great.)

The Stansted Airport Terminal [13.15] is an elegant successor built about twenty-five years later to meet a more demanding brief. Each column there consists of four tubular legs rigidly interconnected at three levels by welded horizontals. It extends about twice as far below the continuous reinforced concrete slab of the main concourse as it does, visibly, above it, and

it is restrained against any lateral movement by this slab. A tubular branch is pinned to the top of each leg, these branches being splayed outwards and held in place by an arrangement of struts and ties that is prestressed to resist varying side loads as well as the dead weight of the roof. Each branched assembly carries one section of roof 18 m. square. These directly supported sections carry others of the same size, simply supported between them and glazed in the centre. There is therefore no flexural continuity in the roof as a whole.[22]

A more frequent use for rigidly based supports has been to carry cantilevered canopies of various kinds over grandstands, spanning only or predominantly in one direction. It has usually been possible to resist any tendency to

of these, also in Turin and built shortly before the Palace of Labour by Morandi, is seen in Figures 13.16 and 13.17(c). The structural requirements were unusual. The whole structure was to be sunk beneath part of a park that was to be reinstated over it when construction was completed. There was thus a heavy super-imposed roof load to provide for, plus inward side loads on the walls. Internally, the overall width of 69 m was to be obstructed as little as possible by columns and it was desirable to keep the roof fairly level in order to achieve the necessary headroom everywhere without dropping the floor more than was essential below the finished ground level of the park.

The roof was therefore built as a slightly arched slab of reinforced brickwork varying

Figure 13.16
Underground exhibition hall, Valentino Park, Turin (Morandi). (Source: Cement and Concrete Association)

overturning at right angles to this direction by linking together a series of parallel canopies. The supports have therefore been called upon chiefly to provide restraints in the direction of the overhang. Since, however, it has usually been desirable to keep the supports well back from the centres of gravity of the canopies, they would be subject to large overturning moments in the absence of further support. To reduce these moments, secondary tie-down supports have been provided at the rear edges. This was the manner of support chosen by Torroja for the cantilevered roofing shells of a grandstand in Madrid [13.17(d)]. The secondary counterbalancing tie was at the rear of the prototype shell as seen in Figure 8.14.

Auxiliary tension support has been used in a few other instances of a less typical nature. One

from 240 to 440 mm in thickness and supported by two intersecting sets of deep but thin prestressed-concrete diagonal ribs. The principal supports were hinged at both top and bottom, their feet were set a little in from the walls, and they were inclined inwards towards one another. The inclination not only prestressed the central parts of the ribs through the compressive arching action that it induced. It also further reduced the effective spans of the ribs and, thereby, the central bending moments.[23]

Short outer hinged ties were provided as secondary supports partly to help stabilize the four-hinged transverse 'arches' thus created, and partly to reduce even further these central bending moments,. The similarity of their action to that of the outer ties provided to stabilize the cantilevered canopies of the Madrid grandstand

250

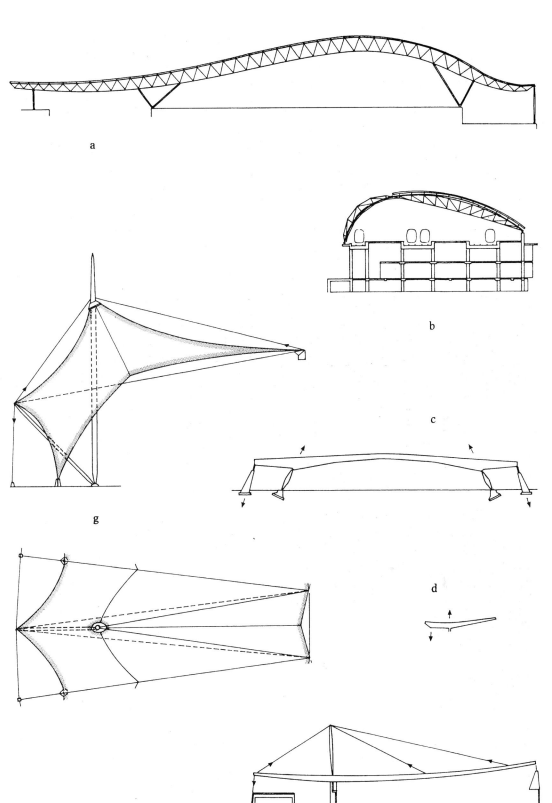

Figure 13.17
Diagrammatic views (to the same scale) of support systems for wide-span roofs: (a) Kansai International Airport Terminal, (b) Waterloo International Rail Terminal, (c) Valentino Park exhibition hall, Turin (Morandi), (d) grandstand roof, Madrid Racecourse (Torroja), (e) Alitalia Hangar, Leonardo da Vinci Airport, Rome (Morandi), (f) INMOS Factory, Newport, (g) grandstand canopy, King Fahd Stadium, Riyadh. (Source: author)

10 m.
1 : 1,000

Figure 13.18 Centre Pompidou, Paris (Piano & Rogers, Arup). (Source: author)

will be apparent from the sketches of both structures in Figure 13.17. Unlike the system in Madrid, that in Turin was, however, statically indeterminate because there were four supports in all for a single continuous rib. This made it possible to prestress the ties for maximum effect. The attentive reader will also have noted a similarity, in the use of the principal inclined supports, to the use of similar supports in the Esbly Bridge. There is an even closer resemblance to the use of such supports in some of Morandi's bridges.[24]

The second instance worth looking at has no such parallel. The Centre Pompidou in Paris [13.18] is, moreover, a multi-storey structure. Nevertheless, each storey is a wide-span one and it is this wide-span aspect that is of greatest interest.[25] The chief relevance of the fact that it is multi-storey is that the principal supports are subject to much heavier load than if they had been called upon to carry only a roof. It was decided to keep their cross-sections to a minimum by using cylindrical steel tubes and ensuring that they were not subject to bending.

252

In the columns of all the structures considered hitherto in this section except the last, some bending did have to be provided for. To avoid it, the loading had to be truly axial. To ensure this, the connections with the spanning trusses were pinned, using for the purpose large steel castings. The joint is seen in close-up in Figure 3.18 and the whole castings are seen towards the right of Figure 13.18 and against the sky at the left. On the side of the main trusses they have only short projections, to which the trusses are hinged. On the other side they project much farther, and the vertical ties are attached at their outer ends. These were prestressed to balance the bending moments due to the loads from the trusses and the total load transmitted to the columns was thereby made axial.

This arrangement alone, with all its articulation, would, however, have left the structure unable to resist a sideways disturbing force. To prevent sidesway, the two end frames were given, in addition, the three inverted-V braces seen between each pair of trusses. In the twelve otherwise similar intermediate transverse frames along the length of the building, similar braces would have caused obstruction. Here, therefore, they are omitted, and the stabilizing effect of the end frames is transferred to them by horizontal diagonal braces in the floors. Secondary diagonal bracing is also provided externally on the long frontages, as seen on the right of the figure.

Although the basic concept for this building was not entirely new, its practical realization was far from straightforward structurally, especially in the pioneering use of large steel castings. In its frank exposure and display of the structure on a prominent Parisian site, it paved the way for similar designs elsewhere, such as the Hongkong and Shanghai Bank described in Chapter 15.

Figure 13.19 Burgo Paper Mill. Mantua (Nervi). (Source: author)

Halls with cable-supported beam, truss, or space-frame roofs

In a further group of single-storey structures with trussed and space frame roofs, the primary support has been partly a tensile one. The roof has been suspended by vertical hangers, like the deck of a suspension bridge, from two or more catenaries. Or it has been supported by a number of straight ties as in a cable-supported bridge. Such support is valuable chiefly where longer spans are required than can economically be provided by other means, particularly where long cantilevered spans are required.

In all such structures there is also a need for tall columns, masts, or braced frames able to carry the weight of the roof down from the high points of attachment of the cables to the ground. If the resultant loads exerted by the cables or ties are not vertical, these supports must either be inclined in line with the resultants or otherwise provided with the ability to resist horizontal loads in the planes of the cables as well as vertical ones. They must also, of course, be able to withstand possible side loads from other directions.

The earliest built examples appeared only towards the middle of the nineteenth century, several decades after the first such bridges, by which they seem to have been inspired.[26] One was the roof of the 42-m central span of a building for making ships' masts at the naval arsenal at Lorient, under construction in about 1840. Here the roof was suspended by vertical hangers from catenaries slung between towers very like those of contemporary suspension bridges. Another, on a circular plan, was the roughly contemporary roof of the Panorama des Champs Elysées in Paris. This was suspended from straight ties passing over masts standing on the interior walls and anchored back to outer walls.

Figure 13.20 Alitalia
Hangar, Leonardo da
Vinci Airport, Rome
(Morandi).
(Source: author)

Most development has taken place only since 1960. In the Burgo Paper Mill, with a central span of more than 160 m, the resemblance to the suspension bridge was closest [13.19]. The trussed roof was suspended by vertical hangers from four parallel main cables. But the frames carrying these cables were inclined outwards to offer better resistance to the greater pulls of the longer lengths of cable carrying the central span.[27]

In the contemporary Alitalia Hangar [13.20] the form is more that of the cable-stayed bridge [14.2(f)]. The hangar proper is to the right of the slightly inclined pairs of struts supporting the cable stays. It is completely separated from the two-storey part of the building to the left by a wall that was required as a fire barrier and serves as a support for the struts. At the right there is a completely open front, free of all fixed supports, to allow aircraft to be moved in and out. (The doors are carried independently.) To allow this open front, the roof as a whole acts in much the same way as the cantilevered roof of the grandstand in Madrid and is similarly tied down along the line of the rear wall. The weight of the rear part of the building was thus used as a counterweight to balance the excess weight of the part of the roof over the hangar [13.17(e)].[28]

Twenty years later, the INMOS factory required a series of 36 m-deep wings on each side of a wide spine corridor. The wings on one side had to be heavily serviced, unobstructed, and to have an unusually high standard of air cleanliness. The layout was also required to facilitate major changes in the production process that could not be foreseen in detail at the time of construction. The requirements were met partly by installing the servicing above the central corridor and partly by hanging the roof from

tubular trusses supported at intermediate points by tension stays attached to cross-braced groups of tubular masts erected over the corridor [13.17 (f) and 13.21]. These trusses were attached at the corridor end to short triangular trussed columns that also served as the feet of the masts, and they are held at their outer extremities by splayed legs acting partly in tension and partly in compression. The stays are solid steel rods for maximum stiffness and minimum deflection.[29]

Resistance to side loads acting transversely to the span was provided in the Burgo Mill by the rigid-jointed portal frames carrying the main cables, and in the Alitalia Hangar partly by analogous portal frames consisting of the interconnected pairs of struts similarly carrying the stays, and partly by slightly splaying out the three pairs of stays carried by each of these frames. In the INMOS factory it is provided partly by both transverse and longitudinal diagonal bracing of the masts above the central corridor and partly by the splayed legs at the extremities of the trusses carrying the roofs.

Halls with cable-net and membrane roofs or canopies

Two types of cable network and membrane roofs acting in direct tension were discussed in Chapter 7. One type – the air-supported membrane – calls for no further consideration here because the only supports it requires are those of the air inside and a fairly continuous tying-down where it meets the ground. Given these, the membrane itself is the whole structure.

That can never be true of the other type – the slung membrane or cable network. These modern counterparts of the tent or of the velarium that was stretched on ropes as an awning to shade some of the seats of Roman theatres and amphitheatres always require support from above. They always carry their loads upwards, so that other elements are required to bring them down again to ground level.

The poles of the tent and the continuous outer wall of the amphitheatre, stabilized respectively by guys and by self-weight against being overturned by the inward pulls of the roof, have continued to be the prototypes for this support, whether so recognized or not. In the structure illustrated in Figure 5.10, the main support was obviously just a large tent-pole, guyed by the cables of the roof itself and other cables attached to the lower boundary cables. In some nineteenth-century unexecuted projects for large circular exhibition halls with roofs carried by cables or wrought-iron chains slung between central towers and the outer walls, the basic concepts of the two types of support were combined. The chief departure from the prototypes was the substitution, for the massive continuous wall, of less massive elements playing the same role. The group of temporary exhibition pavilions built by Suchov in 1896 kept closely to the prototype of the large circus tent, replacing the masts by more substantial ones or by latticed towers spanned by a ridge truss and replacing the fabric, as noted in Chapter 7, by networks of flat iron strips [13.22].[30]

Two later roofs supported by counterparts of the continuous outer wall were also referred to in Chapter 7. The two inclined reinforced concrete arches of the Raleigh Arena [7.31 and 7.32], intersecting one another scissor-fashion at the lowest points of the roof, were the closest counterpart. Since the roof of the Terminal Building at Dulles Airport [7.30] has cables spanning in one direction only, it could be slung between two straight rows of outwardly cantilevered piers which were held apart against the

Figure 13.21 INMOS factory, Gwent (Rogers, HuntYRM). (Source: Anthony Hunt Associates)

pulls of the roof by the transverse beams of the concourse floor acting as struts. They were stabilized against being overturned inwards largely by their outward overhangs.

The roof of the Yale Hockey Rink, built between the Raleigh Arena and the Dulles Terminal, also spans transversely but is stiffened, like the Raleigh roof, by cables running in the other direction and curving upwards rather than sagging downwards [13.23 and 13.24]. As at Dulles, the equivalents of two walls were provided. But they were arched horizontally against the cable pulls, their central sections were also held apart by transverse struts at foundation level, and they carried stiffer continuous flat top sections just over 2 m wide as the main supports for the cables. A third arch in a vertical plane, with the same parabolic curve as the other two for much of its length and carried down to the ground at its two low points, was provided in addition as a central spine. This allowed the outer walls to be kept as low as the required headroom above the tiered seating allowed.[31]

Elsewhere, where the plan of the roof has been more nearly circular, it has been possible to resist the inward pull by means of a continuous ring-member analogous to the continuous tie at the foot of a domical shell but acting throughout in compression rather than tension.

Figure 13.22 Exhibition Hall, Niznij Novgorod (Suchov). (Source: Archive of the Russian Academy of Science, Moscow)

This could be carried on vertical supports at any desired height just as the shell roof of the Market Hall at Algeciras was.

There is, however, no need for the plan of such a roof to be a simple rectangle or a closed curve. In the early 1950s Otto embarked on a wide-ranging exploration of other possible configurations, and he was stimulated further by construction of the Raleigh Arena.[32] Some of them had rigid boundary supports; others had flexible boundaries and one or more intermediate mast supports. Working largely with small models, he was less inhibited than others might have been by the practical limitations that arise from the need to keep stresses fairly uniform and the difficulties of doing so when building full-scale. His ideas had great aesthetic appeal and led to a number of commissions for medium-sized tented pavilions at garden exhibitions. A first opportunity to test them on a large scale came with the construction of the West German pavilion for Expo 67 in Montreal. Here a square mesh of steel cables was suspended from seven masts to cover an area of 10 000 m^2 with continuously undulating tent-like forms, and a PVC membrane was suspended below it. The mesh was stretched between boundary cables at the foot and other cables hanging in elongated pear-shaped loops from the heads of the masts.

Figure 13.23 Interior of the Hockey Rink, Yale University, New Haven (Saarinen, Severud). (Source: author)

A greater challenge was presented by the roofs of the stadium and other structures built for the Munich Olympic Games in 1972 [13.25]. Here the total area to be covered was 22 000 m^2. Like Utzon's original idea for the Sydney Opera House, the original proposal of the architects Behnisch & Partner was impracticable. It was for a single roof suspended from only a few masts with large excessively flat areas between them. To make it feasible, it was remodelled with the additional participation of Otto and the engineers Leonhardt & Andra. The number of masts was increased, the roof was partitioned into smaller areas, and the flat areas were thereby eliminated. Much further work was nevertheless required to define the geometry with sufficient precision and decide upon all the necessary details, and it was made more difficult by the need to start construction while it was still proceeding. As at Sydney, and as again later at Mannheim, current analytical techniques also had to be developed further to allow it to be brought to a successful conclusion. Again nets of steel cables with a square mesh were stretched between boundary cables and other cables that were carried at their high points by the masts. Large prestressing forces were required to maintain the desired shape. But the continuity of the networks and main cables meant that, when these forces were applied in places where this could be done most easily, they were felt throughout. To complete the roofs, panels of acrylic glass were suspended from the joints in the steel mesh.[33]

The fact that it had already been possible at Osaka to construct an almost flat air-supported fabric roof of considerably greater span stemmed from the fact that it was relatively easy to reduce local curvatures (and thereby stresses in the fabric) by adding a cable network *above* it and held in contact with it simply by the internal air pressure. As noted in Chapter 7, membrane roofs on the scale of the Munich roofs, and unsupported by a separate cable network, became feasible only with the availability of stronger fabrics in the later 1970s.

The limitations on feasible geometries, their requirements for support, and the need for prestressing, are essentially the same in practice as for cable networks, but the simplicity of the pure membrane lends a greater elegance if the main supports are not too obtrusive. The fine example at the Denver International Airport Terminal [7.33] has already been described in Chapter 7. An earlier example, more directly comparable with the Munich roofs and designed some ten years after them, is the King Faud Stadium roof, Ryadh [13.26]. Twenty-four identical units arranged in a circle provide shade to 52 000 m^2 of seating and surrounding concourse. The section and plan of one unit [13.17(g)] indicate the way in which the fabric is supported by cables stretched between a tall main mast, a stayed inclined secondary mast, a

Figure 13.24 Exterior of the Hockey Rink, Yale University, New Haven (Saarinen, Severud). (Source: author)

Figure 13.25 (top right) Pavilion for the 1972 Munich Olympic Games (Behnisch, Otto, Leonhardt). (Source: author)

Figure 13.26 (bottom right) King Faud Stadium roof, Ryadh (Fraser Roberts, Geiger Berger, Schlaich, Birdair). (Source: Horst Berger)

ring cable, and two ground anchorages. This support system was erected first. The fabric was then laid out on the ground, hauled up the main mast, attached to the cables, seamed, and finally stressed.[34]

Although the design of cable net and membrane structures can never be completely free, they do offer a freedom that did not exist before and, as these examples show, the necessary disciplines can make a positive contribution.

Chapter 14: Bridges

The problems that a bridge presents to its designer obviously have something in common with those presented by a wide-span building. But there are also, and always have been, significant differences. While the developments of the two forms have impinged on one another from time to time in various ways, that of the bridge has therefore followed its own course – a course that has been particularly interesting in the last two centuries.[1]

The most characteristic difference stems from the different functions of the two forms. The primary and almost the sole function of nearly all bridges has been to provide a way for some sort of traffic, or even a watercourse, over an obstacle such as a river, a deep valley, another traffic route, or open water. This is a simpler and more clearly defined function than that of the great majority of buildings, and has tended to be reflected in basically simpler and more openly expressed structural forms. There have been exceptions, of course. Buildings like airship hangars, aircraft hangars and train sheds are, in this respect, more like most bridges. And there have been bridges that were more akin to buildings in their primary functions, such as the seventeenth-century Khaju Bridge in Isfahan [14.1] which was built more to linger on and enjoy the cool breezes over the water than for

caravans to pass over. But such bridges have never been typical.

The characteristic simplicity of function of the bridge has not meant, however, that the structural requirements to be satisfied have always been less onerous or easier to satisfy efficiently and economically than the corresponding requirements for wide-span buildings. Quite the contrary. With the coming of the railways in the nineteenth century and of fast motor roads in the twentieth, it has, for instance, been necessary to design for large dynamic loads imposed by the traffic that repeatedly moves across the span. These have had no parallel in the loading of most wide-span buildings. Much more attention has also had to be given to wind loading as spans have become longer and relatively lighter, while there have always been problems associated with building under and over water that do not arise on dry land.

Likewise, the basic simplicity of form of most bridges has not always been matched by a corresponding simplicity at a more detailed level. Nor have bridges been any less advanced in their exploitation of new structural techniques. For the last 200 years, the reverse has usually been true. Partly this has been a consequence of increases in span far ahead of those seen in contemporary buildings. Partly also it

Figure 14.1 Khaju Bridge, Isfahan. (source: author)

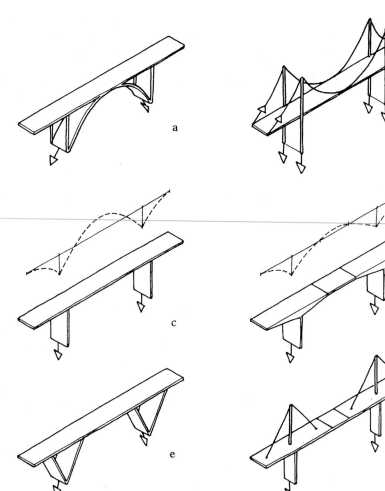

Figure 14.2 Diagrams of the principal types of bridge. Arrows indicate the directions of the loads exerted on the ground. Broken lines above (c) and (d) indicate bending moments for self weight loading. (see text). (Source: author)

has come about because the challenge presented to structural engineers by a proposed new bridge has been of a kind that they have found particularly stimulating. It has stretched their powers to the utmost while, at the same time, allowing these powers to be used fully, unfettered by a need to satisfy also a multitude of non-structural requirements. Much of the development of the structural uses of iron, steel, and reinforced and prestressed concrete has thus taken place initially in bridge building. Also, of the basic one-way spanning systems illustrated diagramatically in Figure 14.2, all but one (or perhaps two, depending on the niceties of definition) have been developed initially for bridging purposes and only later used also for wide-span buildings. The development of spans curved in plan has also taken place largely in bridges.

There is even one respect in which the basic structural forms of many bridges have necessarily been rather more complex than those of their building counterparts. To provide a convenient more or less level route for the traffic, it has usually been necessary, as indicated in Figure 14.2, to associate with primary spanning elements such as arches and catenaries other more nearly horizontal elements. In principle, these need only be propped up by or slung from the primary spanning elements. Some structural

interaction is almost inevitable, though, and much of the interest of some recent bridges lies in the way in which this interaction has been turned to advantage in providing the requisite overall strength and stiffness.

Masonry arch bridges

The first major bridge-building activity was closely associated with the building of a great new network of roads to facilitate communications in the Roman Empire and, to a lesser extent, with the supply of water to its principal cities.[2] Bridges of many kinds were built − timber beam and trestle, timber arched, and stone and concrete arched. But the only ones to have survived for something like 2000 years have been some of the stone and concrete arched ones.

The Pont Julien [6.8] may be taken as representative of the best road bridges though of more modest scale than some, and the Pont du Gard [4.6 and 4.7] of the best elevated aqueducts. Both these structures have excellently constructed stone-voussoir arches of typically Roman form and proportions, with spandrels built up solidly above them. The piers were founded directly on rock and have widths which, characteristically, are between one-quarter and one-fifth of the span. Since the arches spring fairly low down on them, these piers would have been well able to resist an unbalanced thrust from one side only, both during construction and in the event of one or more arches being damaged. On the other hand, such wide piers could greatly impede the flow of a river in flood. It was therefore fairly usual to construct smaller relief arches above them as seen in the Pont Julien. Being constructed above the level of the springings of the main arches, they did not jeopardize their stability. They were particularly necessary when the piers could not be founded directly on rock, because the Romans did not have the means to carry the foundations down under water to a sufficient depth to remove the risk of undermining by scour.[3] Such undermining was probably the most frequent single cause of collapse.

Figure 14.3 Bridge at Sharistan, near Isfahan. (Source: author)

Up to the later Middle Ages, masonry arch bridges generally followed the Roman pattern without, as a rule, being constructed of quite such excellent masonry or approaching the best Roman examples either in span or the refinement and carefully considered unity of their designs.[4] Foundations, in particular, seem often to have presented even more of a problem than they did to the Romans. Extremely massive piers, with widths about equal to the spans of the arches they carried, were common [14.3]. Their very massiveness would have served to compact any poor ground beneath them before the arches were built, and to confer, thereby, some added security. But they inevitably increased further the risk of scour in time of flood and made all the more necessary the provision of relief arches above the springing-levels of the main ones.

The commonest change was the adoption of pointed profiles in place of the semicircular or segmental arcs preferred by the Romans. Their greater rise for a given span sometimes resulted in steeply inclined approaches. But, in multi-span bridges like that at Sharistan, they had the advantage of making it possible to maintain fairly constant levels for both the springings and crowns of the arches (including any smaller

relief arches) simply by varying the sharpness to compensate for the variations in span. Another advantage was a reduced need for centring.

To reduce further the need for centring, other changes were sometimes made. At Sharistan (in a land with little timber) the bricks were not laid voussoir fashion but pitched along the lines of the arches as described in Chapters 4 and 6. Where stone voussoirs were used, each arch might be constructed as a number of parallel ribs spaced a little distance apart with slabs laid above them to close the gaps as in some Roman barrel vaults and their Romanesque successors.[5] Centring sufficient for the temporary support of one rib would then have sufficed for a whole arch if the ribs were constructed one by one and it was repeatedly reused. To reduce weight on the arch rings and save materials and labour, it was also not uncommon to leave voids in the fill that carried the roadway.

The 'chain arch', already referred to in Chapter 6, appears to have been used only in places like Venice and Amsterdam and the Yangtse delta in China where it was difficult to construct pier foundations and abutments that would not tilt or slip in the soft ground. The Venetian example that was illustrated in Figure 6.12 is a very small one, and the one portrayed in a painting by Gentile Bellini in the Accademia in Venice was probably not much larger although it did have three arches.[6] Some Chinese examples are considerably larger, and the multi-span ones are notable for the slenderness of the intermediate piers. This was made possible partly by placing the springings of the arches almost at water level. The most important characteristic of all these bridges is the way in which, as described in Chapter 6, the very slender and flexible arches were stabilized and prevented from collapsing by the fills above the haunches.

It may also have been in China that arches of flat segmental (circular arc) profiles were first used.[7] The earliest extant European example, the fourteenth-century Ponte Vecchio in Florence, has three such arches, of which the centre one rises only 4.2 m to the crown over a span of 28.7 m. The An-chi Bridge in northern China, which may date from the early seventh century, has only a single arch which springs almost from water level and has the proportionately rather greater rise of 7 m over its somewhat greater span of 37.5 m. The arch itself is also proportionately deeper and all its voussoirs are bound to one another by iron cramps. It is notable for the opening-out of each spandrel by the construction of a relief arch directly over the haunch. This was probably essential to reduce obstruction to the river in time of flood. These arches will also have served a useful structural purpose in relieving the haunches of a large weight of masonry, and they may have served as a model for a similar lightening of the haunches in some European bridges from medieval times

Figure 14.4 Maglova
Aqueduct, Belgrade
Forest, near Istanbul
(Sinan). (Source: author)

onwards, though it is equally likely that this happened independently.

In post-medieval times, Roman structures like the Pont du Gard continued to provided the usual model where a roadway, canal, aqueduct, or railway had to be carried, as nearly on the level as possible, over a deep valley. Semicircular arches were usually preferred and the chief changes were various expedients to lighten the structure. Usually this was again done by introducing voids, commonly by supporting a road or railway on a series of parallel walls over the arches rather than on a solid fill.[8] An unusually elegant alternative was adopted in the Maglova Aqueduct [14.4]. Here Sinan tapered the piers as they rose, diminished the widths of the successive tiers of arches accordingly, and penetrated the piers with smaller arches both longitudinally and transversely.[9]

Where there were no steep banks and it was again desired to keep the roadway as level as possible while still causing the least possible obstruction to the waterway, arches with a small rise in relation to the span became the usual choice. Architecturally the finest example is undoubtedly the sixteenth-century Ponte Santa Trinita alongside the Ponte Vecchio in Florence [14.5]. Its spans and the ratios of rise to span are very similar to those of the Ponte Vecchio, but the profiles selected for the arches were subtly different. Instead of being simple circular arcs meeting the piers at sharp angles, they were pairs of curves of constantly varying curvature tangential to the piers at the springings and meeting at a slight angle at the crown. Thus both the thrusts of the arches and the eyes of the stroller by the side of the river are carried smoothly into the piers, while the slightly pointed crowns (masked by carved pendants

surmounted by cartouches) avoid the risk of collapse due to an excessive flatness there and give a greater strength of line than simple continuous ellipses would have done. To absorb the horizontal thrusts, the piers were given widths more than one-quarter of the central span, and great care was taken with the foundations.[10]

In the second half of the eighteenth century, Perronet took the further important step of reducing the widths of the piers to just enough to take the vertical loads of the arches. The opposed horizontal thrusts of adjacent arches were left to cancel one another out as in the sort of columnar arcade seen in Figure 12.7, and heavy abutments were provided to resist the unbalanced thrusts of the two end arches. He did so in a series of bridges starting with one over the Seine at Neuilly, carried the idea further at Pont-Sainte Maxence over the Oise, and applied it finally to the Pont de la Concorde in Paris.[11] This last alone still stands and it has been considerably widened to take modern traffic, but happily without destroying its character [14.6]. As first designed, it was to have had arches with rises less than one-tenth of the spans, carried on piers with widths only a little greater than one-tenth of the spans and each divided into two parts, one on each side of the bridge. Official caution forced Perronet to increase slightly the rise of the arches and to make the piers continuous over the width of the bridge, in spite of the fact that the design had been no more daring than that of the bridge at Pont-Sainte-Maxence which had already been successfully completed.

Another feature of the arches of these bridges was that there was no longer a clear distinction between arch and spandrel. With the intention of making the two work together as fully as possible, the voussoirs were bonded into the spandrels in the manner seen in Figure 4.15. One is thus reminded by them of the reinforced flat arches currently being used in buildings [3.15 and 8.10], just as the treatment of the piers as supports only for the vertical weights of the arches is reminiscent of the similar treatment of the supports immediately beneath the dome of the French Pantheon [12.31 bottom left].

Perronet's bridges, in fact, really marked the end of a road in much the same sense as the French Pantheon had done. Had the material been available, their flat arches could have been constructed more easily in iron, steel, or reinforced concrete, either still as arches or as beams. And had they been so constructed they would have been lighter and would have thrust

less. Thus the need to balance the thrusts in order to reduce the widths of the piers would also have been diminished and it would no longer have been necessary to have all the spans approximately equal and to construct all the arches together and strike all the centers simultaneously.

Large masonry arch bridges nevertheless continued to be built until the beginning of the present century, but rarely with such flat arches. Usually the arches had a considerably greater rise. To reduce the total weights and thrusts, the spandrels above them were opened out, as in the An-chi Bridge, by carrying the road or railway on what were effectively superimposed viaducts of many shorter spans.[12]

Figure 14.5 Ponte S. Trinita, Florence, as reconstructed to the original design after being largely demolished in 1944. (Source: author)

Figure 14.6 Pont de la Concorde, Paris (Perronet). (Source: author)

Figure 14.7 Road bridge at Chepstow (Rastrick). (Source: author)

Timber, iron, steel, and reinforced-concrete arch bridges

When materials able to resist tension have been used for the main spanning elements, the structural distinction between arch spans and beam spans (including spans cantilevered out from the supports among the latter) has not always been sharply defined. Many recent bridges whose spans have arched soffits closely resembling those of the Pont de la Concorde have, structurally speaking, been beam bridges. Others have been intermediate or hybrid forms. While their spans have acted to a considerable extent as beams, they have had enough compression induced in them by horizontal thrusts at the supports for them to be regarded also as arches. Both these true beam bridges and the intermediate or hybrid forms will be considered later.

If we confine our attention here to those bridges in which the principal action of the spans has been an arching one, it is possible to identify two basic forms. On the one hand are those bridges in which the arches are self-sufficient. They have all the stiffness needed to carry the varying imposed loads, and the deck is merely propped from it as sketched in Figure 14.2(a), or suspended from it. On the other hand are those bridges in which a significant part of the requisite stiffness is contributed by the deck or, in a less determinate manner, by multiple interconnections between a fairly large number of elements. Among the masonry arch bridges just considered, the closest parallels would be the open-spandrel An-chi type of bridge for the first group, and the 'chain arch' type of bridge for the second. Further alternative possibilities arise in bridges of the first group. When the deck is capable of resisting tension, its tensile strength can be used to resist the horizontal thrusts of the arch or it may be left to the supports to provide balancing counter-thrusts.

Probably most early timber arch-bridges as well as many early cast-iron ones relied on multiple interconnections of a large number of elements. Two examples – Trajan's Bridge over the Danube and the Coalbrookdale Bridge [3.16 and 6.15] – have been mentioned already in Chapter 6. The fine example at Chepstow illustrated here [14.7] was constructed about thirty-five years after Coalbrookdale Bridge. Though much simplified in its detailing, it is still a highly indeterminate structure. It is not even possible to say, from a purely visual inspection, to what extent the spans as a whole behave as pairs of cantilevers fixed to the piers. Certainly there are no separate self-sufficient arches, and none of the upper arched ribs is even continuous throughout a span. Such arches as there are must be envisaged as extending over the full depth of the spandrels. In this respect they might well be compared with the arches of the Pont de la Concorde. The proportions are also comparable, though the central span of 34 m at Chepstow is substantially longer. But the change to a lighter form of construction did indeed permit the relaxation of some of the restrictions on design that Peronnet had to observe, without calling for any noticeable increase in the widths of the piers. Here, instead of being almost equal to equalize the thrusts, the spans diminish towards the abutments, first to 21 m, then to 9 m.

In bridges like those illustrated in Figures 6.14 and 9.7(a), the arch, usually built up from horizontal laminations, became a clearly defined element. But its depth was insufficient to enable it, unaided, to stand up to moving loads without excessive deflections and vibrations. The span was therefore stiffened by a truss through which the roadway or railway passed. This truss, by itself, would also have been inadequate. But it stiffened the arch on each side by distributing any concentrated loads on the deck more widely over the span.

The very elegant series of bridges built by Maillart in the 1920s and 1930s with thin-slab arches of reinforced concrete, represented here by the Valtschiel Bridge of 1926 [14.8], followed the same principle but with considerably more clarity.[13] The arches were given only the minimum thickness required to resist the direct compressions induced by the largest possible uniformly distributed load from the deck, and they were made slightly polygonal in profile since this load was transmitted to them at a few points only. If the decks had been equally flexible, large local concentrations of load on parts of the spans would have led to gross deflections and consequent collapse of the arches. To prevent this, the decks were given the extra stiffness of deep trough-shaped cross-sections. To further ensure that no bending moments were transmitted to the thin arches, the interconnecting vertical elements were also made as thin as was possible without running the risk of their buckling.

Though the parallel with the masonry 'chain arch' bridge has been mentioned, one significant difference should be noted. This is that the Valtschiel Bridge and others like it were intended to retain their shape under the action of moving loads comparable in magnitude with the weight of the bridge itself. The 'chain arch' bridges, on the other hand, were designed specifically to accept large deformations of the flexible arch consequent on foundation movements, but to stand up only to the action of relatively much smaller moving loads. For this purpose, the heavy spandrel fill between the deck and the arch, capable of deforming only slowly to follow the deformations of the arch, was the appropriate form of stiffening.

A notable feature of another bridge of the group, the Schwandbach Bridge, is that the deck was built on a continuous horizontal curve to allow the road which climbs up one side of the narrow valley to turn back on itself at the other side without excessively sharp bends at the two

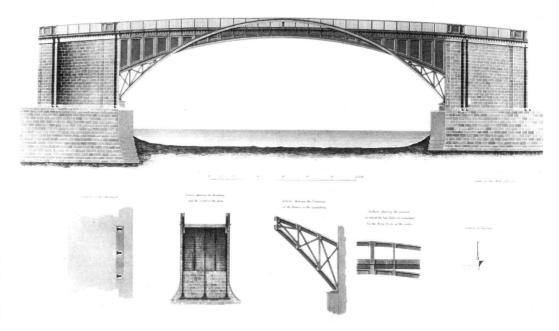

approaches. To allow this, the arch was made curved in plan on the inside of the curve and straight on the outside, and the vertical walls (with battered edges on the outside of the curve) were aligned at right angles to the straight edge to minimize the inevitable torsional moments introduced by the curve.

The first iron bridges with arches clearly distinguishable from the spandrels (and probably not thought of as relying on any stiffening from these to ensure their continued stability under the fairly light moving loads of the time) were those whose arches were assembled as parallel ribs from large numbers of voussoir-like castings as in the bridge over the Wear at Sunderland mentioned in Chapter 6. These castings were considerably deeper than the individual arched ribs used, for instance, at Coalbrookdale and Chepstow. The bridges tended, nevertheless, to be excessively flexible and shaky. More than one collapsed soon after completion.[14]

Rather more interesting is Telford's Buildwas Bridge [14.9] constructed over the

of defence on the indeterminate degree of further stiffening provided by the plating of the spandrels over the flatter ribs.

The design of the Buildwas Bridge was not wholly satisfactory. The coupled arches of different radii had no freedom to 'breathe' – to adapt to changes in temperature, for instance. Something had to give way, and it was only after many years of patching that it was recognized that this was an inevitable and not necessarily harmful act of self-adjustment on the part of the structure.[15] It might now be likened to the cracking (equally innocuous from a purely structural point of view) that may so often be observed in masonry structures with a similarly high degree of statical indeterminacy. At the time, especially with the steady growth of the ability to analyse the forces in statically determinate structures, it naturally led to a preference for a more nearly statically determinate form.

In several of Telford's own later wide-span arch bridges – at Craigellachie [6.16], in the Galton Bridge at Smethwick, and in the Mythe Bridge at Tewkesbury – this may be seen in the

Severn not far downstream from the slightly earlier Coalbrookdale Bridge. Here also the arch was clearly distinguishable. But it consisted, on each side of the bridge, of a pair of ribs of different radii braced together. It is possible that the flatter rib at each side, together with similar ribs under the deck, was thought of as carrying most of the dead weight of the bridge, leaving the other ribs chiefly to assist in carrying concentrated moving loads. The flatter ribs themselves were probably deliberately made so flat to increase their thrust and prevent similar trouble to that which had been experienced at Coalbrookdale on account of the lightness of the span – a tendency of one abutment to be moved inwards by a slight progressive slip of the earth bank. Be this as it may, Telford certainly seems to have aimed at fairly stiff and self-sufficient arches and to have relied only as a second line

adoption of trussed arch ribs connected to the deck by light crossed bracing with no separate uprights.[16] The Eads Bridge in St Louis, and Eiffel's Maria Pia and Garabit Bridges [4.19 and 14.10], went further in the same direction by making the trussed arch completely self-sufficient, though not completely statically determinate.[17] For the latter purpose, three pinned articulations would have been required in each arch, whereas there were none in the Eads Bridge and only two in each of Eiffel's bridges. But these bridges were constructed of steel or wrought iron rather than cast iron, so they were better able to accommodate themselves to changes in temperature and the like without fractures. Also some statical indeterminacy was valuable in conferring extra stiffness to withstand better the heavy railway loadings for which these bridges were designed – loadings

Figure 14.10 Garabit Bridge (Eiffel). (Source: author)

for all but the longest steel spans.[18]

In Maillart's Thur Bridge [6.18] the two parallel open trough-section arches were, for instance, carried up to the deck and united with it over the central half of the span. In addition a third hinge was introduced at the crown, where, in the Garabit Bridge, the bending moments due to unsymmetrical loading from a train on one half of the span called for the maximum depth of truss.

In other longer-span bridges like the Bayonne Bridge [9.9] and the bridge at Sainte-Pierre-du-Vauvray [14.11], it was often found more convenient, or even essential, to suspend much of the length of the deck from the arch instead of carrying it above on struts. It was then in the right position to serve also as a tie to absorb the horizontal thrusts of the arch. But only where these thrusts cannot more easily be absorbed directly into the ground is the complication of the design entailed by giving the deck this dual role justified. This situation is most likely to arise where the arch is carried on fairly tall piers and the span is considerably less than the 504 m of the Bayonne Bridge.

The first major use of the tied (or so-called 'bowstring') form was probably in Robert Stephenson's High Level Bridge in Newcastle completed in 1849. Here there were six 38-m spans high above the Tyne. Strutted up from the arches were railway tracks. Below each of them, a suspended roadway deck served also as the tie. Figure 14.12 shows a recent example with only the suspended roadway deck.[19]

If the tie was fully effective, the tied arch was, of course, equivalent to a beam. In fact, with the addition of diagonal bracing as the Whipple arch-truss, it was one of the many forms of beam-like truss developed around the middle of the nineteenth century. The High Level Bridge and others like it might, therefore, be better regarded as beam bridges, if this term is taken to include bridges with trussed and other non-thrusting analogues of the continuous-web beam as their spanning elements.

much in excess of those for which any of Telford's bridges were intended. The techniques used in the erection of two of the bridges were described in Chapter 4 and the early (though not the first) use of compressed-air caissons to found the piers of the Eads Bridge in the difficult waters of the Mississippi in Chapter 10. Another interesting feature was the splaying-out, in plan, of the arches of the Maria Pia and Garabit Bridges towards the abutments so that they could act as fixed-ended trussed horizontal beams in resisting side winds.

Most other wide-span arch bridges from the mid-nineteenth century onwards may be regarded as variants on the forms of the Eads, Maria Pia, and Garabit Bridges, usually with the substitution of solid-walled box or trough sections for the arches where these have been constructed of reinforced concrete, and likewise

Figure 14.13 Bridge at Bassano del Grappa (Palladio). (Source: author)

Beam (including cantilever) bridges

Probably the earliest of all man-made bridges were also beams, but much simpler ones. With convenient banks, boulders, or other natural supports at each side of a gap that it was desired to cross, it would merely have been necessary to manoeuvre into place between them a suitable length of timber or slab of stone. Similar simple bridges have continued to be made even to the present day, usually with man-made supports. Spans are, however, limited. The granite spans of over 20 m apparently achieved in at least one Chinese bridge must always have been highly exceptional. Often therefore they were reduced by cantilevering out the piers as they rose. Or the beams were enabled to span farther by being propped at intermediate points by inclined struts as in Palladio's bridge seen in Figure 14.13.[20]

From about the beginning of the nineteenth century, beams of iron, steel, reinforced or pre-stressed concrete, or their trussed and other analogues referred to above, have, in turn, been substituted for the simple timber beam or stone slab to further extend the limits of both span and load-carrying capacity.

Where each span has been simply and independently supported, the bridge has still had essentially the same basic structural form and action as its earlier prototypes. Greater care has merely had to go into the design of the supports to ensure this. Not only must they bear all likely loads (both vertical loads and the horizontal ones due, for instance, to wind or the braking action of a train). They must also provide the freedoms of rotation and longitudinal expansion that are necessary at the ends of each span if it is to be free of bending moments there and truly non-thrusting. But, with no horizontal thrusts generated by the vertical loads, this design has still been much simpler than the design of the

supports for an arch bridge. It therefore seems unnecessary to say more about this type of bridge than what has already been said about its spanning elements and supports in Chapters 8 to 10. Bridges with continuous or cantilevered spans or with inclined supports as in Figure 14. 2 (e) or 14.13 do, however, call for some further discussion.

The chief drawback of the simply supported beam and any of its equivalents is that, with no bending-moments at the ends, the maximum moment always occurs near midspan where it is two to three times (according to the loading) what it need be if the ends are fully restrained from rotating. In turn, this increased maximum moment means that the span must be heaviest there, with a consequent further relative increase in the moment. This is true even if there is a partly compensating increase in depth towards midspan.

Making several spans continuous over their common supports as indicated in Figure 14.2(c), or cantilevering them out from the supports as indicated in Figure 14.2(d) with or without short simply supported central spans as shown there, overcomes this drawback by transferring much of the moment to the neighbourhood of the supports where more efficient provision can be made for it. This is indicated in the diagrams above the bridges themselves, in which the broken lines show the variations in moment along the spans. (The dipping of the lines representing the moments below the base lines in the neighbourhood of the supports means that the moments there act in the reverse direction, inducing tension at the top of the beam and compression at the bottom.)

The similarity between the two diagrams makes it clear that, for self-weight loading, the cantilevered form can act in precisely the same way as the form with continuous spans if it has simply supported central spans of appropriate length. It differs in that the moments must always reduce to zero at the ends of the cantilevers, whereas the locations at which they do so in a series of continuous spans vary with the distribution of the loading and, for a given pattern of loading, depend on the variations in bending stiffness along the spans, on the precise manner of construction, and on any relative settlements of the supports. The cantilever form is, in this respect, statically determinate: the continuous-beam form is not.

To take full advantage of the continuous-beam form called, therefore, for a greater analytical ability than to take similar advantage of the cantilevered form. The designer had to be able to analyse the distributions of moment that would arise from particular choices of stiffnesses and manner of construction. For this reason, it is only from about the 1950s onwards that it has been fully exploited. It was, nevertheless, the first of the two forms to be exploited on a large scale. The occasion was the construction, between 1846 and 1850, of the Britannia Bridge over the Menai Straits [14.14]. The tubular beams [5.7] were very similar to those of the single-span Conway Bridge [8.6] constructed only slightly ahead of it. By making the three Menai tubes continuous over the piers, it was possible to increase the central span to 140 m from the already unprecedented 122 m at Conway. Just how they were made fully continuous for the live loads of trains passing over, while arranging that the maximum moments under self-weight load would be about equal over the supports to those at midspan instead of considerably greater than these, was described in Chapter 4.

Despite this initial success with the continuous-beam form, it was the cantilevered form that became dominant for wide-span railway bridges in the latter part of the nineteenth century. Probably its main attraction was still its statical determinacy. This not only simplified the analysis of its behaviour. It also overcame an inherent drawback of the continuous-beam form whenever there was a risk of one support sinking relatively to others. This drawback was that the continuity of the spans allowed the sinking support to shed some of the load that it was intended to carry to adjacent supports, thereby upsetting the whole designed distribution of bending moments in the spans. In addition, of course, the cantilevered form was the ideal one for construction over water by progressive canti-levering-out from the supports. This method can also be used to construct continuous spans, and frequently is now so used. But, to achieve the desired continuity in relation to self-weight loading, it is then necessary to adjust the initial distribution of the internal bending-moments by some sort of prestressing operation.

The outstanding achievement of the time was undoubtedly the Forth Railway Bridge [14.15]. It was a structural masterpiece, with its three great pairs of trussed cantilevers reaching out 207 m from each anchored foot and carrying simply supported central spans of another 106 m to leave two clear waterways of 520 m.[21] Breathtaking in these giant leaps and impressive in its manifest strength, it was also a remarkably clear and legible structure. Only on close inspection does one realize that the two outer pairs of cantilevers had to be tied down at their outer ends to great weights of ballast (concealed in the piers at the ends of the approach viaducts) to prevent their being tipped inwards by the otherwise unbalanced loads of the central spans with a train on one or other of these spans. And only on even closer inspection from beneath the

Figure 14.14 Britannia Bridge, Menai Straits (Robert Stephenson). (Source: author)

railway tracks is one perhaps inclined to criticize the manner in which the much lighter trusses carrying these tracks have been strutted up from some of the transverse bracing-members of the main cantilevers [10.15]. The detailing of the principal joints was, like the structure as a whole, superbly three-dimensional. But neither could really serve as a model for the twentieth century, when economy in fabrication and erection almost invariably demanded something simpler and with less variation in cross-sections.

Two twentieth-century successors of the Forth Bridge are shown in Figures 14.16 and 14.17. The East Bay Bridge, with a main span of 425 m, is fairly representative of a number of steel-truss bridges of comparable span, except that its lines are rather cleaner than most and that it had to be designed for possible earthquake loadings calling for unusually deep underwater foundations. The trusses characteristically are carried on high and relatively slender piers rather than rising directly from water level. Also they have members of more uniform cross-section arranged in a more repetitive and less three-dimensional manner than in the Forth Bridge. But their varying overall depth still reflects the variations in bending moments along the spans.

The Bendorf Bridge, with a main span of 208 m, is similarly representative of a more recent group of prestressed-concrete box-girder bridges including the Medway Bridge illustrated previously [4.20]. It differs from the East Bay Bridge in having no central simply supported span. The two cantilevered arms simply meet in the centre where they are joined in such a way

as to allow slight relative longitudinal movements due to expansion or contraction while also transmitting any vertical loads that may be necessary to prevent one arm from rising in relation to the other. Because of this, the bending moments reduce continuously to the centre. This reduction is reflected in the continuous reduction of depth which gives an arch-like soffit to the span as a whole. A sketch of the arrangement of the prestressing cables was given in Figure 4.19.[22]

A very different recent example is the footbridge shown in Figure 14.18. It consists largely of two identical balanced cantilevers meeting over the river and connected there in essentially the same way as the arms of the

Figure 14.15 Forth Railway Bridge. near Edinburgh (Sir Benjamin Baker). (Source: author)

Figure 14.16 San Franciso Oakland Bridge: East Bay crossing from Yerba Buena Island (Purcell). (Source: author)

Figure 14.17 Bendorf
Bridge (Finsterwalder).
(Source: author)

Figure 14.18 Footbridge,
Durham (Arup).
(Source: author)

Bendorf Bridge. A third section in the fore-ground, aligned at a slight angle to the main span, is independently cantilevered-out from the bank. To avoid the expense of constructing centring over the river, each main cantilever was constructed on its final base but in line with the river bank. It was then swung round into position on a conical bearing which, after both spans had been finally levelled, was grouted solid with liquid cement.[23] Direct comparison with the other bridges is ruled out by the much-reduced scale and loading of this bridge. Its simple elegance is, nevertheless, particularly notable.

The corresponding twentieth-century succes-sors of the Britannia Bridge have, on the whole, been bridges with multiple spans that are individually much shorter than most cantilever spans, though they have included a number of bridges like the present Waterloo Bridge in London and the Zoo Bridge in Cologne [8.7] which are partly of the cantilever form.[24] By far the largest number have been built since the 1950s to carry elevated roads, for which purpose they have often proved more economical than bridges with multiple simply supported spans like the Corso Francia [8.13]. Box-section beams of prestressed concrete, or of steel with a concrete deck, have been the usual choice for the spans, since these have rarely been long enough to justify the use of trusses. Thanks to the high torsional stiffness of such sections, it has been possible to reduce the supports to a single line of columns along the centre with considerable aesthetic gain, as seen in Figures 5.2, 8.15, and 10.10. It has also been relatively easy to build along a continuous curve, as seen in the first and last of these figures. Breaks must be left at intervals in a long straight series of spans to allow for longitudinal expansion, and all but one of the supports of a continuous length must incorporate sliding, roller, or rocker bearings unless they are them-selves sufficiently flexible to allow the expan-sion to take place freely. In a continuously curved series of spans it should be possible to increase the spacing of the breaks by allowing part of the expansion to take place radially rather than longitudinally.[25]

The use of supports that are inclined in the direction of the span, whether single legs or V-shaped pairs of legs, allows a number of choices according to the internal detailing and the final

273

stress distribution resulting from the construction procedure. The simplest beam bridges with inclined supports have single supports at each end of the span, beyond which they continue as cantilevers towards the abutments.

If, as in the footbridge in Figure 14.19, the ends of the cantilevers are left to float freely, with restraint only against longitudinal movement, and there are effective hinges at the tops of the supports, the system is statically determinate. In addition to reducing the effective central span – particularly useful on motorways as a means of doing away with piers in the central reservation without adding excessively to the cost – the inclination of the supports results in an inward thrust on this span which can be enough to eliminate all tension near the centre. The cantilevered overhangs also greatly reduce the bending moment there and permit a very shallow depth. If these conditions are not met the behaviour becomes statically indeterminate.

If the ends of the cantilevers are tied down by means of ties inclined in the opposite direction, it is then possible, by prestressing the ties, to achieve any initial distribution of bending moments along the span that may be desired. It is also possible, by jacking at the supports, to achieve any desired initial inward thrust as Freyssinet did at Esbly [4.22 and 6.20] and Morandi did in Turin [13.16 and 13.17(c)]. If, instead of hinges at the tops of the supports, there is full continuity between beam and support, the behaviour becomes essentially that of an arched portal frame with cantilevered extensions without any need for tie-downs or jacking. Freyssinet's last major work, the St-Michel Bridge in Toulouse [14.20] is an example of this, as may be seen from the variations in depth over the span. The depth is a maximum where the beam meets each support, and it tapers to a minimum at the centre of the span and at the end of each short cantilevered extension towards the adjacent span. This bridge was actually a highly ingenious reconstruction of an earlier cast-iron arched bridge of five main spans, which had to be undertaken around the existing structure while traffic continued to use it and which yet succeeded in paying proper respect to the eighteenth-century bridge seen beyond it.[26]

A series of V-shaped supports as in Figure 14.2(e), with both legs of each support acting simply in compression, has also been used to reduce the spans of a deck continuous over several spans. The beams have then sometimes been of constant depth throughout or, for better appearance on a straight alingment, they have reduced in depth towards the centre of each span, remaining at a maximum over the V-supports. In the fine Kylesku Bridge, constructed more than twenty years after the St Michel Bridge at a very different site on the north-west coast of Scotland, the V-supports carry a continuous prestressed concrete box-section

Figure 14.19 Motorway footbridge near St Albans (Herts. County Engineer). (Source: author)

deck over a single main span and two shorter side spans. But here the alignment is curved, especially towards one end of the main span. Each leg of the V is therefore a pair of struts which meet directly under the deck but then separate and splay outwards to give a broader base and resist the considerable side loads due to the curvature and the extreme exposure to wind. The deck itself is of constant depth and is reinforced to resist the torsion that results from the curvature. It was constructed by first cantilevering from the two piers and then lifting into place a prefabricated central section. Continuity throughout was established by a final prestressing operation.[27]

Simple cable and stressed-ribbon bridges

The simplest alternative to the beam or the arch as a means of crossing a gap is a rope slung across it and securely attached at each end. It was probably the last of these three forms to appear, and crossing would be possible only by tediously moving along it hand-over-hand. In due course, crossing was made easier by adding further ropes for the hands to assist in maintaining balance when walking across the main rope and then by linking several main ropes alongside one another to create a wider walkway. Stay ropes could be added at points along the span to limit the tendency to sway sideways.[28]

Examples are still found in various parts of the world, using ropes made from locally available natural fibres. A straight substitution of iron chains for ropes of natural fibre occurred almost two millennia ago in China. With the further substitution of steel ropes, similar walkways are still used for access during spinning of the main cables of large suspension bridges [2.1 and 14.25].

In other modern counterparts, stiffer catenaries made of laminated timber and less prone to sidesway have been slung side-by-side to carry a timber deck. Or concrete decks have been directly supported by steel cables or steel flats. When the latter is done, there is a close similarity to the slung roof of the Dulles terminal building [7.30]. The chief difference, other than a reduced relative width, is that the cables may be prestressed for greater stiffness and to avoid cracking under live load. The structure is then known as a stressed-ribbon.

As yet, however, most built stressed-ribbons have been footbridges for which an appreciable sag in the centre is acceptable and the tensions to be resisted by the abutments are not excessive even for spans of 100 m or more.[29] A flatter profile for road bridges can be obtained only at the expense of much greater tensions. For a given weight and traffic loading, each halving of the sag will roughly double the tensions to be resisted. This complicates the design of the abutments and cancels out the saving in cost that might otherwise be achieved by simplifying the spanning element. Moreover the profile and the corresponding tensions will be much more sensitive to changes in temperature of the cables. For such reasons, and perhaps also on account of a deep-seated, even if irrational, prejudice against a sagging crossing, the most ambitious designs hitherto, both by Finsterwalder in the early 1960s, were never built. The first, for the Zoo Bridge in Cologne, had a suspended central section of 176 m between shorter cantilevers and central sag over this distance of about 1 m. The second, to cross the Bosphorus, had three main spans with the longest suspended section almost twice as great. It was claimed that wind-tunnel tests showed that the wide decks, also reinforced transversely to distribute traffic loads but a mere 25 cm or so in thickness, would experience no aerodynamic instability.[30]

Bridges with cable-supported decks

Only a separate deck can provide a level or slightly rising passage without the penalty of greatly increased tensions in the suspension system. When the deck is suspended by vertical hangers from slung catenaries it becomes what is usually referred to as a suspension bridge [14.2(b)]. This term may be extended to include also the less common form in which it is propped up from catenaries slung below it but independently anchored. When it is suspended directly from one or more pylons by a series of straight inclined stays it becomes what will be referred to as a cable-stayed bridge [14.2(f)].

As largely tension structures, both these types of bridge are inherently lighter than arch and beam bridges. With modern high-strength catenaries and stays, they are therefore capable of much greater spans before reaching the limit at which they would break under their own weight. But this lightness has the drawback of making them more susceptible to sway and vibration under wind loads. With one exception, they cope with this hazard in different ways and will therefore be dealt with separately in what follows. That exception is the use, in long-spanning recent examples of both of them, of mechanical damping systems such as tuned-mass dampers. These are essentially masses that are so mounted that they are set in motion by the structure's own movements, and tuned to move out-of-step with these and absorb energy thereby.

Figure 14.20 Pont St Michel, Toulouse (Freyssinet). (Source: author)

275

Suspension bridges

A bridge with a horizontal deck propped up from five parallel chains anchored in the abutments was proposed by Stephenson in 1821.[31] But when the deck was propped in this way it was more often made strong enough to resist the chain tension by developing an equal internal compression. The chains were then anchored to it at each end to make what is better regarded as a trussed beam, and the same is done today when the deck is of steel or reinforced concrete and wire ropes or prestressing tendons take the place of the chains.

It is unclear when a level separate walkway or roadway suspended from the slung ropes or chains first appeared. In the West it is first seen in an engraving published in about 1615.[32] This depicts ropes slung as catenaries from tall posts and carrying pulleys over which other ropes are looped to serve as vertical hangers supporting a light timber deck [14.21]. But there is no evidence that anything like it was built at the time. Real development began only at the start of the nineteenth century.

In 1801, Finley built a bridge over Jacob's Creek in Pennsylvania that had all the essential elements of the modern suspension-bridge, and he subsequently patented the design. It had a span of 21 m, two wrought-iron chains suspended between towers rising about 3 m. above the roadway and anchored at their outer ends in the ground, and a timber deck hung from these chains by vertical chains evenly spaced along its length. This deck was stiffened partly by lapping the joints of the longitudinal beams and partly by light diagonal bracing of the parapet.[33] He then went on to build a number of other bridges of somewhat greater span but of the same general form and designed in the same wholly empirical manner.

Finley's bridges received wide publicity through Thomas Pope's *Treatise on bridge architecture*,[34] and it is possible that this was partly responsible for stimulating Brown and Telford to design a number of similar bridges in Scotland and England in the 1810s and 1820s. In 1814, Telford prepared designs for a span of over 300 m at Runcorn. This bridge, probably fortunately, was never built. His main achievements were the Conway and Menai Road Bridges [6.21 and 14.22], the first alongside and the latter not far from the sites later chosen for Stephenson's tubular railway bridges. The larger of the two, the Menai Bridge, was designed in 1817-18 and completed in 1826.[35] It has a central span of 177m carried by flat eye-bar chains of a form developed originally by Brown and used in his Union Bridge over the Tweed. (The photograph shows a fresh set of chains which, in a reconstruction in 1940, replaced the original ones without departing substantially from their form.) The chains were securely anchored in tunnels driven into the rock at the ends of the approach spans, though the original

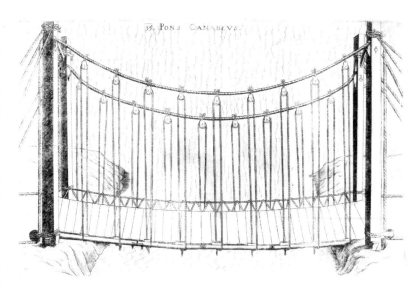

Figure 14.21 Design for a
rope suspension bridge,
from Verantius, F.,
Machinae novae, Venice,
c.1615. (Source: author)

intention – changed only when construction was
well under way – had been to fan them out over
the towers and carry them down into the springings of the masonry arches of the approach
viaducts. To have done so would have helped to
limit any tendency of the chains to develop
wave-like motions in the main span under the
action of moving loads. To achieve this objective in the revised design, they were tied down to
the viaducts by otherwise unnecessary 'hangers'
in the side spans.

The timber deck was both lighter than that in
Finley's patent and lacked its stiffening parapets. These changes were no doubt thought to be
justified by the tying-down of the chain in the
side spans and by a reduction in their sag in the
main span from about one seventh of the span in
Finley's design to little more than half this. At
the expense of considerably increasing the chain

tension, this reduction would indeed have
reduced the changes in its profile due to changes
in the loading – thereby making the suspension
system stiffer in much the same way as a slender
flat arch is stiffer and less likely to collapse
under a moving load than an equally slender
arch that is not so flat. But, as events soon proved, it did not make enough difference to allow
the use of quite such a light deck. This had to be
repaired within a week of the opening of the
bridge. Ten years later a vertical oscillation of
up to 5 m was observed during a severe gale.
Finally, after another three years and during an
even worse gale, much of it collapsed, calling
for its complete replacement to a different
design. The trouble was that the analogy with
the masonry arch bridge is only partly valid.
Even the worst wind-loading on such an arch is
usually insignificant in relation to other loads.

Figure 14.22 Menai
Suspension Bridge
(Telford).
(Source: author)

Figure 14.23 Brooklyn
Bridge, New York
(Roebling).
(Source: author)

On a much lighter suspension bridge it is far from insignificant.

At the time two possible ways of improving stiffness presented themselves. One was to stiffen the deck itself, by the use of either trusses or deep beams. This led to a structural system closely analogous to that later adopted by Maillart in his stiffened slender-arch bridges [14.8]. The other was to add to the simple suspension system of catenaries plus vertical hangers a number of inclined stays which would limit deflections in much the same way as diagonal tensile braces in a truss or the braces that had often been used already to limit oscillations when the deck or walkway was laid directly on the catenaries. Rendel, when reconstructing a bridge in Montrose whose deck had been destroyed in a hurricane in 1838, chose to stiffen the deck by means of a truss 3 m deep. The bridge then remained in use for almost a century until it had to be replaced to carry greatly increased traffic. John A. Roebling – following, whether he knew it or not, Brown's choice in reconstructing a much smaller bridge at Dryburgh Abbey in 1818 – chose to combine both alternatives in a series of outstanding bridges built from 1845 onwards.[36] As described in Chapter 6, he also adopted Vicat's recommended use of parallel-wire cables spun *in situ* in place of chains.

One of Roebling's bridges, that over the Niagara River, is particularly interesting in relation to Stephenson's Britannia Bridge, since it was constructed only a little later and was likewise designed for the railway traffic of its day. Stephenson had originally thought of a suspension bridge with a stiffened deck, and this was the reason for the subsequently needless height of the towers [14.14]. But it was then realized that the amount of stiffening required for this heavy loading was so great that the tubular beams developed to provide it had no need of further support over the shorter spans of the Menai crossing. Roebling, by using large numbers of inclined stays in addition to stiffening-trusses, was able successfully to retain the basic suspension form without making the trusses excessively heavy or even bracing them very efficiently. Later he built the Cincinnati Bridge on the same lines and designed his masterpiece, the Brooklyn Bridge [14.23], but had to leave his son to build it. As well as being again designed for heavy loading, this had the then unprecedented span of 486 m. It remains in use today, little changed apart from a modification of the deck trusses between 1950 and 1952.

The achievements of the two Roeblings helped to spur on the building of a succession of even greater spans. Between the 1920s and 1960s these were all in the United States, notably the George Washington Bridge over the Hudson River in New York (1927 to 1931 with a central span of 1070 m), the Golden Gate Bridge in San Francisco (1933 to 1937 with a central span of 1280 m – Figure 1.4) and the Verrazano Narrows Bridge at the entrance to New York Harbour (1959 to 1965 with a central span of 1300 m – Figure 14.24).[37] However, the only two characteristic features of the Brooklyn Bridge to be carried over were the use of parallel-wire cables for the catenaries and the method of spinning these [2.1, 6.22 and 14.25]. The more immediately obvious features of the heavy and stiff masonry towers and the combined use of a stiffened deck and large numbers of inclined stays to overcome the inherent flexibility of the basic suspension form were, for good reasons, abandoned.

Although it was originally intended to clothe the towers of the George Washington Bridge in masonry, this cladding was never added and masonry towers subsequently gave way completely to lighter and more flexible steel ones [1.4, 4.21, 10.16, 14.24 and 14.25]. This not only proved more economical both in itself and in relation to the foundations: it also did away with one of the most troublesome details, not only of the Brooklyn Bridge but also of its

Figure 14.25 (right) Forth Road Bridge, near Edinburgh, temporary catwalks and overhead hauling-ropes for spinning the main cables (Freeman Fox). (Source: author)

Figure 14.24 (below) Verrazano Narrows Bridge New York Harbour (Ammann and Whitney). (Source: author)

278

the bearings (or saddles as they are called) to move slightly on the towers. For this reason they were usually placed on rollers. But the rollers never proved to be effective for long, so the only small alleviation was that provided by such limited flexibility as the towers themselves possessed.[38] Steel towers, being more flexible, are able to move sufficiently in the directions of the cables to equalize the pulls.

Inclined stays were rejected on account of the very high degree of statical indeterminacy that they introduced into the suspension system. In the Brooklyn Bridge it was made all the greater by the fact that the deck was continuous past each tower and rigidly fixed to it, being effectively hinged at the centres of the main span and of each outer span and at the shore abutments. Over the main length of the bridge there were thus virtually two independent systems. Superimposed on that of the basic suspension form with its catenaries, vertical hangers, and deck was another consisting of the inclined stays and the deck. This second system was tantamount to a couple of cantilevers whose river arms were directly connected at midspan and whose shoreward arms were linked, via the hinges in the outer spans, to the abutments. Somehow loads had to be shared between the two systems. In the cantilever system, moreover, locally concentrated traffic loads could be shared between a reasonable number of the inclined stays serving as the tension members only if the deck was stiff enough, by itself, to distribute these loads between them. This led to the paradoxical situation, when the two systems were combined, that the inclined stays could be really efficient as stiffeners – acting in concert and not just one or two at a time – only if the deck was already stiff enough for there to be no need of them. A further factor that does not seem to have been fully realized was that, in the cantilever system, the deck had to serve as the compression member and was thus subject to an increasing overall compression towards each tower.

The normal form of stiffening therefore became the deck truss alone. Initially very stiff trusses were used, but as experience was gained, as improved methods of analysing the interaction of the truss and the suspension system were devised, and as spans increased (making the varying traffic-load an ever-smaller proportion of the unchanging self weight of the deck and cables), the ratio of depth of truss to span was progressively reduced from an earlier value of about 1/40 to 1/168 at the Golden Gate [1.4]. At the George Washington Bridge it was even found possible to dispense with the truss, thanks to the unusual width of the deck and unusual weight of the cables (intended as these were to carry a second deck which was added only later).

Eventually the reduction in depth was carried too far. Then the lesson that had been learnt in the first half of the nineteenth century

predecessors like the Menai Bridge. In any suspension bridge it is necessary to provide bearings at the tops of the towers for the catenaries, around which they change direction and through which the entire weights of the suspended spans have to be evenly transmitted to the towers [14.26]. The trouble arose in the earlier bridges from the inevitable tendency of the pulls in the catenaries at the two sides to change slightly relative to one another with changes in temperature and in the distribution of the traffic load. With stiff masonry towers, the only way of alleviating the changes in the pulls was to allow

Figure 14.27 Severn Suspension Bridge (Freeman Fox). (Source: author)

from a large number of collapses in high winds of inadequately stiffened decks was re-learnt from the excessive movements experienced by some recently built decks followed by the dramatic collapse of one of them – Tacoma Narrows Bridge – in what was only a fairly moderate wind. This bridge, completed in 1940, had a span of 850 m. But it was designed for much lighter traffic than the George Washington Bridge, was itself correspondingly lighter, and had a deck little more than a third of the width. Yet the truss was again dispensed with, being replaced by a pair of solid-web girders whose

depth was only 1/350th of the span. They gave to the whole deck a cross-section that had little torsional stiffness and proved to be particularly susceptible to movements akin to the fluttering of a leaf in a breeze. These movements, continued with progressively increasing amplitude over a period of about two hours and changing finally from a simple up-and-down wave-like motion to a more serious twisting one [2.3], broke it to pieces.

Unlike the earlier collapses however, this one led to a painstaking scientific inquiry into its precise causes.[39] The outcome therefore was more than just another decision to provide more stiffness in future: the basic requirements were greatly clarified, and a radical reassessment began of the best ways of meeting them. Two results of that reassessment in actual construction are illustrated in Figures 4.21 and 14.27.

The Forth Road Bridge was provided with a fairly conventional stiffening-truss (seen under construction in Figure 4.21) with a depth of 1/120th of the span. But the design of this truss was guided by extensive wind-tunnel tests that indicated, among other things, the value of introducing longitudinal gaps in the roadway to break up eddies.[40] In construction, more extensive use was also made of welding and of bolted *in-situ* connections than had been usual previously.

In the Severn Bridge more radical changes were made [14.27].[41] An entirely new form of deck was adopted, similar to one recently proposed by Leonhardt in an alternative design for the Emmerich Bridge. Although this was little deeper than the Tacoma Narrows deck, it was a torsionally stiff closed box that was shaped almost as an airfoil [8.9] instead of a torsionally weak section shaped with no regard to its aerodynamic characteristics. This first change led to the next, since the sections of the box were continuously welded together throughout the length of the main span and therefore lacked the ability of earlier types of deck to absorb the energy of wind-induced movements by very slight play in the joints. To make good this loss of 'damping' capacity, all the hangers were slightly inclined along the length of the bridge, alternately one way and the other way, so that they looked in profile like the alternate inclined struts and ties of a simple Warren truss [9.7(f) without the verticals] much elongated in the vertical direction. The deck was freely supported on rocker arms at the towers and was restrained from moving longitudinally to and fro relative to the catenaries only by the alternate lengthening and shortening of each hanger that any such movement entailed. It was then necessary only to select for the hangers a suitable type of cable whose repeated lengthening and shortening would itself absorb enough energy to provide the requisite damping. A third change was that the legs of the towers were single boxes, stiffened internally by welded ribs instead of being multi-cellular as were those of the Golden Gate and Forth Bridges – a feature that allowed the later insertion of tubular columns in the corners to increase the strength when this was made necessary by an increase in traffic loading much beyond what had been anticipated.

Since the two bridges were very similar in width, span (1020 m for the Forth and 990 m for the Severn), and design loading, these changes in the design of the second provide, incidentally, a good illustration of the freedom of choice of form open to the really skilful designer even when working close to the limits of practic-

ability set by the available materials and the functional requirements.

Subsequent bridges of the pure suspension type have broadly followed one or other of these alternative possibilities with progressive increases in span. Apart from designers' preferences, the chief factors influencing choice have been the span and the loading, including the type and anticipated magnitude of the traffic load. Important aspects of the latter are the width of deck in relation to the span and whether or not provision is required for heavy rail traffic. The first of these greatly affects the stiffness of the deck and the second affects both the acceptable defections under load and the strength requirements, including resistance to fatigue failure due to repeated changes in stress. To put different designs in perspective it may be noted that most bridges have been designed for road traffic only with either four lanes or six lanes on a single deck but that, at the other extreme, the Verrazano Narrows Bridge has as many as twelve lanes on two decks. The heaviest loading for which provision has hitherto been made has probably been that of four road lanes plus four rail tracks on separate decks on the Shimotsui-Seto and twin Besan-Seto Bridges between the Japanese islands of Honshu and Shikoku.

Aerodynamically shaped box decks certainly lead to the most elegant designs and have generally been preferred in Europe. Sequels to the first Severn Bridge have included two over the Bosphorus, the Humber Bridge (whose central span of 1300 m was the first to exceed that of the Verrazano Narrows Bridge), and designs for crossings of Great Belt between Denmark's largest islands (with a central span of 1624 m) and the even wider Messina Strait (with a proposed span of 3300 m and towers 25 per cent taller than the Eiffel tower).[42] All are single-deck and all but the last are for road traffic only. But this last deck has a planned width of 60 m, considerably greater than any of the others, and resembles that of the Forth Bridge to the extent of having two longitudinal gaps that are closed only by grids. Between them are three separate aerodynamically shaped closed boxes. Without these, it would probably not be feasible to reduce its sway to within tolerable limits even with additional stays. Some, but not all, of these later bridges have the inclined hangers of the Severn Bridge. These proved to be a mixed blessing since their alternate lengthening and shortening led to fatigue problems at the connections that do not arise with vertical hangers. Alternative means of absorbing energy have been the permanent incorporation of tuned-mass dampers in the decks and hydraulic buffers at the supports that permit some longitudinal movement of the deck.

Elsewhere, notably in Japan where (as at Messina) it is necessary to provide also for earthquakes, more conservatively designed trussed decks have been usual. In the Shimotsui-

Seto and Besan-Seto Bridges this choice was the obvious one to accommodate also the second deck for the rail tracks within its depth. But it was also the choice for the single-deck Akashi-Kaikyo Bridge – not due for completion until 1998, but with a record-breaking designed central span of 1990 m that was increased to 1990.8 m by a horizontal displacement of the foundation of one of the already completed towers in the 1995 Osaka earthquake.[43]

In nearly all these bridges the catenary cables were fanned out near the ends of the side spans and anchored there in massive anchorage blocks, usually of reinforced concrete. The chief exceptions arose where (as in the second Bosphorus Bridge) there were no side spans or where (as in the twin Besan-Seto Bridges) two side spans met end-to-end at a single shared anchorage block.

An alternative to thus carrying the catenaries down to ground anchorages capable of resisting their horizontal pulls was always available in principle, since the deck was always at the right level to absorb them as a continuous horizontal

through improvements in design and the introduction of stronger steels. Indeed it is unlikely that any more such bridges will be built, since the cable-stayed bridge is now more economical for spans like that of the Cologne-Mulheim Bridge.

Cable-stayed bridges

Cable-stayed bridges [14.2(f)] differ from suspension bridges in having the deck as the primary spanning element. Since only the outermost stays can have the horizontal components of their tensions resisted by a ground anchorage, the deck must also anchor the feet of all the others in the same way as the deck of the self-anchored suspension bridge anchors its catenaries. It will therefore be subject over all or most of its length to direct compression or tension according to how continuous it is and where it is restrained against longitudinal movement. When it is restrained only at one pylon and without any ground anchorage and is continuous throughout, it acts in much the same

Figure 14.28 Albert Bridge, Battersea, London (Ordish).
(Source: author)

strut. This alternative – a counterpart to the tied arch – was occasionally adopted where the nature of the ground would have made it difficult to provide the usual anchorages and the deck could be made stiff and strong enough to resist the resulting compressions. To do the latter, however, necessitates considerable increases in weight in any long deck. Thus the longest self-anchored suspension bridge ever built was the first Cologne-Mulheim Bridge, with a span of 315 m.[44] Today the economic limit is probably a good deal lower on account of the lighter decks that have become possible

way as a balanced cantilever [14.2(d)]. Even if it is not continuous throughout, flexural continuity is essential in each section.

Cable-stayed bridges also differ from suspension bridges in that their stiffness in the vertical plane resides in the triangulation of the whole system of pylons, stays and deck and not largely in the deck alone as with the suspension bridge. This has two consequences. One is that this triangulation has a potential weakness in the necessary inclination of the stays. Not being vertical, they tend to sag. This sag reduces their stiffness, all the more so as their horizontal span

Figure 14.30 Severin
Bridge, Cologne (Lohmer,
Leonhardt).
(Source: author)

from the pylon increases. To limit the sag, cables should be stressed to a high tension under all states of load on the bridge. The other consequence is that moving loads cause much larger variations in stress in the stays than occur in the hangers of a suspension bridge. This brings risks of fatigue failure in the stays or their anchorages.

Since, nevertheless, the basic principle of support is such a simple one, it is possible that it was thought of at quite an early date. Certainly it is implicit in the typical drawbridge and (to look a little further) in much ships' rigging. Another of Verantius' five bridge exemplars had a deck held up partly by straight inclined stays and partly by a single vertical hanger from a higher catenary. But it is the least practicable of the five since the main members of the deck were formed of the same eyebar chains as the stays and catenary.[45] More realistic purely stayed projects, mostly unexecuted, began to appear in the late eighteenth century and became more numerous in the early nineteenth, but the full requirements where not yet recognized and no purely stayed bridge actually built then seems to have stood for long.

In Roebling's bridges the stays provided only secondary support to the deck, as has been seen. Though the Albert Bridge at Battersea [14.28] was also a hybrid, it was the closest approximation to a successful purely stayed bridge at that time. Its deck was better able to provide the necessary anchorage for the stays; they do not span as far horizontally as in Roebling's bridges; and they seems to play a more important role than the vertical hangers. A few purely stayed bridges followed. But it was only in the 1950s that the requirements became better understood and the bridge as now known appeared and rapidly established itself for spans of up to a few hundred metres.

At this time, continuous steel plate-girder decks and prestressed concrete decks were also becoming established. They had the necessary characteristics, and the addition of stays usefully extended the range of spans over which they could be employed. Newly developed prestressing techniques allowed more efficient

use to be made of the stays and newly developed analytical techniques made it possible for the first time to cope with the high degrees of statical indeterminacy that had to be faced.

In the early bridges, a few stays only were used. This simplified design, and they were perhaps regarded as taking the place of similarly spaced intermediate piers. The first to be completed was the Strömsund Bridge in Sweden with a main span of only 182 m, a rather heavy-looking plate-girder deck, and pairs of stays fanned out in each direction on each side of the deck from the goal-post towers. Two larger and more elegant German bridges across the Rhine followed shortly afterwards.

The first of these Rhine bridges, the Theodor Heuss Bridge in Düsseldorf, has a 260 m. main span, a shallower steel deck and three parallel stays in each direction attached at different heights to each of the four single pylons. [14.29].[46] Aesthetically pleasing though this harp-like arrangement is, it is less efficient than

the fan adopted previously because all the stays are at the maximum angle to the vertical. In the second, the Severin Bridge in Cologne,[47] the more efficient fan arrangement was adopted again [14.30], and it was adopted once more for the delightful Schillerstrasse Footbridge in Stuttgart [14.31].[48]

Because of its location near to the cathedral, the Severin bridge was given a single A-shaped pylon straddling the deck and echoing the silhouette of the cathedral spires, so that here the fan arrangement was also better visually. The stays were attached as close together as possible near the top of the pylon and splayed out laterally towards the two sides of the deck. Although they fan symetrically towards each span, these spans are not equal and the shorter span helps to stiffen the longer 302 m span seen to the left of the photograph. The footbridge, built almost at the same time, has a span of a mere 92.5 m but demonstrates very well the versatility of stayed support. Its single pylon is

Figure 14.33 Brotonne
Bridge (Muller).
(Source: author)

situated towards one end of the span, where the deck divides into two outwardly curving arms which pass on each side of it without making direct contact.

These German bridges were all constructed of steel. Morandi, as in his other works, used reinforced and prestressed concrete in his roughly contemporary Maracaibo Bridge [14.32].[49] Here the resemblance to the balanced cantilever form of Figure 14.2(d) is very close, extending even to the inclusion of suspended sections in the centre of each span. Only the stays were of steel. They soon suffered from corrosion. In a later similar bridge, the steel was encased in concrete and was prestressed not only against the weight of the deck but also against this concrete so as to precompress it. Besides giving protection to the stays, the encasement was thus made to participate fully in any subsequent changes in stress due to variations in the load on the bridge. Changes in stress in the steel were thereby reduced. The overall design appears, in comparison with others, to be unnecessarily heavy-handed with its X-shaped piers below the deck as well as the stays above it, and it was not emulated by other designers.

In later stayed bridges, and especially as spans increased, the numbers of stays have tended to increase. Larger numbers allow lighter stays and simpler anchorages, reduce bending moments in the deck so that these can also be lighter, and facilitate the construction of spans over water by progressive cantilevering. Fan arrangements or modifications of the fan with a small vertical spacing of the upper anchorages have also been usual. But so many designs of pylon, types of deck, and ways of supporting the deck at the pylons and ends of the spans have been explored that only a broad review with a few further examples is possible. Indeed it is tempting to say that all reasonable possibilities – and even some less reasonable ones – have been tried.[50]

Pylons have usually followed one or other of the patterns already mentioned, though the opportunity for architectural variation has been widely exploited. The greater transverse stiffness of the A shape makes it valuable for exposed sites, and its drawback of a very wide base when the deck has to be high above water has been overcome by tapering the legs inward below it to give an overall diamond shape. Early bridges had steel pylons. Reinforced concrete then became more usual.

To take advantage of the inherently greater stiffness of stayed support, the stays themselves have had to be as stiff as possible. Although they should ideally be of solid cross-section from this point of view, wire cables of higher strength and greater resistance to fatigue have been preferred in one of the forms that reduces stretch under load and yet allows easy handling. Pretensioning to reduce sag and consequent loss of stiffness has assumed greater importance as spans have become longer. Various types of sheathed protection have been adopted to minimize the risk of corrosion.

Stays on each side of the deck have been usual, sometimes in vertical planes and sometimes splayed out transversely from central upper anchorages to improve the lateral stiffness of the whole system. Single sets of stays along the centre of the deck have also been used. This calls for a deeper torsionally stiff deck and was first done at the not entirely successful Norderelbe Bridge in Hamburg. A more successful example is the Brotonne Bridge where, 15 years later, the deck was not a steel one but a stiff prestressed concrete box with a central span of 320 m [14.33].[51]

More three-dimensional arrangements, impossible with simple catenary suspensions, have also been tried. Mostly they have been limited to lightweight structures like the Schillerstrasse Footbridge. But these have included particularly attractive continuously curved spans like the projected Ruck-a-Chucky Bridge to cross a new reservoir in California.[52] This was to have been supported by arrays of cables attached at close intervals to each side of the deck and attached at their high points to rock anchors spread over the steeply sloping banks. More recently road bridges designed for heavier traffic and following S-shaped curves have been supported from centrally placed single pylons, and a tentative beginning has been made in the use of newer materials with the construction of a footbridge of reinforced plastic.[53]

The Normandie Bridge [14.34] took the greatest leap forwards in span – from a previous 602 m to 856 m. It has a streamlined steel box girder for the central 624 m of the main span and reinforced concrete box beams to carry the deck of the side spans and the sections of the main span nearest to the pylons.[54] The side spans were constructed first and rigidly connected to the pylons so that, as the main spans were progressively cantilevered towards a meeting point in the centre, they provided the maximum restraint to sidesway. To further limit this and the resulting relative movement of the two free ends in high winds, tuned-mass dampers were attached to the advancing deck, although in the event they proved to be an unnecessary precaution. Provision also had to be made to limit vibration of the long stays in the wind. This had already become a problem at lesser spans like that of the Brotonne Bridge. They were reduced there by installing dampers at the attachments of the stays to the deck as seen in Figure 14.33. In the Normandie Bridge they were further reduced by careful attention to airflow over the cables and by interconnecting the stays by the ropes seen as darker lines in Figure 14.34. As well as absorbing energy, these ropes reduce the vibration periods of the stays well below that of the deck.

Designs have been made for spans approaching 2000 m. But they are never likely to match the 5000 m or so that may one day be achieved with the catenary suspension system.

Figure 14.34 Normandie Bridge (Virolgeux). (Source: Professor Michel Virolgeux)

Chapter 15: Multi-storey buildings and towers

The multi-storey building has been left to the end of this review of the developments of complete structural forms because it is, in an important sense, the most complex. It is the chief representative of the second of the broad categories of such forms referred to at the end of Chapter 5 – the category of forms that cannot, unless we ignore much that is essential, be regarded as analogous to one of the simple elemental forms plus the necessary support. As a whole, it can only be likened to the single column or wall. But it is the very essence of the form that most of this notional single column or wall is hollowed out into a large number of usable internal spaces. In relation to these spaces, the floors are the primary structural elements. Columns, walls, or equivalent elements are required only to support them, though they may, of course, serve also as part of the enclosure.

From this point of view, the entire building might be regarded more as a series of single-storey buildings with beam or slab roofs, these buildings being stacked one on top of another.

There is more to this, though, than simply imposing increased vertical loads on the supports of the buildings that are lower down in the stack. It is also necessary to ensure lateral stability. The Panch Mahall at Fatehpur Sikri [15.1] must have almost the maximum number of storeys at which this can be done by relying solely on accurate cutting and assembly of monolithic columns and beams and on gravitational resistance to any tilting of the columns.[1] In taller buildings it must be ensured by adequate stiffness in bending and shear throughout the height.

As, since the latter part of the nineteenth century, numbers of storeys increased under a variety of commercial and other pressures, this need to ensure lateral stability increasingly came to dominate the purely structural aspect of design and to distinguish these multi-storey forms most sharply from those discussed in Chapter 13. Lateral stability under the action of horizontal thrusts generated by the spanning elements rarely presented a problem, since the use of arches and vaults was limited to short spans. Over longer spans they would have been too deep. The main problem has been almost entirely that of stability under the combined action of vertical loads and externally imposed horizontal ones such as wind loads.

The tower, as exemplified by the structures illustrated in Figures 15.2 to 15.5, is structurally much closer to the single elemental column. The provision of the usable floor space is not a primary objective and a much smaller area in relation to the total volume has normally been considered adequate. Since, for comparable heights, this has greatly simplified the problems of design, it will be convenient to look at the tower first.

Towers

Probably the most numerous class of tower is that which we have seen already in the masonry defence towers of Figures 10.7 and 11.13 and in the ancient lighthouse of Alexandria referred to in Chapter 11. It continued to be used for defence purposes until late medieval times (a secondary role of many church towers being defensive), and for observation or communication (the primary role of the lighthouse) until much more recently. It was characterized by substantial outer walls and gained all its lateral

Figure 15.1 Panch Mahall, Fatehpur Sikri. (Source: author)

Figure 15.2 Campanile and Doge's Palace, Venice. (Source: author)

stability from these. Usually they diminished in thickness with height. At the base, the masonry might even be solid throughout the entire width of the structure as, for instance, in the Eddystone Lighthouse.[2] Above it, the internal voids might be no larger than were needed to accommodate access stairways or ramps. Even in the rather broader defence-towers they were still small enough to be spanned by timber beams or stone slabs or by vaults whose outward thrusts were well within the capacity of the walls to resist. In the Campanile of St Mark's in Venice (shown as reconstructed early in the present century in Figure 15.2 but originally built in the same form largely between the tenth

and twelfth centuries), access ramps climbed up inside around a central well to the open belfry 54 m above the ground, but ramps and central well together occupied considerably less than half the total volume [Figure 15.9(b)].[3] Inside the typical Islamic minaret, there was nothing but a spiral stairway, again occupying much less than half the total volume. The platforms from which the call to prayer was made were cantilevered-out externally.

With fairly constant external dimensions throughout the height, much of this thickness was necessary to ensure that there was always enough compression in the masonry due to gravity to avoid the development of net tensions on the windward face under a strong wind. However, the need could be reduced in a tall tower by building in stages of successively diminishing external width. This, as we have seen, was done in the lighthouse at Alexandria and it was also frequently done in church towers. A further advantage was that it distributed the loads more widely on the foundations.

The most efficient form of all, and the only one that permitted really thin walls, was the spire. Given an octagonal, circular, or some intermediate polygonal plan, and an adequate continuous slope of the outer faces (say between 5° and 10° to the vertical), the continuously diminishing thickness could be reduced to not much more than one-twentieth of the local diameter with little risk of the vertical tensions induced by a strong wind exceeding the compressions due to gravity.

No tower seems to have been built as a spire for its full height. Instead, many spires were later additions to church towers with more or less vertical faces. The outstanding fourteenth-century spire of Salisbury Cathedral [7.22] has been referred to already in Chapter 7. It began 62 m above the ground and continued upwards to a total height above ground of 115 m. Unlike some less daring examples, however, it was stabilized to some extent by leaving within it the timber framework around which it was built. This consisted of a central post and radiating struts and other members, rather like the frames of some lead- or shingle-covered spires and spire-like timber roofs but without the rafters that were an essential part of these.[4]

Towards the end of the Gothic period, there were also a few masonry spires constructed in a purely skeletal manner. That surmounting the west front at Ulm reached the greatest overall height – 161 m – though it was completed only between 1880 and 1890 to the fifteenth-century design. It was provided with an open central spiral stairway to which the outer ribs were braced in a manner closely analogous to that seen in the internal frame at Salisbury [15.3]. Elsewhere, as at Freiburg im Breisgau, the interior was completely open.

In the first half of the nineteenth century, all these masonry forms served as models for towers constructed, or proposed for construction, in cast iron.[5] Advantage was naturally taken of the greater compressive strength to reduce considerably the bulk of the walls. One of the most interesting of the proposals was that published by Trevithick in 1832 for a monument to commemorate the passing of the Reform Act in that year. This was to be a spire, circular in plan, 30 m in diameter at the base, 305 m high, and constructed throughout of flanged panels of cast iron bolted together through the flanges. The panels were to be only 50 mm thick and were to be pierced by holes to lessen the wind loading. Up the centre was to run a continuous tube 3 m in diameter containing a 'steam lift' operated like a piston by air pressure on its under side. Some rough calculations show that, structurally, the proposal was not altogether unreasonable. The mean stresses, even under the wind loading to be expected in the London area, would not have been excessive. Nevertheless, had the tower been built, trouble would almost certainly have arisen from the virtual impossibility of achieving even an approximately uniform distribution of stress through the flanged joints and from fatigue of the bolted connections under fluctuating loads. Shortly after publication, however, Trevithick died and little more was heard of the project.

Equally ambitious in terms of height, but far less practicable, was one of the proposals made

Figure 15.3 Looking up into the spire, Ulm Minster. (Source: author)

in 1852 for the re-use of the components of the 1851 Crystal Palace. This amounted to stacking the basic portal-frame elements of the structure (seen to the left of Figure 4.10) above one another to form five concentric and inter-connected towers. The highest was to be in the centre, and each of the others was to be of a lesser height to give the stepped external profile characteristic of many of the tallest masonry towers. No details were given of any modifications to the columns or the connections. Had the proposal been carried out without substantial modifications to both, it is almost certain that the lower columns would have buckled long before the intended height was reached even without any great load from the wind.[6]

As might be expected, all the towers actually constructed were much lower. Up to 1851, 30 m was the limit. Several lighthouses and fire towers were constructed up to this height, some following the general pattern of Trevithick's proposal and some with simple column-and-beam framing with the individual column about 5 m high. Two towers built shortly afterwards in New York by Bogardus were of greater interest in relation to later developments in the construction of multi-storey buildings. These were both for the making of lead-shot and had heights of 53 m and 66 m. They were of octagonal plan, were slightly tapered in profile with maximum widths only about one seventh of the height, and had column-and-beam framing like some of the earlier towers. But they differed significantly from these in that the beams now carried relatively thin brick infill panels. Whether or not this was taken into account in design, these would have helped considerably in ensuring lateral stability.[7]

The introduction of wrought iron with its greater ductility and tensile strength, together with a better ability to analyse the forces in large structures, further stimulated the development of framed forms in the latter part of the century. If we now leave unexecuted projects on one side, the main achievements were undoubtedly those of Eiffel, culminating in the tower built for the 1889 Paris Exhibition [15.4].[8]

Here, at last, the height of 305 m – the 1000-ft target of several earlier projects – was triumphantly achieved as the culmination of the systematic and continued development of similar framed tower-like piers for large iron bridges such as those illustrated in Figures 4.19 and 14.10. The tower might also be regarded, from another point of view, as a counterpart, much more appropriate to the material used, of the skeletal masonry spire. Both the stiffness required to guard against buckling under self-weight loading, and the further stiffness required to stand up to high winds, were provided in the third and principal stage of the tower by diagonal bracing between the riveted box-section corner members that carry the vertical loads. This was not feasible in the lower stages. Had these been similar, the necessary braces

Figure 15.4 Looking up at the two lower stages of the Eiffel Tower, Paris (Eiffel). (Source: author)

would have had to be excessively long on account of the continued widening of the tower towards the base that was desirable for overall stability. Here, therefore, four separate legs were provided, each consisting, like the single leg of the third stage, of box-section corner members diagonally braced together. These lower legs were splayed out and connected to one another by arches and deep trusses supporting a first-stage platform. The splay was sufficient to ensure, by a wide margin, that the loads in the corner members and the ground reactions would remain compressive even under the worst likely wind. Numerous similar towers, mostly smaller, were built later in emulation.

With two exceptions, the design of later tall steel towers has not departed radically from the form established by Eiffel. In Tokyo, a slightly taller but considerably lighter tower for television transmission was closely modelled on Eiffel's design. Most towers for high-voltage transmission purposes have also usually been broadly similar.

Figure 15.5 Television mast, Berlin (Frost). (Source: author)

The tallest, a radio transmission tower in Moscow built between 1919 and 1922 rose 150 m. but an earlier project had more than twice this height and yet called for less than a quarter of the weight of steel used in the Eiffel Tower.[9]

The other exception has been the use of straight slender masts stabilized by guy-ropes. Sometimes these have been tubular, usually of constant external diameter. Elsewhere they have been braced like the individual legs of the Eiffel tower, but more slender than these and reaching up to twice the height.[10]

The use of reinforced concrete permitted, however, a considerable parallel development of a minaret- or chimney-like form up to heights approaching those of the tallest braced masts. Much thinner walls could be used than in the masonry prototype on account of the higher compressive strength that can be attained, the possibility of resisting very high wind loads partly by tension on the windward face, and the possibility of 'rooting' the tower much more effectively in the ground by modern spread footings or inverted-cone foundations. Such a tower behaves exactly as a very tall cantilevered hollow column. The height of the taller tele-communication towers has meant, however, that it has not been sufficient merely to provide enough stiffness to prevent buckling or collapse under wind load. The more critical requirement has frequently been the prevention of excessive sway in the wind. Since the wind can blow from almost any direction, a cylindrical form, as well as being the most efficient structurally, is the best to achieve a favourable air flow and reduce the resultant loads.

The form was established by the first Stuttgart tower, designed by Leonhardt in 1953. It further resembles the minaret in having only enough space inside to allow access and accommodate service runs. Working floors are therefore cantilevered out. The Berlin tower [15.5] was the first (at 365 m) to exceed in height the Eiffel tower. Built fifteen years after the Stuttgart one. it followed essentially the same design, as most later towers have done.[11]

At lesser heights the reinforced concrete cylinder has also been used since the mid-twentieth century, both for lighthouses and, more widely, as the most elegant support for elevated water tanks constructed of the same material [1.14].

Pre-twentieth-century bearing-wall buildings

The typical Roman tenement of late Republican times was built of necessity to a height of several storeys to accommodate the increasingly crowded population, as Vitruvius remarked.[12] It probably differed from the Egyptian house illustrated in Figure 11.2 only in having a narrower frontage and being able, therefore, to dispense with internal columns to support the timber floors. These could have spanned directly from

One exception was the use of a hyperbolic paraboloid form pioneered by Suchov for water towers and lighthouses of modest height at the end of the nineteenth century and then developed further for telecommunications and power transmission in the early twentieth century. Typically these towers, or each stage of the taller ones, began and ended in a stiff circular ring. In each stage, these rings were connected and the upper one supported by two families of relatively slender straight struts. While one family spiralled upwards in a clockwise direction, the other spiralled anticlockwise. Lighter horizontal rings connected and stiffened them at intermediate levels When there was only one stage (of 68 m in one instance), the resulting form was a simple hyperbolic paraboloid, rather more slender than the later typical cooling tower. When multiple stages were used to reach a greater height, the reduction in diameter of successive main rings resulted in a more spire-like form and proportions, differing chiefly from the spire in being gently waisted in each stage.

one party wall to the other. Initially these party walls were of mud-brick. Later, by the time Vitruvius was writing, concrete was at least recognized as being superior. It permitted a reduction in the thickness of the walls as well as being more durable and less likely to allow the whole structure to collapse on the inhabitants – apparently a not infrequent occurrence in the older tenements. But it did little to reduce the serious fire risk introduced by the timber floors and stairs and the frequent overcrowding. Only after the great fire of A.D. 64 were really effective measures taken to reduce the fire risk, both by planning legislation and by the substitution of concrete vaults at least for the floors between dwellings.

The resulting structures were the real Roman contribution to multi-storey construction. Today the most extensive remains are to be found in the abandoned port of Ostia. Unfortunately though, all the upper storeys there have disappeared. A better idea of the form can therefore be gained from the better preserved remains of Trajan's Market in Rome itself [15.6] and, as far as architectural form and appearance are concerned, from the older tenements of many once-Roman towns.[13]

At street level there was frequently a shop or *taberna* open to the street. A typical row of such shops is seen at the left in the detail of the Market. Above the shop fronts and below cantilevered balconies (whose outline alone is now visible there), smaller windows gave light and air to timber-floored mezzanines or garrets approached from within the shops by timber ladders. The main horizontal divisions were, however, concrete barrel vaults spanning between the party walls, and access to the upper levels was provided by concrete stairways. In the Market, the arrangement differed from that in the normal tenement blocks chiefly in that the upper storeys there were also shops, but shops that faced away from the street seen in the photograph and opened on to a large inner hall at a higher level at the other side. The party walls were commonly spaced only about 5 m apart, but extended back considerably farther than this from the street front. With a thickness of only about 0.5 m, they must have relied to some extent on the stabilizing effect of the returns at the front and rear in resisting any unbalanced thrusts of the vaults and any other tendency to instability in the transverse direction. It is possible also that the spans of the vaults were short enough, and the concrete fill over their haunches deep enough, for the development, if need be, of some portal-frame action in this direction.

There was no very significant improvement on this Roman form until about the middle of the eighteenth century. The sort of achievement represented by the Gothic cathedrals showed that this was not entirely for want of technical skill. Indeed such technical skill is sometimes evident in multi-storey construction in details

Figure 15.6 Via Biberatica, Trajan's Market, Rome. (Source: author)

like the carrying (in the late fourteenth century) of the heavy outer walls of the second-floor Sala del Maggior' Consiglio of the Doge's Palace in Venice on comparatively light two-tiered arcades, as seen at the right of Figure 15.2.

Economy in first cost was probably the chief reason (at least from the thirteenth century onwards) for a continued willingness to make do, once again, with combustible non-thrusting timber-beam floors. In most tall tenements and other tall buildings in the centres of towns, these spanned, as before, between fairly closely spaced party walls, though a broader plan with substantial walls in both directions was necessary to ensure the lateral stability of free-standing buildings. The walls were of brick or stone where these materials were readily available, and were comparable in thickness to those of the Roman tenements. Elsewhere timber-framed walls were substituted. But this substitution was a long way from giving a self-sufficient column-and-beam frame of the kind

Figure 15.7 North-south section of the War, Marine, and Foreign offices, Versailles (Berthier). (Source: Bibliothèque Nationale, Paris)

subsequently developed in iron, steel, and reinforced concrete. The walls remained walls and functioned as walls particularly in the role of guarantors of lateral stability. To enable them to do so, the principal upright and horizontal members of the frames were heavily braced in their own planes either by brick infillings (following the much earlier practice that can be observed, for instance, at Herculaneum), by large numbers of secondary timbers, both upright, horizontal, and diagonal (as was common particularly in France and Germany), or by a very close spacing of the uprights combined with light forms of timber lining and siding (as in the 'balloon frame' house developed in Chicago in the mid-nineteenth century).[14]

To reduce the fire risk to the level attained by the later Roman form without the loss of economy that resulted, particularly in taller buildings, from the use of fairly deep and heavy vaults to carry the floors, a lighter and shallower incombustible alternative to these vaults was needed. The first such alternative was the shallow tile vault introduced into France at the start of the eighteenth century through Catalonia and Roussillion and first used extensively in a large multi-storey building in 1761.[15] The section reproduced in Figure 15.7 is drawn at the left through the two staircase towers and the corridors of the War Office (which was built round a large courtyard) and at the right through the main wing of the adjacent Marine and Foreign Offices. Rather inconsistently, the staircase towers and corridors had timber floors. All other floors were carried by shallow vaults in every storey. These vaults may be seen in transverse section in the narrow rooms adjacent to the right-hand corridor of the War Office, and in

longitudinal section through their crowns in the larger rooms to the right of these. Care was taken in the planning of the building to contain the thrusts of the vaults without excessive thickening of the walls. This was done by such means as placing smaller rooms alongside larger ones so that their vaults could absorb along their lengths the thrust of the larger vaults, much as in the system shown in Figure 12.1(e) but with all vaults at the same level.

Until the introduction of the reinforced-concrete slab as a second and structurally superior alternative in the latter part of the nineteenth century, similar shallow vaults were thereafter increasingly widely used. One improvement was the use of hollow rather than solid bricks or tiles to reduce the weight and permit even flatter spans within acceptable limits of outward thrust. Much more significant though, in relation to the further development of the structural form of the multi-storey building as a whole, was a different improvement in the design of large mill buildings. This was the introduction of cast-iron columns, and shortly afterwards of cast-iron beams, in place of the timber columns and beams that had previously served as intermediate supports for the floors.[16]

A drawing of the interior of a mill completed in 1801 and having both beams and columns of cast iron is reproduced in Figure 15.8. Having been made just before the brick vaults were added (spanning between the lower flanges of adjacent beams), it shows the internal framing very clearly, though it shows only the bottom three of a total of seven storeys. It also shows clearly the thick outer walls on which the outer ends of the beams rested and on which the whole lateral stability of the structure depended.

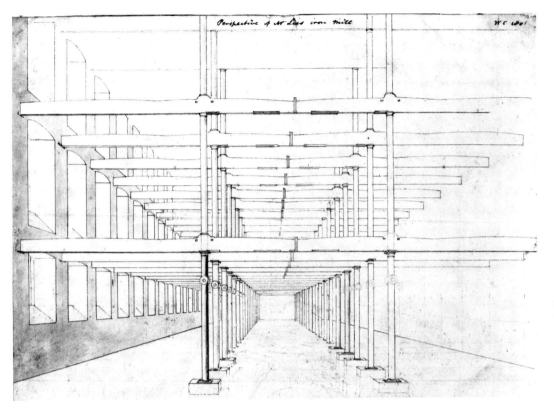

In fact the drawing gives the impression that each frame-bent (i.e. each assembly of columns and beams on a single transverse line) was too dependent for its own stability in the direction of the longitudinal axis of the building on the bracing-action from both sides that the floor vaults would provide.[17] Thus the structure, with others like it, while marking a first step towards a fully framed form, nevertheless remained primarily a bearing-wall one. Its internal framing was still as far from being self-sufficient as it had been in the Egyptian house and in the great audience-halls at Persepolis.

The next major development was to carry this internal iron framing of free-standing columns and beams into the outer walls and to make it completely self-sufficient so that the walls could become no more than a skin. But this took time.

During the period of rapid commercial expansion in the latter part of the nineteenth century, the form with outer bearing-walls of solid masonry and internal frame continued to be used and developed further alongside the fully framed form.[18] The limit was reached with the seventeen-storey Monadnock Building in Chicago, completed in 1891. This was both a structure of considerable architectural merit and one in which the internal framing was much more nearly self-sufficient than it had been in the earlier mill buildings. It nevertheless marked the limit in terms of height because the outer stabilising walls had to be almost 2 m thick at the base to ensure their independent stability [15.9(a)]. This meant that they both occupied too much space in the lower storeys and, through their weight, added considerably to the difficulties of providing adequate foundations.

Nineteenth- and early-twentieth century fully-framed buildings

Already, however, there had been other developments. The most important of these will be considered next, after noting briefly another roughly contemporary one that was of more symbolic than practical significance. This was the use of cast iron columns and beams for the principal facade or facades while still leaving stability largely dependent on masonry walls elsewhere in the structure. The most prolific innovators in this field were Badger and Bogardus working in New York around the middle of the century, but the finest surviving example is in Glasgow.[19] Gardner's Store [15.10] has cast iron facades on both street sides and cast-iron framing inside also. Only the two rear walls were of masonry.

The first more significant steps towards self-sufficient multi-storey frames were made in two buildings of moderate height to which some reference has already been made in Chapters 3, 4, and 10 – the Crystal Palace of 1850, and the Sheerness Boat Store of 1859 – although there were partial anticipations in a few smaller structures.

Paxton's Crystal Palace, built for the Great Exhibition of 1851, was a large structure consisting of tall single-storey nave and transepts all flanked by three-storey aisles and galleries [15.11]. The transept was roofed by a timber-framed barrel vault. Elsewhere, and most significantly from the present point of view, it was framed entirely in iron and glazed throughout. The basic element, repeated many times over on two different scales, was the portal frame. This is seen most clearly in the view of erection in

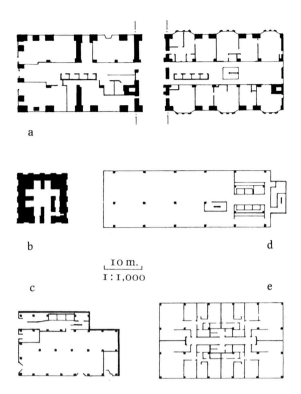

a

b   d

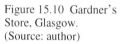
1:1,000

c   e

progress reproduced earlier [4.10]. The columns were hollow iron castings, the trussed cross-beams were also iron castings or were fabricated from wrought iron, and the joint between them was the one illustrated in Figure 4.11.

Stability was ensured primarily by rigid connections between both the longitudinal and transverse cross-beams and the columns. Only 'as a matter of extra precaution' were diagonal braces formed of wrought-iron rods also provided to a limited extent.[20] In Figure 15.11, these diagonal braces can just be distinguished in some of the bays on each side of the transept. Similar braces will also be seen as a continuous network in the transept roof. The remarkable feat of completing the construction in less than six months has already been commented upon. It seems all the more remarkable when it is remembered that the size of the building vastly exceeded that of anything built before and that there was no precedent for many of the techniques adopted to rationalize the whole process of fabrication and erection.

The Sheerness Boat Store, completed within ten years of the Crystal Palace, marked a major further advance [15.12]. Its overall length was little more than a tenth of that of the Crystal Palace. The spans and heights of the constituent frames were little more than half. But, with a central 'nave' open for the full height of the building (seen to the right of the figure) and side aisles and galleries of the same overall width as the nave, its general form was very similar, and the spans in the longitudinal direction were somewhat greater than in the earlier building. There were, moreover, three gallery floors on each side, each liable to be subjected to heavy loads, so that the vertical loading was correspondingly more severe. To allow for free manoeuvring of the boats, no diagonal bracing whatever was allowed between the internal columns:

Figure 15.10 Gardner's Store, Glasgow. (Source: author)

nor was any provided externally, where all the cladding was glass and corrugated iron. Longitudinal and transverse stability were therefore ensured entirely by the portal-frame actions developed by rigid interconnections between each column and the longitudinal and transverse beams that it carried.[21]

Apart from the secondary floor beams not in line with the columns, all columns and beams were of an H or I section. The columns and the shorter beams were of cast iron. The longer (longitudinal) beams were built-up sections of wrought iron. To facilitate making the connections, both types of beam had end plates, either cast-on or riveted. They could thus simply be bolted through, top and bottom at each side, either to the web or to one of the flanges of the column. To obtain the required moment-resisting rigidity, no more was needed save some local stiffening of the web of the column. To have left the bolts also to transmit the vertical loads carried by the beams would, however, have subjected them to large shear stresses. To avoid this, the beams were also seated on

brackets cast on the columns for this purpose – the brackets cast on the webs to carry the longitudinal beams also serving as the necessary local stiffeners. The practical superiority of the connection will be clearly seen if it is compared with the corresponding detail from the Crystal Palace.

The chief feature that these structures lacked was the fireproof floor, though it could not be said that any of them would have completely attained the fireproof ideal even with such floors. In neither of them was the iron frame protected against the damaging effects of the heat of a possible fire.

The first building that overcame this shortcoming was a large six-storey warehouse constructed in 1865 at the St Ouen Docks in Paris. Everywhere but on the relatively short principal facade, the iron frame was carried through to the outer walls which consisted only of brick panels of the same uniform thickness from top to bottom carried, storey by storey, on the outer beams of the frame. Internally now, all the floors were shallow vaults of hollow bricks topped with concrete.[22]

The further development was greatly stimulated by the intensive rebuilding of the commercial centre of Chicago after the fire of 1871.[23] Within little more than a decade, the Home Insurance Building was erected. This was framed throughout and all members of the frame were given the protection of a fireproof casing, although the rather heterogeneous choice of sections and materials made it inferior, in this respect, to the frame of the Sheerness Boat Store. This choice included cylindrical and box-section columns, I-section beams, cast iron, wrought iron, and the first use of steel in a tall building. The floors were again made from hollow tiles topped with concrete, but the bricks were now shaped so that they could be arched with a flat soffit. By the early 1890s, buildings having many of the characteristics of their modern steel-framed counterparts of similar height were becoming common. They lacked chiefly the welded or high-strength bolted joint and the reinforced-concrete floor slab.

The finest example, the Reliance Building [15.13], was completed in 1895. It was fourteen storeys and 60 m high and all loads were taken by the slender frame. Even allowing for the fact that the Monadnock Building was 6 m higher, a comparison of the plans of the two buildings [15.9 (a) and (c)] will show the huge reduction in space taken up by the supporting structure that this full transference of load made possible. The building was particularly notable both for the lightweight terracotta and glass cladding and for the way in which lateral stability was ensured.[24] In principle this was done by using fairly deep solid- or open-web beams to obtain rigid connections with the columns, and by making the columns themselves effectively continuous throughout the height of the building. Of course, such continuity could not

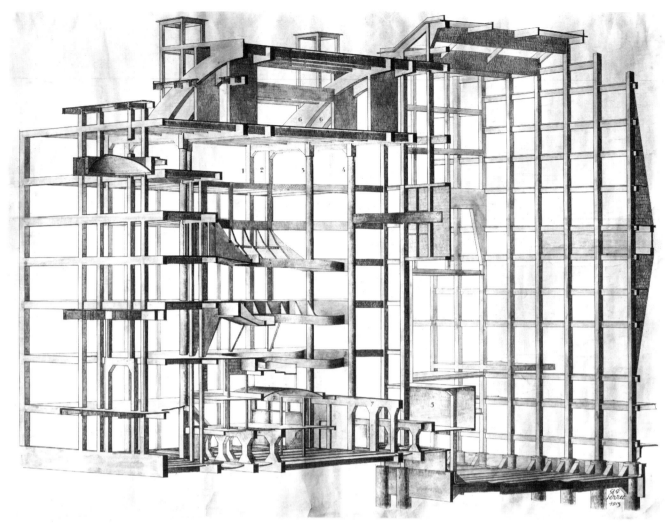

be obtained without splicing. But great care was taken over the splices. They were staggered in adjacent columns, and each column length as delivered to the site was already two storeys high. Both transversely and longitudinally, the whole frame was thus tantamount to a large number of rigid-jointed portal frames continuous with one another both vertically and horizontally. Continuity was interrupted only within the building, in the transverse direction, by the spacing of one row of interior columns more widely than the corresponding exterior columns. This defect was made good by having the flooring beams spaced much more closely than the columns. Only limited assistance was given by a low masonry party walls near the foot on parts of two sides.

By this time, reinforced concrete was becoming more widely used. Once the reinforced concrete beam and slab and the reinforced concrete column had been developed as separate elements for other uses, it was a relatively simple matter to employ them in a similar manner in multi-storey construction. They even made it easier in two ways to achieve the necessary lateral stability through a multiple portal-frame type of action. First (if the concrete was cast *in situ*) it was easier to obtain structural continuity between element and element. Then there was the considerable structural superiority

of a reinforced-concrete slab over the earlier forms of fireproof floor. Being very stiff in its own plane, such a slab could readily transmit any horizontal loads throughout the width of a building and thereby share them among a large number of columns.

The 64 m-high and sixteen-storey Ingalls Building in Cincinnati constructed between 1902 and 1903 was the first to match the structural achievement of the Reliance Building in the new material, though it did so with less architectural distinction.[25] Shortly afterwards, somewhat lower buildings were being constructed in the United States and in Switzerland with beamless flat-slab floors and mushroom-headed columns, and Perret, in France, was giving the earlier column-and-beam form the architectural refinement it had hitherto lacked.[26] The technical mastery achieved early in the next decade is well illustrated by Perret's drawing [15.14] of part of the framing provided to meet the complex functional requirements of the Théâtre des Champs-Élysées. In this drawing, of 1913, the structure is cut on its longitudinal centreline and the foyer and smaller theatre above it are omitted at the left. The view looks through the auditorium and past the proscenium to the rear wall of the back-stage area.

In the meantime, the focus of development of the steel-framed structure had moved from

Figure 15.14 Reinforced concrete framing of the Théâtre des Champs-Élysées. Paris (Perret). (Source: Chevojon)

Figure 15.15 York University (Robert Matthew, Johnson Marshall and Partners, Scott, Wilson, Kirkpatrick and Partners). (Source: author)

Chicago to New York. Contemporary with the Théâtre des Champs-Élysées was the Woolworth Tower in New York [15.27]. This was 242 m high and was carried throughout its sixty storeys by a steel column-and-beam frame. In the early 1930s the Empire State Building went even higher with a frame that did not differ greatly in general form. Both buildings (in common with similar buildings of the time) had frames that were much heavier than that of the Reliance Building on account of the greater heights. Columns and beams were built up from rolled sections and flats riveted together as sketched in Figure 10.14(d), and they were joined to one another with heavy riveted arched portal braces and knee braces (Woolworth

Figure 15.16 Habitat 67, Montreal (Safdie, Kommendant). (Source: Central Mortgage and Housing Corporation)

Building) or riveted semi-rigid joints supplemented by knee braces in interior bays (Empire State Building) to give the frames the lateral strength needed to resist wind loads.[27] Measurements on the Empire State Building during and after its completion nevertheless showed that its exterior cladding and internal partitions made substantial contributions to its overall stiffness and were really necessary to avoid excessive side sway.

Nothing comparable was attempted at this time in reinforced concrete on account of the large dimensions that would have been called for in the lower lengths of the columns and the difficulties that would then have been experienced in casting the concrete *in situ* at great heights.

Later twentieth-century framed and bearing-wall buildings of medium height

Higher strength concretes were nevertheless becoming available and precasting was soon being practised on a larger scale. Steel strengths were likewise increasing; the more efficient broad-flanged sections referred to in Chapter 10 were being introduced; and welding was making possible more compact and fully rigid joints. A decade or so later, prefabrication (of which an early example in concrete was seen in Chapter 4) was becoming commoner, especially in the production of other steel components and new forms of lightweight cladding. At the same time there was a continued growth in servicing possibilities and requirements; there were growing demands from business clients for larger floor-spaces unobstructed by columns to permit freer choices in the arrangement of internal partitioning; and there were soon pressing demands from public authorities for new housing and schools for the masses. These changing possibilities and demands led to the emergence of a variety of new forms in both steel and concrete and even to a transformation of the masonry bearing-wall structure. The range of choice rapidly expanded, initially largely for buildings of heights at which the more serious problems of standing up to lateral loads did not arise, though even at these heights there was no escape from the need to ensure lateral stability.

To meet the demand for mass housing and for new schools and similar buildings, prefabrication of all components of the structure, became the usual ideal from the later 1940s until the 1960s. A particularly interesting example of its use for school buildings was the steel-framed English CLASP system developed partly to meet the special requirements of an area subject to subsidence on account of old mining works. The frames had effectively pinned joints and their stability was ensured by diagonal tensile braces that yielded at preset loads when subsidence occurred under some columns. In their total prefabrication and this reliance on

diagonal bracing, they returned almost to the basic form of the aisles and galleries of the Crystal Palace, but with the bracing as the sole guarantor of lateral stability rather than just as an extra precaution. Figure 15.15 shows an adaptation of this system for the construction of university students' hostels on a site where there was less risk of subsidence. Here the cladding was heavier than was usual in schools in order to give better fire resistance and, incidentally, a more substantial and permanent appearance.[28]

For housing, the almost universal practice was simply to precast floor and wall panels – and to a lesser extent beams – rather than casting them *in situ*, but to adopt essentially the same structural forms as for construction *in situ*. This use of prefabrication will therefore be mentioned later. But there were also numerous projects to prefabricate complete rooms and even complete dwellings and then either to stack

Figure 15.17 Lever Building, New York (Bunshaft, Weiskopf & Pickworth). (Source: author)

Figure 15.18 Lakeshore Drive and East Chestnut Street, Chicago. On the right are the Lakeshore Drive Apartments (Mies Van der Rohe). Behind them to the left are first the De Witt Place Apartments and then the Hancock Centre. (Source: author)

these on one another or to plug them into larger structures that would serve simply as supports and allow them to be periodically removed and replaced.

The only realized project on a substantial scale was the 'Habitat 67' cluster built as part of Expo 67 at Montreal. The support here was a conventional one consisting of vertical access and service towers spanned at appropriate levels by horizontal beams. It was clearly revealed at the rear, but only glimpsed occasionally through gaps in the receding tiers of seemingly randomly arranged interlocking precast-concrete boxes that make up the individual dwellings when these were viewed from the front [15.16]. The beams carried both horizontal service-runs and partly glazed pedestrian streets. Because they occurred only at two of the twelve levels of the cluster, the tiers of boxes had to be made self-supporting above, below, and between these levels. For this purpose each box had to be stiffened by top and bottom ring-beams and vertical pilasters, and the boxes when stacked above one another were tied together by prestressed vertical steel rods to permit some cantilevering of the upper ones. Adjacent boxes at the same level were also bolted together. Thus, while the support structure carried some of the vertical load of the upper tiers and made a large contribution to the overall lateral strength, stiffness, and stability, it did not confer on the cluster as a whole the freedom of piecemeal replacement and rearrangement of the individual dwellings often claimed for such projects. Making the boxes bear other loads in addition to their own weights also called for far more variation in detail according to their precise intended locations in the cluster than would otherwise have been necessary. This must have contributed significantly to the very high unit cost in relation, for instance, to that of the much simpler alternative of merely stacking similar boxes vertically above one another to form compact slab-like blocks as in some of the heavy concrete systems developed for the post-war reconstruction of Eastern Europe.[31]

For other purposes, and at heights more like that of the Reliance Building or a little higher, the steel-framed form dependent on rigid joints for its lateral stiffness and stability was simply made more efficient and economical. This was done by using the more efficient broad-flanged sections, by rigidly welding them together to exploit fully their potential stiffnesses and strengths rather than relying on additional bracing, and by adopting more regular column grids.[32]

Two early examples were Bunshaft's Lever Building in New York [15.9(d) and 15.17] and Mies Van der Rohe's apartment blocks at 860 Lakeshore Drive in Chicago [15.9(e) and right foreground of 15.18].[33] Architecturally, both were highly influential and the apartment blocks, in particular, introduced a new refinement of proportion and detail to the multistorey

form, just as Crown Hall [13.11] had done earlier for a related wide-span form. (Note that only the heavier-looking verticals of the facades of the apartment blocks that are seen almost face-on in Figure 15.18 and only the corresponding ones of their other facades are load-bearing columns. Between them are just light window mullions.) Structurally each building would have been stiffened to some extent by the walls enclosing the lift shafts and stairs and by the concrete fireproofing of the steel. But their stability was largely ensured by the rigid joints of the steel frame in spite of the fact that the reduced cross sections of the higher strength steel tended to make the individual beams and columns more flexible.

A more radical departure from the earlier forms was made in the thirteen-storey IBM Office Building in Pittsburgh [15.19].[34] On each facade a close diagonal grid was substituted for the usual vertical and horizontal grid of beams and columns. Its high natural shear stiffness ensured stability. It also provided a very neat way of removing the obstruction to free access that can be presented by closely spaced vertical columns at ground level. Instead of calling for

Figure 15.19 IBM Building, Pittsburgh (Curtis and Davis, Worthington, Skilling, Helle and Jackson). (Source: author)

heavy beams at first floor level to bridge between more widely spaced columns there, all that was necessary was to stop short most of the diagonals at first-floor level and to make those that continued to the ground from a higher-strength steel capable of carrying the increased load. On the other hand, the grid was almost excessively stiff, and difficult to connect to the floor system. It had no direct progeny. But diagonal framing was adopted in modified form in buildings such as Bush Lane House in London, and it was used later in association with normal framing in taller structures.[35]

In brick masonry and reinforced concrete, the principal developments were parallel transformations of the traditional masonry bearing-wall form. These exploited the much wider ranges of compressive and shear strengths that could now be obtained in quite thin walls of either brickwork or concrete. With appropriate choices of concrete mixes or bricks and mortar, it became possible to achieve the compressive strengths necessary to resist the gravity loads of twenty or more storeys with walls no thicker than were previously usual only for two or three storeys. Those developments that kept closest to the usual multi-cellular masonry prototype also exploited the great structural superiority of the reinforced-concrete slab over earlier types of floor – in particular its ability to distribute wind loads, to tie the walls together, and to stiffen them against buckling. Indeed it was only with its help that such slender walls could be used in this way for buildings of more than a few storeys.

An early example of this new bearing-wall form in reinforced concrete was Highpoint One,

whose thin external concrete walls are shown still under construction (in 1934) in Figure 4.5. Two later, taller, examples are shown in Figures 15.20 and 15.21.[36] In the first of these the walls are all of brick, never more than two bricks (220 mm) thick. In the second they are of concrete and were precast in storey-height panels as seen in Figure 10.11. Since the assembly of such precast units does not naturally give the tensile continuity that is easily obtained with *in-situ* construction by continuing the reinforcement over the joints, additional reinforcement was called for to remove the risk of a collapse such as that of an otherwise similar building seen in Figure 5.14.

In all these buildings, the floor slabs are capable of acting simultaneously in all the three roles just referred to as well as in their primary role of carrying over to the walls the vertical loads imposed by the occupants. Since, moreover, the walls were disposed in two directions at right angles, the structures as a whole may be likened to tall cellular cantilevers whose flanges and webs are the walls and whose transverse stiffening diaphragms are the floors. To a first approximation, the walls on the windward face are placed in tension by the wind, those on the leeward face in compression, and all those parallel to the wind in shear [5.5(c)], though there should be little if any net tension on account of the compressions induced everywhere at the same time by gravity loads. In practice, the necessary presence of openings in the walls (especially in those that are placed in shear by any particular wind-loading) results in rather less concerted actions characterized by a certain amount of what is known as 'shear lag' across

Figure 15.20 (left) Residential towers, Essex University, (Architects' Co Partnership, Harris and Sutherland). (Source: author)

Figure 15.21 (right) Residential tower, Thamesmead, London (GLC Architects, Balency and Schuhl). (Source: author)

the building. But, by careful planning of the layouts of the walls and openings, this was kept small enough for most of the potential structural economy of the form to be realized. Comparison of the floor plans with those of the Monadnock Building would show almost as dramatic a saving in the proportion of space occupied by the bearing structure as was achieved by the frames of the Reliance Building and its more recent steel-framed counterparts.

A related development that was of wide significance for the later development of taller buildings was that of the tower-like reinforced concrete core. This usually contained only access stairs, lifts, and perhaps service runs. But it was an inherently stiff form if openings were kept to a minimum, and was capable of giving to the building as a whole some or all of its lateral stiffness as well as carrying some or all of the weight of the floors and the vertical load on them down to ground level. One example of such a core has been seen already [4.12].

Wright was the first to use it as the sole support in his Research Tower for the Johnson Wax Company at Racine, Wisconsin.[37] Fourteen floors, alternately square and circular, were cast integrally with the hollow core, and individually cantilevered out from it. The circular floors projected a little less than the square ones, and the absence of peripheral columns was emphasized by running windows, two storeys in height, round the entire exterior between horizontal bands of brickwork at the levels of the square floors. In the twenty-three-storey Knights of Columbus Building in New Haven five such reinforced concrete towers were built ahead of the floor construction and were then spanned by steel beams to carry the floors [15.22, 15.25(e)]. The outer circular ones contain stairs and were post-tensioned to avoid tension in the concrete under wind loading. The central square tower contains the lifts.[38]

When using the core or cores as sole support, the alternative to cantilevering the floors individually or carrying them by direct bridging was to construct stronger cantilevers at the top, and perhaps at one or more intermediate levels, and then to suspend the floors from these transfer structures, using vertical hangers. The cantilevers then had to be the depth of a full storey for maximum efficiency. Figure 15.23 shows two early examples of this alternative under construction.[39] The nearer of the two, almost complete structurally, has only ten suspended floors carried by a single set of cantilevers at the top, walled in as a plant room. The farther one is just over twice the height and has similar steel trussed cantilevers at both mid-height and the top, carrying respectively thirteen and twelve suspended floors. In the photograph, construction of the floors in the upper section is only just commencing, and both the reinforced-concrete core which carries all the loads to the ground and some of the first lengths of the peripheral hangers can be clearly seen. The

inner ends of all the floor beams are, of course, carried directly by the core. As in the forms in which each floor is independently cantilevered-out or independently bridges between several cores, all the lateral stiffness must be provided by the core, acting as a relatively narrow vertical cantilever. The roundabout path of the peripheral loads does, however, have the merit of automatically inducing a considerable compression in this cantilever throughout its height and thereby reducing the need for tensile reinforcement as compared with that in the other forms.

One advantage of these forms is the absence of any need for peripheral vertical supports at ground level. Whether or not the area around the core or cores is left completely open, this can both contribute to easy access to the completed building and facilitate the construction of both foundations and superstructure on certain cramped or otherwise restricted sites.

Figure 15.22 Knights of Columbus Building, New Haven (Roche Dinkeloo, Pfisterer Tor). (Source: author)

Taller framed and core-stiffened buildings of the 1960s and 1970s

From the late 1950s onwards the demand for taller buildings spread and soon became world-wide. It remained chiefly a commercial one as it had been in New York, with office use as the main requirement and further uses for apartments and hotels. For offices, vertical access and service runs were required to be concentrated in the centre with the maximum unobstructed floor spaces elsewhere to permit later free partitioning. For hotels and apartments, access was preferably less concentrated and there was less need for partitioning at the occupants' choice. In the 1960s and early 1970s when the main advances were made, changes in taste and planning legislation led, however, to a preference for more cubic building envelopes rather than the numerous setbacks and overall tapered profiles of the Woolworth and Empire State Buildings.

Of the forms just considered, only those relying largely on a stiff reinforced-concrete core or cores for their lateral stability and resistance to wind could be heightened with little change, just as the pure cylindrical tower could be. The twin Marina City Towers in Chicago [15.24 and 15.25(a)] were an early example of this heightening. Each tower relied on a central cylindrical core almost as much as did the Commercial Union and P&O Buildings and the building illustrated in Figure 4.12. To make this core as stiff as possible, it had its necessary access doorways staggered to left or right from floor to floor. Around it, from foundation level upwards, thirty-two columns in two concentric rings of sixteen assist in carrying the floors, thereby avoiding the need for massive cantilevers while also offering some resistance to wind. At the time of completion, these towers were the tallest multi-storey buildings yet built in reinforced concrete, with a height of 180 m. They were notable also as being part of a new type of multi-use development comprising a yacht marina in the basement (with direct access from the Chicago River), offices, shops, and restaurants above this, then space for car parking up to the twentieth floor, and finally forty floors of apartments.[40]

In the National Westminster Tower in London, completed almost two decades later and only slightly taller but designed for heavier floor loads, a similar reinforced concrete core is the sole support [15.26]. In plan, both core and building envelope are clover-leaf in shape. The core has the same internal dimensions throughout, but reduces in thickness at three intermediate levels from a maximum of 1500 mm at the foot to a minimum of 300 mm rather more than half-way up. Within it are three separate sets of five lift shafts to serve the three leaves of the clover. Around it the floors are carried on the outside by steel columns which are supported well above ground level by three separate cantilevers. These are evenly spaced around it and separated from one another by two-storey intervals. Each is 9 m deep to bear the weight of up to forty-three normal floors and three more plant floors.[41]

The simple rigid-jointed frame as seen in the Lever Building and Lakeshore Drive Apartments was less appropriate for greater heights because its overall lateral stiffness and resistance to wind are too dependent on the bending stiffness of its columns and beams. Whatever the structural system, increased height inevitably carries some penalty on account of the greater weight to be supported at the lower levels. Hence the considerably increased thickness of the core of the National Westminster Tower at these levels. But in the column-and-beam frame dependent on rigid joints for its lateral stiffness and stability, there is a further penalty. This arises because the large shear forces near the

Figure 15.23 Commercial Union and P & O Buildings, City of London, under construction (Collins, Melvin, Ward and Partners, Scott, Wilson, Kirkpatrick and Partners). (Source: author)

Figure 15.24 Marina
City, Chicago
(Goldberg, Severud).
(Source: author)

base due to wind can be resisted only by the beams and columns acting primarily in bending while the columns are simultaneously carrying high axial compressions. This they are poorly fitted to do because their individual cross-sections are small on account of their large number. In order to keep the sidesway of the at reducing the stiffness required of the frame by supplementing it with that of a reinforced concrete core or separate shear walls or vertical trusses. The most radical were those that placed most of the frame on the exterior and increased its stiffness by transforming it into something more like the walls of a continuous tube so that

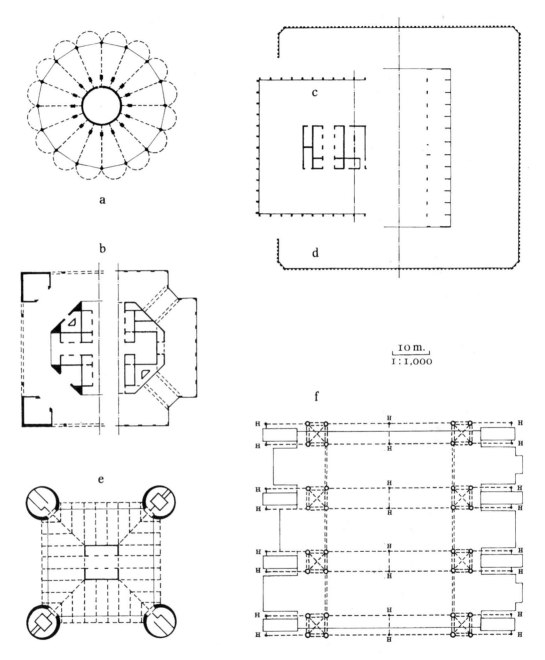

Figure 15.25 Plans (to the same scale) of taller buildings constructed since 1960: (a) Marina City, Chicago, 1963, 180 m.; (b) Messeturm, Frankfurt, 1994, 250 m.; (c) Brunswick Building, Chicago, 1962, 145 m.; (d) World Trade Centre, New York, 1972, 410 m.; (e) Knights of Columbus Building, Newhaven, 1970, 100 m.; (f) Hong-kong and Shanghai Bank, Hongkong, 1985, 180 m. Vertical supports are shown in heavy full line, with hangers lettered 'H'. Principal beams in (a), (b), and (c) and principal beams and trusses in (f) are shown in heavy broken line. (Source: author)

Woolworth and Empire State Buildings and others like them within tolerable limits, they had to be greatly increased beyond what was needed simply to resist gravity even with the added stiffening of claddings and partitions. Ways had to be devised to reduce this necessary increase.

Several approaches were developed in the 1960 and 1970s. Some were directed towards increasing the lateral stiffness of the frame itself without the same 'penalty for height' and some

it would resist the wind, as far as possible, by tension in the columns on and near the windward face and compression in those on and near the leeward face. This entailed either changing the proportions of the individual beams and columns to reduce their bending or adding diagonal members to create a trussed vertical cantilever.[42]

The principal first steps in these directions were made in Chicago in two buildings on

Washington Boulevard and East Chestnut Street that are seen in Figure 15.13 beyond the Reliance Building and in Figure 15.18 beyond the Lakeshore Drive apartments. Indeed the earlier framed form – in which much of the actual lateral stiffness came not from the frame itself but as a free bonus from the heavy masonry cladding and infill partitions – is also seen to the left of the second figure, and a further step is seen across the street in the distance.

In the Brunswick Building on Washington Boulevard [15.25(c)] and the DeWitt Apartments on East Chestnut Street, the excessive weakness in shear of the normal frame as seen in the Lakeshore Drive Apartments (with wide column spacings and beams with structural depths are little more than a twentieth of the span) was partly overcome by changing the proportions of the beams and columns. The columns were much more closely spaced and the beams were given depths of between one-half and one-quarter of the span.[43] In effect, the non-load-bearing mullions of the Lakeshore Drive Apartments were turned into load-bearing columns and its beams deepened. Since the close spacings of the columns were not acceptable at ground level, fewer columns were used here and they were bridged by deeper transfer beams. In the Brunswick Building, these were a whole storey in depth with recessed panels as seen at the bottom right of Figure 15.13.

These changes were not enough, however, to eliminate all shearing or racking distortions of the outer 'walls' in line with the wind, so that the behaviour of both buildings under wind or other lateral load still falls short of that of the corresponding ideal solid-walled tube. In this ideal tube, with no such shear distortion or 'shear lag', most of the resistance would be provided by tension in the whole windward wall and compression in the whole leeward wall and a uniform variation from tension to compression in the walls parallel to the wind – though the windward tensions should do no more than reduce the compressions due to gravity loading, unless the building is unusually narrow. It was estimated that in the DeWitt Apartments, where internally there are only free-standing columns instead of a continuous stiff core to give greater freedom in planning, the lateral stiffness is about a quarter of that of the corresponding ideal tube. This means that about three-quarters of the sidesway must still be attributed to shearing or racking distortion of the 'walls' parallel to the wind. To increase the stiffness of the Brunswick Building, this was also provided with a stiff solid-walled tubular core to create a 'tube-within-tube' structure [15.25(c)].

In the twin towers of the New York World Trade Center, each more than three times as high as the DeWitt Apartments, the approach using an external framed tube was taken towards its limit [15.25(d) and 15.27-29].[44] With columns set only 1 m apart and with beams 1.3 m deep, the whole exterior above a transition zone

at the foot became as much a perforated wall as a frame and a very close approximation to the ideal solid-walled tube was claimed. To simplify erection, these 'walls' were prefabricated in sections three storeys high and three columns wide, and the sections were assembled with staggered horizontal joints. This procedure and

Figure 15.26 National Westminster Tower, London (Seifert, Pell Frischmann). (Source: author)

the novel proportions are seen in Figure 15.28, where the upper parts of the last two rows of prefabricated sections to be erected are sharply silhouetted against the sky.

Internally again, there are only free-standing columns, designed just to take their share of the vertical load from the lightweight wide-spanning floors [15.25(d) and 15.29] without playing any

Figure 15.27 World Trade Center, New York, nearing completion (Yamasaki and Roth, Skilling, Helle, Christiansen and Robertson). The tall tower in the right background is the Woolworth Building. (Source: author)

part in resisting the wind. However another interesting and novel feature was introduced: the trussed floor beams – which, like the internal columns, play no direct role in resisting the wind – are only connected to the outer walls through viscoelastic damping sandwiches. These allow slight relative movements and, as the towers sway slightly in the wind, some of the energy of these movements is absorbed, and the amplitude of the sway is thereby reduced.

A further development of this type of framed tube occurred in Chicago again with the construction of the Sears Tower. Instead of a single tube, nine such tubes were bundled together, some rising higher than others and only the topmost pair being taller than the World Trade Centre. Without calling for diagonal bracing, a close approximation to ideal tubular behaviour was obtained by thus linking together a number of 'tubes' of smaller plan dimensions – or, looking at it another way, by introducing continuous internal diaphragms to stiffen the exterior.[45] The system also provided an easy means of obtaining a tapered overall form.

It is, of course, easier to give the desirable stiffness in shear to framed 'bearing walls' by bracing the frames diagonally and turning them, virtually, into giant vertical trusses. In the first example – the IBM Office Building in Pittsburgh already referred to above – the diagonals provided not only shear stiffness but also the only vertical support. This was inefficient because it left the vertical loads to be supported by inclined struts. In the hundred-storey Hancock Building in Chicago, not much lower than the World Trade Centre, diagonal braces were added to a fairly normal column-and-beam frame [15.18, left background].[46] These braces not only take all the wind loads but also redistribute the vertical loads from column to column as needed, so that these could all be of the same cross-section. The diminution in overall width with height above the ground, as well as being structurally advantageous, was also

Figure 15.28 World Trade Center, New York, under construction. (Source: author)

Figure 15.29 World Trade Center, New York, typical interior before subdivision by non-load-bearing partitions to the requirements of individual tenants. The light trusses of the floor system (with sprayed-on fire protection) span 18 m. from the perimeter 'wall' to the central core. (Source: author)

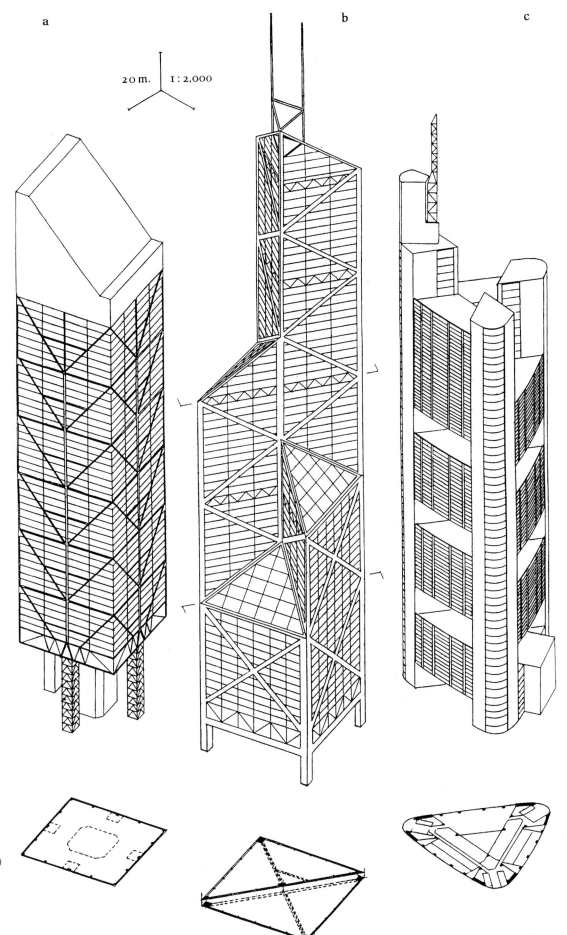

a   b   c

20 m. | 1 : 2,000

Figure 15.30 Isometric
views (to the same scale)
of the primary structural
systems of: (a) CityCorp
Centre, New York
(Stubbins, Le Messurier);
(b) Bank of China, Hong
Kong (Pei, Robertson); (c)
Commerzbank, Frankfurt
(Foster, Arup). The
CityCorp Centre is shown
with its facing largely
stripped away.
(Source: author)

justified by a change in use from an office building below to apartments with lower floor-to-floor heights above.

Numerous examples of the diagonally braced tube followed on the success of the Hancock Building, some, like it, in steel, and a smaller number in concrete. The most interesting example from the later 1970s is the Citicorp Centre in New York.[47] Its unusual design [15.30(a)] grew partly out of a desire to have the building square in plan and so sited that one corner would have to bridge over an existing low building on the site. Partly for architectural reasons, it was decided to leave all four corners similarly open up to the height needed to clear the existing building. This entailed chanelling the weights on each of the four sides partly to the reinforced concrete core and partly down a single trussed pier. For this purpose a system of diagonals forming inverted Vs was devised, with some of the load-transfer characteristics of the exterior framing of the IBM Building and some of those of the Hancock Building. At the foot, however, this system alone would have thrown more wind load on the four piers than their small width and depth could withstand. Here, therefore, a storey-deep truss transferred the excess to the core. To further limit movements at the top in a strong wind, a damping system again was installed, now of a different kind. It was probably the first tuned-mass damper in a tall building. A large concrete mass weighing about 2 per cent of the weight of the whole building was mounted on a smooth surface near the top and connected to it by spring and shock absorber. These were tuned to give the mass the same natural frequency as that of the building when it is made to float by pumping a thin bed of oil beneath it. As the building starts to sway, the mass stays put on this film until pulled by the spring and then it counters the force of the wind by always moving in the opposite direction.

Later twentieth-century taller framed and core-stiffened buildings

From the 1980s onwards tastes changed again and a continual search for variety in geometric massing and silhouette ensued, coupled sometimes with a striving for height for its own sake and not just as a way of increasing floor area on a restricted site. But the underlying structural forms have largely remained those established in the 1970s. These have been exploited further, combined in new ways with the help of new computer-aided design procedures, and assisted more frequently by mechanical damping of wind-induced movements.[48] Economy in the costs of construction seems sometimes to have been considered less important than in the previous period. One example of the trend is the multi-faceted Bank of China Building in Hong Kong [15.30(b)]. This is a novel exploitation of the architectural potential of three-dimensional

truss support, with successive lateral contractions giving an irregular spire-like form.[49]

A further development has been the use of concrete at heights where only steel was previously used for the main supporting structure. An early example of this was the Frankfurt Messeturm [15.31]. Structurally this is closest to the Brunswick Building among those considered hitherto. A stiff core is linked to an external perforated wall, though the proportions are somewhat different and the linkage is here more positive [15.25(b)].[50] Again a transfer structure was called for above the entrance level shown on the left side of this figure. The Messeturm is more than 100 m higher than the Brunswick Building. It has been followed by the linked

Figure 15.31 Messeturm, Frankfurt (Murphy Jahn, Nötzold). (Source: author)

twin Petronas Towers in Kuala Lumpur which are almost 150 m higher, again excluding slender spires at the top. Here, however, steel was used for the floors to speed construction.[51]

A later Frankfurt structure, the Commerzbank, is less easy to classify. But it has strong affinities with the column-and-beam framed tube and will therefore be considered here. The chief differences are that the tubes have wide openings at intervals that extend over four storeys and include the principal walls of three core-like structures.[52] The piled foundations were considered already in Chapter 10. The main vertical elements of the structure rise directly, as seen at the top of Figure 10.22, from the cellular substructure of rounded triangular shape that caps the piles. In each rounded corner of the triangle they consist of a pair of diagonally braced composite steel-and-concrete curved walls and a number of associated columns and lighter walls. Together, these walls and columns form three stiff towers which, like core structures elsewhere, accommodate lifts and services. The affinity with the framed tube arises from the interconnections of the towers around the remainder of the perimeter. These consist of steel walls, perforated by windows, which are continuous vertically for six, eight, twelve or fourteen floors between the four-storey openings [15.30(c)]. They are described as Vierendeel trusses (that is, trusses composed only of rigidly connected vertical and horizontal members) since they not only link the towers to assist them in resisting wind loads but also partly carry the floors around a central triangular atrium that is open for the whole height of the building. Without sacrificing structural efficiency, they allow natural ventilation at all levels and views of sky gardens from all upper floors.

Alternative forms

We have now traced the principal ways in which stability has been maintained, even under considerable side load, when heights increased beyond the point at which storeys consisting of simple columns and beams or walls and floors could simply be piled on one another as in Figure 15.1. Up to heights of twenty to thirty storeys this was done by merely exploiting further the stiffness in shear of stronger enclosing and dividing walls and the stiffness in bending developed by new rigid joints between columns and beams. No other departure from the forms and proportions appropriate to lower buildings was called for. At greater heights, three approaches have been followed. In the first, the same sources of stiffness were drawn upon, but a stricter planning discipline was imposed, most columns were closely spaced around one or more closed peripheries, and beams were made proportionately deeper to create, in effect, perforated tall tubes. In the second, a compact core containing lifts, stairs, and vertical service runs was made to resist much or all of the side load

and much or all of the vertical load. In the third, the column-and-beam structure was stiffened by diagonal bracing, or most of the loads were passed on to a more self-sufficient vertical truss.

In the second and third of these approaches the core or vertical truss can become a megastructure into which the individual floors and rooms are simply filed, to which they were attached, or from which they were hung. This was already the way in which the core of the Johnson Wax Research Tower was treated, and likewise the four corner towers of the Knights of Columbus Building and the cores of all buildings in which the floors were suspended from cantilevers at the top.

Such structures share a great advantage with structures like deep-sea oil rigs and the towers considered at the outset. Their material is better concentrated to carry large loads over considerable heights than is that of the structure in which it is distributed between large numbers of more slender columns and beams. The possibility therefore exists of building structurally highly efficient larger megastructures simply as supports which could then be used, like giant coat hangers, to carry floors and walls arranged with as much freedom as the internal partitions of a large open floor. The idea of doing so has been around for a considerable time and received wide publicity from several groups of architects in the late 1950s and early 1960s.[53] Habitat 67 was an attempt to put the idea into practice on a very modest scale. The Pompidou Centre – described in Chapter 13 as an example of a wide-span structure – also went part-way towards realizing the ideal of separating basic structure and internal subdivision and thereby paved the way for further experiment. Although it had only six storeys, its basic structure was initially conceived as a support for a highly adaptable interior. Other projects have continued to appear, but usually without going beyond exploring possible geometries.

A most successful recent structure of this kind is the Hong Kong headquarters building of the Hongkong and Shanghai Bank [15.32]. The brief called for both an ability to accommodate possible extensive future changes in banking requirements and a design that would allow construction around and over the existing bank while, for a time, this continued in use. This wish to be able to build around the existing bank led to the choice of a suspension system, and this choice was retained even when it was later decided to start by clearing the whole site. For a number of reasons including the cramped nature of the site, it was also thought desirable to adopt a considerable measure of prefabrication.

As built, the basic supporting structure consists of four main supporting frames of three different heights, each 5.1 m broad and separated by clear spaces of 11.1 m.[54] Each frame consists of two square towers spaced 33.6 m apart and carrying, according to its height, four or five pin-jointed suspension trusses, each of

Figure 15.32 Hongkong and Shanghai Bank, Hongkong (Foster, Arup). (Source: Foster Associates, Ian Lambot)

these trusses being two storeys deep and having an overall span of 64.8 m. In the photograph, the outer halves of the north towers and trusses are seen fully exposed on the exterior; also the tops of the adjacent interior towers and trusses. The towers each consist of four large-diameter cylindrical steel columns that diminish in wall thickness towards the top and are joined ladder-wise by Vierendeel braces to form highly efficient larger composite columns. These are then braced in their own plane by the suspension trusses and, out of sight in the transverse direction, by diagonal cross-braces between adjacent towers at the level of each truss and one lower level. So braced they form a completely self-sufficient megaframe capable of standing up to the worst typhoon.

The floors that accommodate all the activities of the bank are suspended from the three low points of the suspension trusses by thick-walled steel tubular hangers [15.25f]. As origin-ally completed, they filled varying proportions of the available space, leaving the height of two storeys completely open at the foot, an open banking-hall atrium extending through another nine storeys in the centre, and stepped-back openings on the east side to respect the then current planning restrictions. Outside the composite columns are hung prefabricated service modules.

The megaframe was designed with the intention that it should be possible to change these modules and the floors at will, including filling most of the spaces originally left open. As at Habitat 67 however, nominally identical modules differ too much to be readily interchanged, although there should be no real difficulty in adding to the floor areas. Again achievement fell somewhat short of intention. Compromises had to be made. But the result was a fine and memorable building that came closer to the ideal megastructure than anything else had yet done.

Part 4: Design

Chapter 16: Structural understanding and design

We have now reviewed some of the more significant steps in the development of structural forms up to the present day by looking – largely in the light of our present understanding of structural actions as outlined in Chapter 2 – at the consequences of the choices made by particular designers in the situations in which they found themselves. How were these choices made? How has the process of structural design itself developed to permit the construction of a structure like that illustrated in Figure 1.2 little more than a hundred years after the first tentative experiments in reinforcing concrete with iron when the far simpler form illustrated in Figure 1.1 was probably arrived at only after a much longer period of repeated trial and error? How has the even more rapid development in recent years of new wide-span and tall multistorey forms in steel, prestressed concrete and newer materials been possible?

We can hardly hope to answer these questions fully in a single concluding chapter. It seems worth while, nevertheless, to attempt a partial answer in fairly broad and general terms, while recognizing that there must, in fact, be something unique about the genesis of almost every new form if only because different people are involved or the circumstances are different. In particular, as was pointed out in Chapter 1, the structural requirement is rarely, if ever, the only one to be met, and others have frequently been more important in leading to the choices finally made. Also designers are hardly ever in a position to act entirely on their own initiative, as sculptors or painters sometimes may. They must both work through others and for others if their designs are to be realized. Thus the changing patterns of patronage have probably been at least as important as the wayward flowerings of structurally and architecturally creative genius in determining the course of development, and the individual patron has sometimes played a far from negligible role in the total creative process.

Briefly, it is clear that the major change that has permitted the accelerated pace of development over the past 200 years or so has been a much-improved and continually growing ability to understand, observe, and predict the structural actions brought into play. The developments could not have taken place without the new materials and techniques that they exploited. But this exploitation would not have been possible without the deeper understanding and analytical ability. Much was achieved with more limited means in earlier centuries, however, and design procedures today still share some fundamental characteristics with what was done then.

Understanding and design up to the Renaissance

In the beginning, there was probably a large measure of simple trial and error in which pure chance could have played a major part. Nevertheless it would have been possible to conceive prototypes of forms like those illustrated in Figures 1.1 and 7.1 *ab initio* on the basis of no more than the simplest of the spatial or geometric requirements for stable equilibrium discussed in Chapter 2, and it seems virtually certain that they would have been thought of in these terms once built. This is not to say that these requirements could then, or would then, have been explicitly formulated, any more than they are formulated today by the average child playing with small blocks on the floor. They would rather have been intuitively apprehended in a dual recognition of the tendency of every object to fall and the possibility of preventing the fall by placing an obstacle in its way.

Coupled with this intuitive apprehension, or intuition, are likely to have been two others.[1] The more fundamental of them would be a closely related but quite distinct physical or muscular intuition derived from direct bodily experience of the pushes and pulls associated with extension and contraction and with different ways of supporting a weight. From these bodily sensations, it would be only a short step to inferring analogous pushes and pulls in built structures. The other, based on a recognition of the different ways in which different materials and structural elements respond and eventually succumb to the pushes and pulls and to other actions such as bending, would have been some association of these observed responses with those other characteristics of these materials and elements by which they were normally identified.

Drawing further on just the first two of these intuitions – which will be called the spatial and muscular ones – it would also have been possible to conceive all the elemental forms employed in later timber and masonry structures. Given enough imagination, it would have been possible not only to see the potential use of a log as column, strut, or beam, but to see, for instance, the possibility of constructing a freestanding voussoir arch or a simple roof truss whose coupled rafters were prevented from

would lead no further than the knowledge acquired by the typical craftsman that this is what to do. But for others with a more questioning approach, the whole experience of placing stone on stone, maintaining stability as the structure rose, and always observing its behaviour and noting the circumstances in which any failures occurred, would lead to a fuller qualitative understanding of the pressures and the pulls and thrusts and other structural actions brought into play. This would be an important secondary type of intuition, powerful within its limits but a less reliable guide in situations different from those in which it was acquired.

This was the usual position up to the eighteenth century. The normal pattern of development was one of small variations from one structure or group of structures to the next. So small were these variations that – even today when the vast majority of the structures once

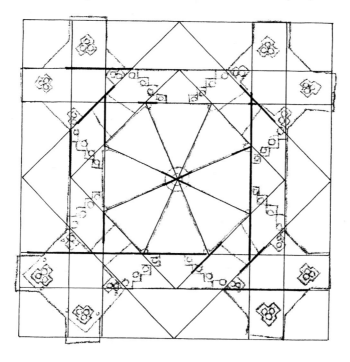

Figure 16.1. Villard de Honnecourt's plan of one of the towers of Laon Cathedral with the scribed geometric setting-out added. Heavier lines denote scribed depressions in the vellum that can still be observed in a glancing light. Lighter lines complete the pattern. Several different levels of the tower are superimposed. (Source: author based on folio 9v of Bibliothèque Nationale MS Français 19093)

spreading by a tie at their feet. It would also have been possible to conceive potentially stable arrangements of larger numbers of elements than in the simplest hut forms. Sometimes the imagination may have been triggered by observing chance occurrences of analogous forms like that of the potential voussoir arch in the defence wall at Segni [6.6]. At other times the trigger may have been no more than pondering on a perceived need.

The basic intuitions would, however, have been poorer guides to selecting, from among the many possibilities, a desirable profile for the arch or, more generally, for selecting appropriate materials and cross-sections for any element or assembly of elements. Here only practical experience could help. Over time, such experience in building many broadly similar structures would suggest that some choices were better than others. For some of the builders, this

standing have disappeared or been extensively rebuilt– we can fairly easily assign relative dates to those that survive from before the time when the perceived needs were fully met and development ceased. In such a process of gradual change, experience was carried over from building to building by the secondary intuition just referred to. And when problems arose they were dealt with in the light of the primary intuitions.

These intuitions must, for instance, have been the chief guides for the continued step-by-step development of the Gothic structural system from its Romanesque beginnings that was outlined in Chapter 12. We know relatively little of the failures that punctuated the advance. But they certainly occurred, as at Beauvais in 1284. On such occasions and when other difficulties arose, it seems to have been the usual practice to convene a panel of experts and seek their opinions.[2] Had these experts made any attempt at precise analysis, it would surely have been mentioned in the surviving records of their expertises. Since there is no such mention, they must have relied largely on the basic primary intuitions in deciding whether, for instance, to underpin, add or strengthen a buttress, or add a tie.

As for the simple carrying over of experience from one building to the next, we have only very limited evidence of how this was done prior to the early thirteenth century. By that time at least, it is clear from a unique survival that very simple geometric disciplines played a considerable part. This survival is the so-called notebook of a French master mason, Villard de Honnecourt.[3] One page from his manuscript was reproduced in Figure 4.3. The most revealing one from the present point of view is reproduced here [16.1]. The chief use of the very simple geometry seen in both these figures seems to have been to assist in arriving at, specifying, and setting out leading dimensions without using the scale drawings that became usual later. But it also provided a means for codifying proportions that had been proved to be adequate and there is evidence that it was sometimes so used. In some circles, particular disciplines based on it even seem to have been seen as guarantors of stability on the basis of associations with Platonic ideas about the role of the five regular solids as guarantors of the stability of the universe. Clearly these disciplines could do no such thing, because they ignored much that was important and they left the designer far too much freedom of interpretation – as the variety of executed designs bears witness.[4]

If there was an exception to this use of the simplest geometry, we might expect to find it in the design, centuries earlier, of Justinian's Hagia Sophia [12.10 and 12.11]. Several centuries earlier still, Archimedes had set down the laws of the balance in a work entitled 'On the equilibrium of planes'.[5] This was the only successful attempt until much later to give quantitative precision to some early intuitive ideas of the conditions of balance between forces. Hagia Sophia's principal designers were, unusually, leading academics who were not only familiar with Archimedes but also masters of all the other relevant theory of the time. But even here there is no evidence that they drew upon more than their fuller knowledge of geometry. Although their whole design was a superb exercise in applied geometry, neither Archimedes nor this geometry could make good the limitations of the basic intuitions when they had to choose particular profiles and cross-sections to meet the requirements of a new combination of forms on an unprecedented scale. The circular arc was adopted throughout for arch profiles and the crucial cross-sections seem to have been based fairly directly on ones that had been successfully adopted in the nearest comparable earlier structures, notably the Basilica Nova

[12.4]. But to allow for continuous broad galleries over the aisles [12.11], a major departure had to be made from the transverse buttressing system that had been adopted there. The openings for these galleries so weakened the interconnections between the main piers and the buttress piers (lettered A,A in the figure) that they had to be strengthened and stiffened by underpinning before the construction of the main vaulting system could proceed. The basic intuitions showed what remedial action was needed, but could not anticipate the need.[6]

Growing understanding and its impact on design from the Renaissance to the eighteenth century

During the Renaissance and the following two centuries, there were slow changes in the design procedures inherited from the previous centuries – changes which laid the ground for more substantial changes in the nineteenth and twentieth centuries. They stemmed partly from a supplanting of the master mason or master carpenter brought up on the building site by others with a different background and hence a different approach, and partly from completely independent attempts to carry further the limited statics inherited from the Greeks and from Archimedes in particular. Initially the approach to design remained rooted in geometry and the two most basic intuitions. Key figures then were the humanist Alberti and the artist Brunelleschi.

Vitruvius had written the first known manual of building design in late Republican Rome, at a time when the use of concrete was only just beginning.[7] Alberti was the first to write a new counterpart, concerned rather more with general theory. For instance, he offered precise guidance, argued from first principles, on matters such as the best profile of an arch.[8] But, to do so, he envisaged the internal forces in purely geometric terms. He thus concluded that the full semicircle gave the strongest arch and saw in the uniform curvature of the hemispherical dome at all points and in all directions, a similar uniformity of internal compressions:

'. . . if the Work were inclined to ruinate, where should it begin, when the Joints of every Stone are directed to one Centre with equal Force and pressure?'

Brunelleschi was not a theorist like Alberti and the professional geometers of his day. But he did bring a fresh enquiring mind and a keen imagination to the problem of constructing the dome over the crossing of Florence Cathedral. His answer, as finally developed, was described in Chapter 7. In its own way, it was just as novel as the whole design of Hagia Sophia. How did he arrive at it?

Observation and the two basic intuitions would have shown him that a circular dome could be constructed, ring by ring, without needing support from centering as does an arch. Relatively simple and perfectly valid geometric reasoning would then have shown that a dome of octagonal plan could be similarly constructed without centering if there was always a strong enough continuous circular ring within the thickness at the working level and if the masonry between the inside octagonal face and this ring was adequately keyed into the ring to prevent it from dropping. The muscular intuition would have shown also that the ring at the working level would always be in compression as it played the part of keystone for the as-yet incomplete radial arches. Even the idea for the keying-in of the masonry between the inside face and the ring [7.16] could, like the concept of the voussoir arch, have been arrived at *ab initio* by purely spatial (or geometric) reasoning. But here I think that it is more likely that a hint was taken from another existing form which Brunelleschi could have observed in the ruins of ancient Rome.[9] This is the flat arch as constructed in brick-faced concrete from the early second century if not earlier. In this, most of the facing bricks were broken ones. But where they had fallen away or been stripped it can be seen that one in every four or five is a complete brick. It penetrated the amorphous concrete mass behind the facing and divided it into voussoir-shaped sections that were much less likely to fall away below the effective arch contained within the overall depth [3.10 and 16.2]. Brunelleschi's spiralling bands of bricks set on edge did just the same, and if the Roman system was the trigger for his imagination he was not just playing a variation on it but intuitively seeing in it a new possibility.

A revealing further light is thrown on the whole design procedure by the unusually full surviving record of the progress of construction. An initial detailed specification significantly ended by leaving open the decision about how to proceed at the upper levels 'because in building only practical experience will teach what should be done', and there were two substantial variations two years and six years later in the light of experience hitherto. The more closely one looks at the development of other innovative designs, the more typical does this pattern appear to have been. The chief way in which the progress in Florence was perhaps less typical is that no changes were enforced by unforeseen shortcomings in need of correction.

Structurally speaking, most design from the Renaissance to the seventeenth century did little more than play variations, usually being content to remain well below the ceiling of past achievement. The chief skill of the artist-designer, apart from a lively formal and plastic imagination, was in that aspect of design called, in Italian, *disegno* – the ability to convey form graphically, even three-dimensional form, with a new precision. While it produced some fine architecture it did not lead to many structural

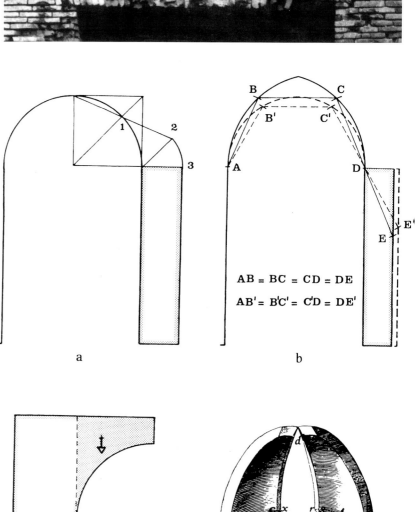

$$AB = BC = CD = DE$$
$$AB' = B'C' = C'D = DE'$$

a

b

c

d

Figure 16.2. Roman concrete flat arch from which many of the broken facing bricks have fallen to expose the full bricks that penetrate the concrete mass. Baths of Trajan, Rome. (Source: author)

Figure 16.3. (a) to (c) Geometric constructions to determine the size of pier required to support the thrusts of an arch or vault, based on drawings in S. Garcia, *Compendio de arquitectura*, c.1681; E. Derand, *L'architecture des voûtes ou l'art des traits et coupe des voûtes*, Paris, 1643; and S. Wren, *Parentalia or memoirs of the family of the Wrens*, London, 1750; (d) The likely mode of collapse of the dome of St Peter's deduced by the three mathematicians as interpreted by G. Poleni in *Memorie istoriche della gran cupola del Tempio Vaticano*, Padua, 1748. (Source: author)

advances. The usual way of dealing with structural problems like determining the dimensions of a pier adequate to support an arch without being overturned by its thrust continued to be to look directly at what had been done before. At most, attempts were made to deduce from this experience certain general rules of safe proportioning. Thus the observed need of flatter arches for more substantial piers was expressed in ingeniously devised geometric constructions like those illustrated at (a) and (b) in Figure 16.3.[10]

Wren criticized this approach and sought a more rational one in his unfinished and posthumously published *Second tract on architecture*.[11] But, lacking a better alternative, he was forced to fall back on Archimedes. The serious limitation here was that Archimedes dealt only with vertically acting weights. Thus Wren was able to determine the required depth of the pier only by restricting his analysis to the stability of half the arch or vault and its supporting pier. He considered these together to be a single rigid body that could fall only by pivoting about the inner edge of the pier [16.3 (c)] and that would be stable if the line of action of its total weight (passing through its centre of gravity) fell within the base of the pier. He therefore calculated the position of the combined centre of gravity and argued that, if each half arch plus its pier was stable, the whole must also be stable. The basic fallacy in this reasoning was, of course, that the halves do not, in normal masonry construction, behave as single rigid bodies and could not pivot in the manner assumed without immediately encountering the resistance of the opposite half, known as the crown thrust. If the arch voussoirs do not slip on one another, their joints actually tend to open at the haunches. This creates at least two bodies from each half, which are prevented from falling by the horizontal thrust. About such horizontal forces, the Archimedean theory had nothing to say and there was, as yet, no means of quantifying such forces.

Much more significant in the long run were a number of more disinterested enquiries that led to a progressive sharpening of the previous qualitative understandings of forces acting in any direction and the conditions of balance between them. Leonardo, for instance, tackled the problem by imagining weighted cords passing over frictionless pulleys to apply inclined forces to a balance-arm [16.4]. He also seems to be envisaging thrust-lines like those sketched in Figure 2.11 in some of his sketches of arches and in one of his more detailed sketches of a possible vault over the crossing of Milan Cathedral.[12] Wren, despite his earlier Archimedean aberration, seems to have been doing likewise in one set of sketches for the dome of St Paul's [16.5]. At about the same time, those responsible for the novel reinforcement of the architraves of the east colonnade of the Louvre [8.10] had tested a model to satisfy themselves of its adequacy, though no details are known of this test.[13]

The real breakthrough came at the end of the seventeenth century with Newton's completely general definition of a force acting in any direction and the formulation of equally general laws of static equilibrium between such forces. Throughout the eighteenth century, practice in the main was probably little touched by it. But the new laws were soon applied in fresh enquiries into the stability of the arch and it supports, supported not long afterwards by physical tests on small models.[14] Then, in 1742 and 1743, came the first recorded theoretic analyses of the stability of a standing structure – the dome of St Peter's. This had suffered extensive radial cracking of the usual kind which extended into the drum and had already been the subject of concern and dispute considerably earlier. Expert mathematicians were called in by Pope Benedict XIV. They examined the cracks [16.6], deduced from them a possible mode of collapse, and calculated from a simplified model of it [16.6, top right,

Figure 16.4. Studies of static equilibrium by Leonardo da Vinci copied from sketches in the Codice Atlantico. (Source: author)

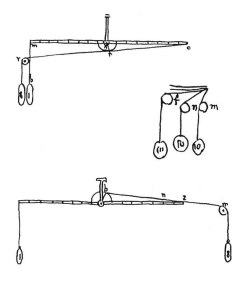

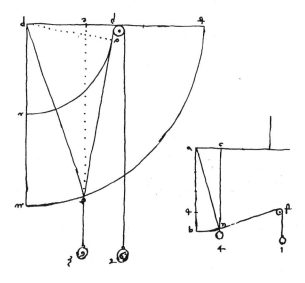

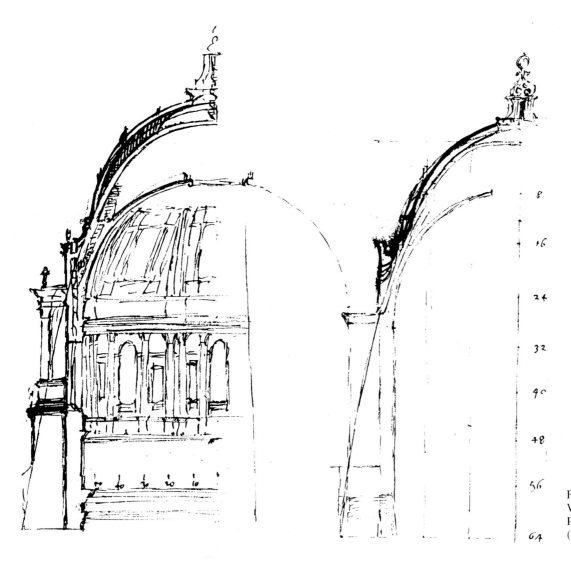

Figure 16.5. Studies by
Wren for the dome of St
Paul's, London.
(Source: British Museum)

and 16.3(d)] the circumferential tie force that
would be needed to prevent it.[15] A second
opinion was obtained from another mathe-
matician who analysed an alternative model, and
further ties were added in 1743 and 1744.[16]

It is doubtful whether there was a full understanding of the significance of the simplifying
assumptions made in either of these analyses.
But a highly significant step had been taken
towards providing a broader basis for future
design. Tests on the characteristic strengths of
iron and other materials had already been made
for a considerable time in pursuit of a more
disinterested curiosity, and these provided the
basis needed for translating calculated strength
requirements into appropriate cross-sections.[17]

Less than two decades later, similar analysis
was pressed into service for the first time to
justify Soufflot's highly innovative design of
what became the French Panthéon while it was
still in the early stages of construction. The
design had been criticized for the slenderness of
the piers that were to support the dome. It was
alleged that they departed too far from proven
practice in recent domed structures [12.31].[18]
The engineer Gauthey came to the rescue with a

calculation that demonstrated the stability of the
dome [16.7, top] and later, on the basis of
further analyses, he suggested that an even more
daring design would have been feasible [16.7,
bottom].[19] Problems had been experienced with
the piers, but these were shown to be the result
of poor construction which threw too high a
proportion of the load onto narrow bands of the
facings where the joints were much thinner than
in the interior.

Further developments
from the nineteenth century
to the mid-twentieth century

In the design of some arched and vaulted
masonry structures, it had thus become possible,
by the end of the eighteenth century, to depart
further from proven practice with a greater
assurance of success. Understanding of the
conditions of stability had grown enough to
provide a basis for design that was less closely
tied to knowledge of what had already been
built. This new freedom was nevertheless
limited to structures that had chiefly to support
their own dead weights and did so largely by a

Figure 16.6. The dome of
St Peter's, Rome, showing
cracking observed in 1742
and the basis of the
statical analysis of its
stability by the three
mathematicians. From T.
Le Seur, F. Jacquier, and
R. G. Boscovich, *Parere
di tre mattematici sopra i
danni, che si sono trovati
nella cupola de S. Pietro*,
Rome, 1743.
(Source: author)

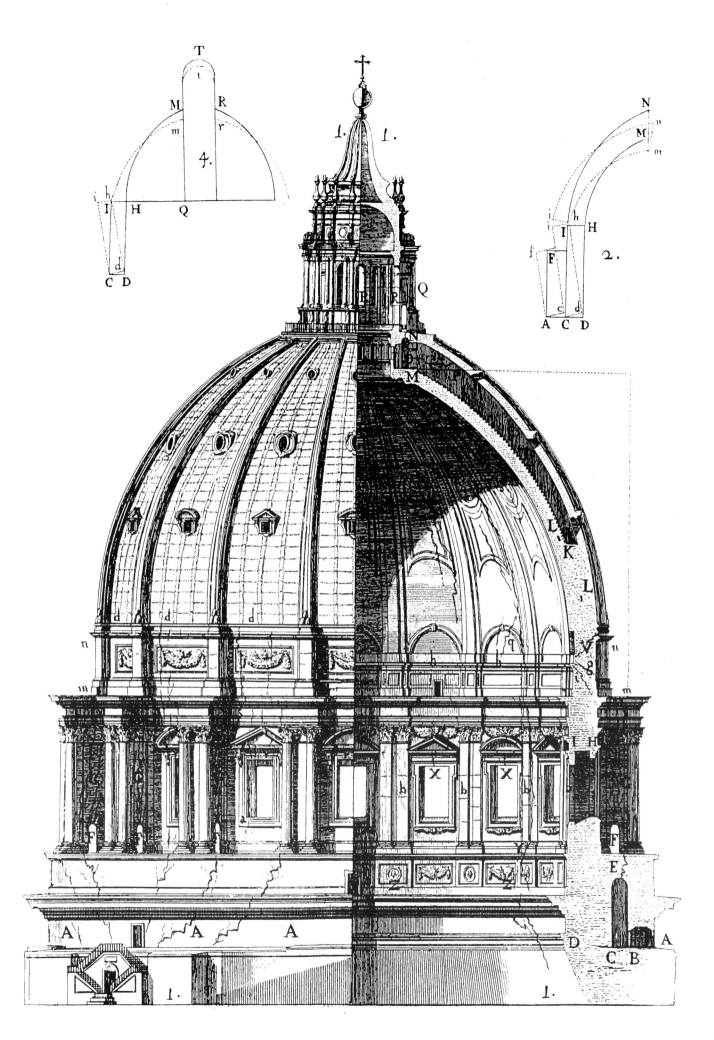

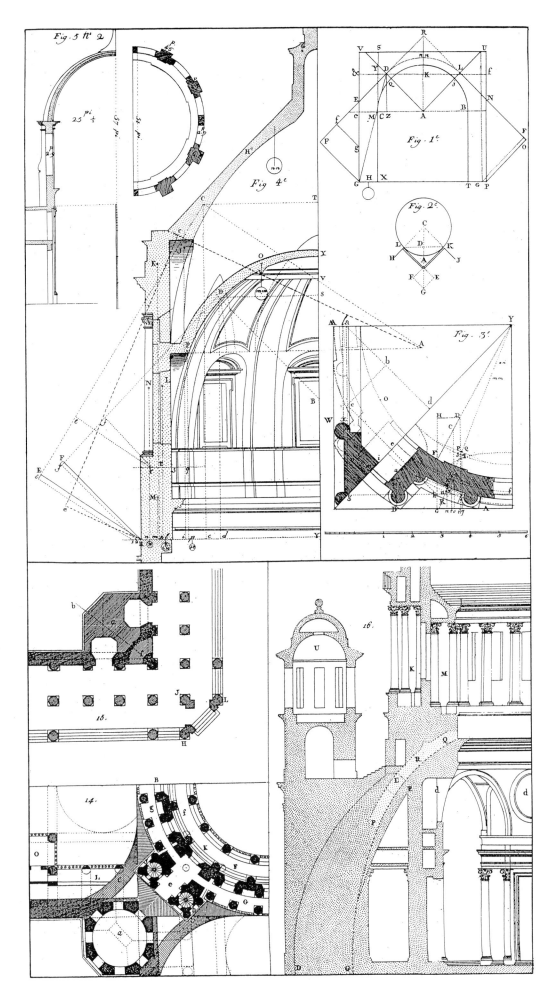

Figure 16.7. (top) Statical analysis of the dome of the French Pantheon. From E. M. Gauthey, *Mémoire sur l'application des principes de la méchanique à la construction des voûtes et des dômes*, Dijon, 1771; (bottom) a later more ambitious project based on a further analysis from idem, *Dissertation sur les degradations survenues aux piliers du dôme du Panthéon Françoise, et sur les moyens d'y remédier* Paris, 1799. (Source: author)

balanced opposition of internal compressions, aided at only a few discrete points by ties. As yet it did not extend to the design of other types of structure or to the proportioning of structural elements like beams which called for consideration of the internal stresses.

New forms such as iron beams were emerging, however, and loads other than self-weight were soon to become more important with the coming of railways and lighter types of bridge-span such as trusses and suspended spans. Especially in Britain and America, the early stages of development of these forms were largely empirical in much the same sense as most earlier development had been, but with a greater incentive to make the best use of material when more costly iron replaced timber or masonry. This called for efficient cross-sections. To check that these were adequate for the intended use, it became common practice to apply proof-loads to columns and beams and the links of suspension chains and the like, before they left the foundry. Even so, there were numerous and often costly failures in service where appreciation of the requirements was less precise than the measured strengths. The demands of industrial, and particularly railway, expansion made it imperative, if at all possible, to cut down these failures without slowing the rate of progress. This called for more precise understanding of the required strengths and stiffnesses and some ability to predict them in advance of manufacture and test.

Achieving this was not as easy as had been the application of simple statical theory to arches and domes. The ground had been prepared by earlier enquiries, particularly by people like Galileo into the bending of simple rectangular-section beams and by people like Hooke into the relationship between load and deformation or, as we should now say, between stress and strain.[20] But the new qualitative formulations of the conditions of equilibrium and the relationships between stress and strain had to be applied afresh to each new form and the results tested. This called for the replacement of the earlier step-by-step empiricism in actual construction by an approach in which deliberate testing played a central role, the tests being devised with the clear objective of carrying understanding further.[21]

The power of this new approach was triumphantly demonstrated by the design of the Conway and Britannia Tubular Bridges [5.7, 8.6 and 14.14]. These not only introduced a new form of built-up beam whose modern steel counterpart is of considerable importance today; they also, by this means and by means of the continuity given to the spans of the Britannia Bridge, achieved spans many times greater than those of any previous beam-bridges.

The crucial contribution to the design was the extensive series of tests by Fairbairn and Hodgkinson that have already been referred to in Chapter 8.[22] They began with a series of exploratory tests on small tubes of various cross-sections. These confirmed predictions that a rectangular cross-section was best, and they provided enough guidance, in the light of theoretical analyses of the results, for the design of a larger model tube, one-sixth of the intended final size. This model was tested six times to failure, being repaired after each of the earlier tests and modified in the light of the results to eliminate the principal weakness that had been disclosed. Figure 16.8 shows three cross-sections of the model after the second test, in which failure occurred by buckling of one of the as-yet-unstiffened webs. By the time that the last of these tests had been made, enough had been learnt to proceed to the detailed design of the single-span Conway tubes. Measurements on these when they had been erected provided confirmation of the calculated deflection. This was of considerable value when it came to designing the Britannia ones, since the procedure described in Chapter 4 for making all the spans there continuous made it necessary not only to give these tubes (like the Conway ones) initial cambers to compensate for the deflections but also to predetermine the amounts by which the outer ends of the outer ones should initially be raised above their final bearings.

Testing thus began with what amounted to a basic investigation into the structural characteristics of the proposed tubular-beam form. This was necessary partly because the tubular form was itself novel and partly also because wrought iron was to be used whereas most previous experience had been with cast iron. It proceeded to an investigation much more specifically directed to finding the best form for the spans of the two bridges, and concluded with measurements on the completed spans. Throughout, it was guided by Hodgkinson's theoretical analyses, as was the application of the results to the final designs.

In the later nineteenth century, purely theoretical developments began to influence design more widely. One of the more significant influences was that of the better qualitative insights gained into the behaviour of pin-jointed trusses. These led to the supplanting of a multiplicity of highly redundant designs of bridge truss such as those seen in Figures 9.7 (a) to (e) and 9.8 by simpler forms such as those in Figures 9.7 (f) and 9.9. A growing understanding of the effects of flexural continuity had less immediate impact because of the difficulties of computation when taking continuity into account. Indeed this growing understanding initially had the negative result of leading some designers to go to considerable trouble to avoid it by making joints truly pinned as noted in Chapter 9. Only when computation was simplified in the 1930s by ingenious procedures of successive approximation was it more fully exploited in the design of multi-storey frames.[23]

When reinforced concrete was introduced, calculations soon became essential to determine

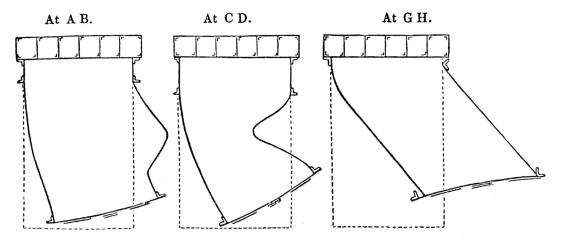

| At A B. | At C D. | At G H. |

Figure 16.8. Transverse sections of the one-sixth scale model of one of the tubes for the Conway and Britannia Bridges after the second test. From F. Clark and R. Stephenson, *The Britannia and Conway Tubular Bridges*, London, 1850. (Source: author)

the reinforcement needed. While the new material was regarded as little more than a straight substitute for timber or iron in beams and columns, they now presented little difficulty in themselves. But some assumptions had to be made about the relative deformations under load of the concrete and steel. Initially it was wrongly assumed, on the basis of too limited tests, that both behaved elastically and that a single constant relationship was adequate. However, a few designers, notably Maillart and Freyssinet, acquired from their own observations better intuitive understandings and saw that the calculations were open to question. Maillart undertook his own tests to substantiate his ideas for slabs [8.18 (c) and (d)], and had a long series of tests made on his bridges as he progressively developed his ideas for trough-shaped-arch forms [6.18] and deck-stiffened forms [14.8].[24] Freyssinet was led by his observations to question the assumed constant relat-

ionship between elasticities and, guided also by numerous tests in his own laboratories, conceived and developed the idea of prestressing.[25]

More recently, this pattern has tended to be repeated with some shifts in the emphases placed on different kinds of testing, on the one hand, and on the purely analytical application of basic insights to more complex modes of structural action, on the other. Some of these shifts have reflected improvements in observational and other experimental techniques or in methods of computation. Others have stemmed from a recognition that much of the necessary investigation could more profitably be carried out as a continuing programme of research than as an *ad-hoc* means of solving the problems thrown up either by new developments or in the course of designing specific structures. This recognition has led, for instance, not only to extensive programmes of theoretical research and associated laboratory tests on different types of

Figure 16.9 Loading test on the Moat Street Flyover, Coventry. (Source: author)

structural element and different kinds of structural behaviour [2.18] but also to tests on a wider range of completed structures in service, or under loadings simulating those of normal service [16.9].[26] Though tests of this latter kind come too late to contribute to the designs of the structures on which they are made, they can play, and have played, an important role in disclosing any shortcomings in the understanding on which the designs were based.

Other advances have come, as in the past, through failures caused by shortcomings in the current design procedures. Of course, some failures occur because of human shortcomings, and failures due to shortcomings in design procedures have become less frequent than they were in the past. But, as designers have continued to exploit new possibilities or to carry existing ones closer to the limit than before, some such failures have continued to occur, and, regrettable though they have been, their value as indicators of a need for change in future designs has been greatly increased by the detailed investigations into their causes that have followed. These have elucidated far more effectively than before the kind of change that is needed and have made further additions to

Figure 16.10 Collapse in a high wind of a reinforced-concrete cooling-tower at Ferrybridge. (Source: West Park Studio)

general understanding. The failure illustrated in Figure 16.10 was not alone in proving to have resulted from ignoring an aspect of behaviour that had not hitherto proved critical in similar structures. The progressive collapse illustrated in Figure 5.14 and the collapse that followed on the severe oscillations of the deck of the bridge seen in Figure 2.3 were found to have resulted from ignoring other critical aspects, as have numerous earlier failures.[27] Elucidation of their causes led, each time, to revised procedures.

The use of physical models and computers as aids to design in the later twentieth century

By mid-century the pace of innovation was still growing and had come to be limited chiefly by the complexities of the calculations needed to explore new possibilities and assess the safety of new designs. Tests on physical models and prototypes provided an alternative to calculation – simple models in the earlier stages of design and more detailed ones to check or supplement the analysis or provide basic date for it such as likely wind loadings.[28]

Gaudi's even earlier use of hanging-chain models [5.3] has already been referred to. These established a pattern of purely tensile equilibrium which could be translated directly into a design for a purely compressive structure of much larger size because stresses were not a critical factor. With later structures in which stresses and deformations under load are more important – structures like shell roofs and arched dams, and multi-storey ones like the National Westminster Tower – it has been less easy to design models whose behaviour would be adequately representative of the full-scale behaviour and it has become necessary to take into account the greatly reduced scale when interpreting stresses and deformations. Even more allowance for the consequences of a reduced scale has been necessary when assessing the dynamic behaviour of a model subjected to a simulated earthquake on a shaking table or simulated wind in a wind tunnel, and when estimating full-scale wind loads from tests of the latter kind.

The most extensive recent use of models in the conceptual stages of design (as distinct from the later stages of detailed dimensioning and assessment) has possibly been in the choice of form for three-dimensional tension structures and some shell and timber-grid counterparts. In his early explorations, as noted in chapter 7, Otto used soap films, which have the useful property of automatically assuming ideal configurations with uniform surface tension for any chosen boundary.[29] For the further development of designs like those for the Munich Olympic tents [13.25] and the Mannheim Bundesgartenschau [7.29] suitably loaded wire models, not very different from Gaudi's, were used. Exploratory models are now commonly made of stretch fabric which has similar characteristics to the fabrics that have largely superseded cable nets in full-scale construction.

Physical models nevertheless have their limitations. Apart from the time and cost involved in making them, their reduced scale inevitably reduces accuracy. This can be particularly important where precise dimensioning of the full-scale structure is important, as it already was with the large cable nets at Munich and continues to be with all large fabric canopies. Without computer modelling to assist in the final development of the Munich and Mannheim designs and those of the large membrane roofs described earlier, sufficiently precise dimensioning would not have been possible.[30] Before this it had already been equally necessary in the detailed design of the Sydney Opera House roofs.

When the Sydney and Munich structures were being designed, computer programs to analyse the proposed forms and provide sufficiently precise geometric coordinates had to be written for the purpose. A decade later there was already a wide choice of programs for many purposes, mostly using the analytical fiction of a division of the real continuous structure (or the part of interest) into small discrete 'finite elements' connected to one another at only a few points or nodes. Elements of different shapes and properties had been devised and their number was growing. On the face of it, these programs reduced the main tasks of the user to selecting suitable elements and a suitable initial mesh to represent the real structure for the purpose in hand and then ascribing suitable properties to the elements to represent the intended materials. The program then calculated the displacements of the nodes for the loading or other conditions imposed on the structure. From these displacements, overall deflections, internal stresses and other types of response that were of interest were then calculated and displayed, different levels of stress (averaged over each element) often being displayed as different colours. But it had to be remembered that these were valid only for the model that was analysed and that this was only a simplified representation of the real structure. Considerable skill was needed, both in devising a model that was adequately representative and in assessing the relevance of its calculated behaviour to that to be expected of the real structure. This skill could be based only on prior understanding of the likely behaviour, and that remains the position.[31]

Of course, no analysis can create a design: only the designer's choices can do this. Theoretical analysis, whether by computer or other means, is merely a way of testing some of the consequences of the choices made – either as an alternative to physical testing or as a complement to it. To test the adequacy of the crucial joint between the individual cylindrical legs of the main columns of the Hongkong and Shanghai Bank and the Vierendeel cross-braces, two typical joints were loaded back-to-back in a test frame until they failed. Loads, stresses and deflections were measured. To corroborate and amplify the results of the test, a finite-element model was subjected to similar load in a computer. Figure 16.11 shows the chosen mesh of elements and the computed deflection. To test the overall behaviour of the support system, one half of the complete system of masts and trusses was also modelled as in Figure 16.12(b) and notionally subjected to side loads, leading to the deflected profiles seen at (a) and (c) in the same figure.[32]

The speed of modern computers does, however, make them valuable aids also in the earlier stages of design. It allows early concepts to be tested without the delays that arise with physical testing or those that previously arose with calculation by hand if similar accuracy was sought. Thus they have made it relatively easy to modify a first design and, without starting afresh, to explore alternatives in order to converge on a better one. This has proved particularly valuable where the limits of feasibility are tightest, as they are, for instance, with doubly curved tensile membranes. The computer has

Figure 16.11. Finite-element model of the joint between the legs of the masts of the Hongkong and Shanghai Bank and their cross braces, showing (greatly magnified) its computed deformation under side load. (Source: Ove Arup & Partners, *Arup Journal*)

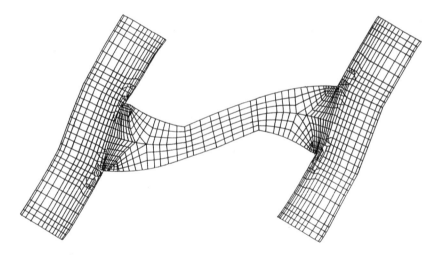

the further advantage, which is particularly valuable with such three-dimensional forms, of being able to present an instantaneous image of the form from any desired viewpoint [16.13(c)]. Although the human mind can be even more versatile in this respect (as in other respects) and can juggle more readily with different views, it is not capable of the same precision unaided; and not all designers have the same powers of visualization.

Computer-aided conceptual design has, indeed, become indispensable for all but the simplest tensile membranes and has proved valuable also in exploring different configurations of space frame.[33] Yet even here it is not, and never can be, totally automatic. For instance, the designer of the membrane must first choose the boundaries and intermediate points or lines of support and any desired limitations on stresses and must specify the types of finite element to be used in the analysis. The program can then display possible membrane surfaces and the stresses within the membrane for the anticipated loadings. By changing some of the boundaries or other supports, it is possible to arrive at an acceptable overall geometry with acceptable stress levels and at corresponding cutting patterns for the fabric, taking its weave and different responses to stress in different directions into account. Figure 16.13(a) shows the mesh of elements selected for analysing the roof that is shown at (c) after several adjustments of the mast positions, inclinations and heights, while (b) shows a computed distribution of stresses over half the surface due to wind uplift. Here the principal stresses are displayed within each element as arrows that indicate both magnitude and direction.

Changes and underlying continuities

As just outlined, these more recent nineteenth- and twentieth-century procedures must seem to bear little relationship to the procedures followed in designing earlier structures. They did indeed differ considerably in important ways.

Earlier procedures aimed directly at finding a form that would be stable under its own weight and a few other (usually far less important) loads such as the wind. Previous experience with similar forms showed that certain proportions had been adequate. Direct comparison with these usually furnished the best criterion for assessing the design. In the approach that became usual during the nineteenth century and continues to be followed today, the strengths, deflections, and similar characteristics measured in proof tests, estimated from tests on models, or calculated by whatever means, furnish in themselves no such criterion. Acceptable values have to be established in some other way. Moreover, they have to be established at levels that allow not only for any doubts about the conformity of what gets built to what is specified but also for inevitable inaccuracies in the results of the tests or calculations, for all ways in which these tests or calculations may not truly mirror the behaviour of the real structure, and for the many uncertainties about such things as actual future loads.[34]

To these changes we might add that design of any major project has now become a much more collaborative matter. The days of the

Figure 16.12. (b) Finite element model of one half of the support system of braced masts and suspension trusses of the Hongkong and Shanghai Bank; (a) and (c): Computed deflections (again greatly magnified) in the two principal directions under maximum anticipated wind load. (Source: Ove Arup & Partners, *Arup Journal*)

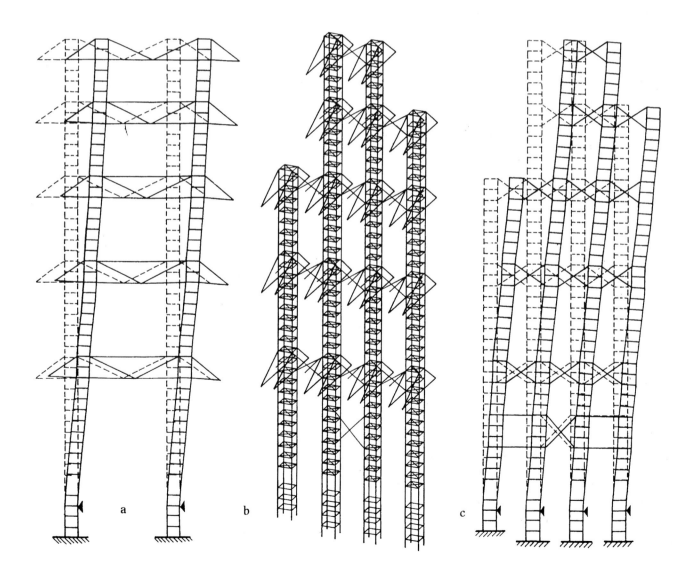

a   b   c

a

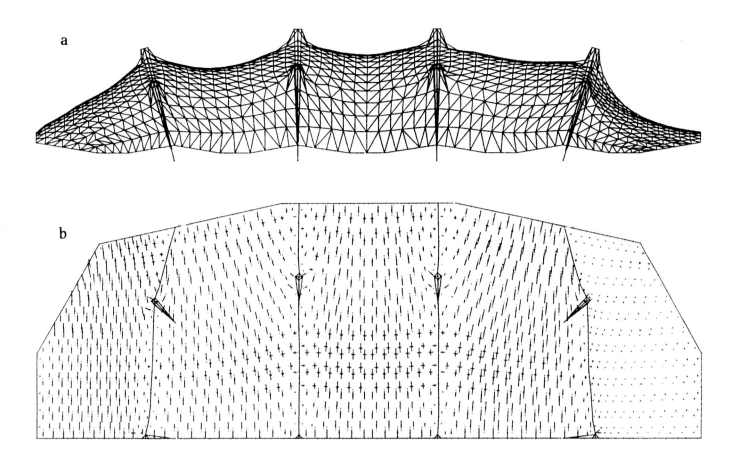

b

c

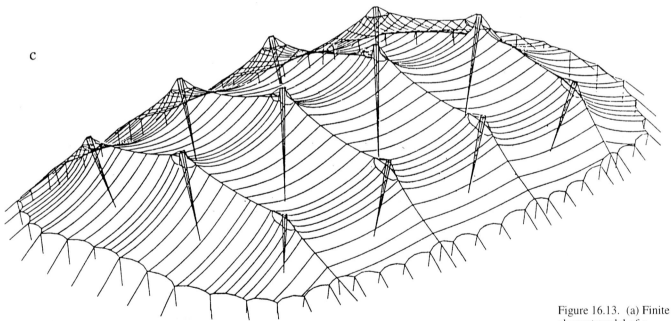

Figure 16.13. (a) Finite element model of a transportable tent covering an area of 95 × 130 m.; (b) Computed stresses under wind uplift; (c) General view of the final design. (Source: Professor Michael Barnes)

single controlling master are largely past, except for some projects that can be clearly defined in almost wholly structural terms. The skilled engineer may remain in control for large bridges, with collaborating experts for some details and possibly a collaborating architect to advise on other aspects of detailing that can make or mar the final appearance. For most buildings the architect is likely to be in control. Herein lie certain dangers if the part played by the structure and the range of structural possibilities are not fully appreciated and preconceived ideas based on inadequate structural understanding are pressed too hard. Examples of unfortunate consequences are not difficult to find. Several of Saarinen's buildings and Utzon's Sydney Opera House could have been among them, had they not been redeemed by Saarinen's and Utzon's mastery as architects and the ways in which their engineer consultants accepted and rose to the challenges.[35]

Underlying these changes, however, there are fundamental continuities. Two of them may be summed up as continuing needs for judgement and for imaginative input.

In the earlier procedures, judgement was called for in the choice of suitable precedents to guide the new design and in deciding which characteristics were important and where changes might be possible or advisable. The counterpart in the later procedures is a need to decide which aspects of the structure's behaviour may be critical and against what criterion each should be checked. Its importance is underlined by the failures just referred to that resulted largely from ignoring some critical aspect. It likewise calls for judgement, for which the tests and calculations themselves give no guidance. In the nineteenth century, this

judgement was usually left to the individual designer. More recently, in an environment of a much reduced tolerance of failures, individual judgement has increasingly been replaced for much design by the collective judgement of a number of experienced designers and other experts as set down in Codes of Practice, Building Codes, and similar directives and advisory documents. But when breaking new ground, the judgement must still be made afresh.

The imaginative input of some new concept, however vague, has always been the essential starting point of new design. Without it, development would be as much at the mercy of pure chance as, over countless generations, we may assume the development of birds' nests and similar forms to have been.

In the past, this concept had to arise directly as a three-dimensional form to be built in the particular way from the particular materials with which the designer was familiar. This had the advantage that there was no more differentiation between what we might now call the structure and its visible embodiment than there was in the slow evolution of the works of the traditional craftsman – something that was possible because of the close relationship between the two in actuality. The geometric disciplines of some Gothic designs were one expression of this [16.14]. The works themselves, illustrated on earlier pages, were its fruits.

Now, with wide choices of materials, ways of using them and construction methods, and with different possible imposed loads and different acceptable levels of response to them, the concept must pay much more attention to all of these aspects. To do this, it must be partly an abstract concept of a system of forces in equilibrium under the varying future circumstances. Sometimes, as with many modern tensile structures, shells and space frames, this abstract concept will imply a specific physical embodiment. At others it will leave this to be developed more freely. The earlier unity of design is then less easily attained. But the bridges of Maillart, Freyssinet, Leonhardt and Arup and other more recent works show that it still can be attained, just as it was in Candela's shells and Otto's tents.

A further continuity underlies the continuing need for judgement and imaginative input. Neither would be possible without the intuitive understandings referred to at the start of this chapter. Those described as the basic spatial and muscular intuitions have remained basic throughout. They have not been superseded but deepened and made more precise by growths in understanding and wider experience. The skilled designer can now visualize, with some feeling for relative magnitudes, not only the thrusts and counter-thrusts in a vaulted masonry structure – as the best designers of some earlier periods would have done – but also patterns of internal stresses in equilibrium with the acting loads such as those sketched in Figure 5.5 and more

Figure 16.14. Transverse section of Milan Cathedral showing an elaboration of one proposed geometric basis for the design. From the Cesariano edition of Vitruvius, *De architectura*, Como, 1521. (Source: author)

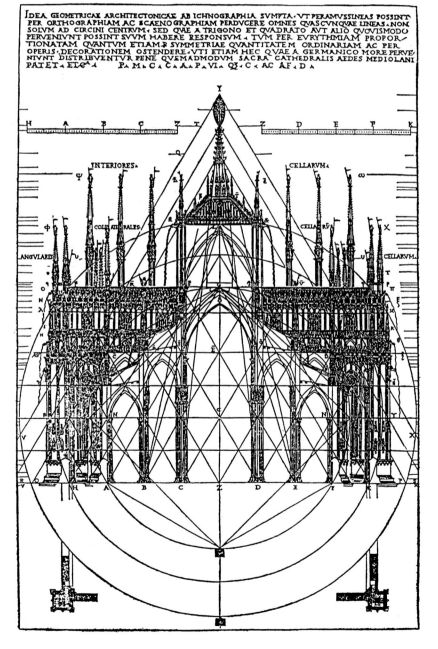

complex patterns of compression, tension, bending and perhaps torsion, in complete structures of many kinds. He or she can also visualize the associated deformations and ways of modifying behaviour by applying internal forces or changing relative stiffnesses. At the same time, what have been called the secondary intuitions of the characteristic behaviours of particular types of structure have also become deeper and more precise, as well as far more wide-ranging, in step with the continued increase in the range of structural types. But the essential role of all these intuitive understandings has remained unchanged. It is they, in some mysterious way, that generate the ideas, while the designer necessarily focuses his conscious attention on the requirements to be met.[36]

Thus the best ideas have always come to designers who, at a deep intuitive level accessible without conscious concentration, have understood best the structural possibilities, appreciated best the characteristics of the available materials and other resources, and appreciated best the achievements of others who have faced similar or related challenges. Much of that understanding and appreciation can come only from personal experience and critical observation. But it has been one objective of this book to convey some of it, and to provide a basis for aspiring designers to build upon.

Glossary of structural terms

Technical terms have usually been defined or illustrated as they were introduced. For convenience of reference the commoner ones relating to structural forms and structural actions are redefined here. For terms relating to the materials of construction the reader is referred to Chapter 3, and for purely architectural terms to one of the many existing glossaries or dictionaries, such as Fleming, J., Honour, H., and Pevsner, N., *The Penguin Dictionary of Architecture*, Harmondsworth, 4th edn 1991.

ACTION (or ACTIVE LOAD). That LOAD or set of loads acting on a STRUCTURE or STRUCTURAL ELEMENT that is regarded as the primary one. Commonly the DEAD LOAD and IMPOSED LOADS that the structure or element is designed to carry as distinct from the counterbalancing set of loads exerted by the supports or other adjacent elements. For a wider sense see STRUCTURAL ACTION

AMPLITUDE. The maximum value of a PERIODICally varying quantity. Usually the maximum displacement to either side of a mean position.

ANTICLASTIC (referring to a surface). Having curvatures in opposite senses (concave and convex) in different directions at all points.

ARCADE. A sequence of ARCHES spanning between a series of PIERS or COLUMNS.

ARCH. A STRUCTURAL ELEMENT capable of spanning a horizontal gap and of carrying its own weight and other loads wholly or largely by internal COMPRESSION. Usually it has a curved SOFFIT [5.5(j)]. See also CHAIN ARCH, FALSE ARCH, VOUSSOIR ARCH.

ARCHITRAVE. The lowest member of an ENTABLATURE, consisting of the principal BEAMS (or LINTELS) spanning between the COLUMNS.

ASHLAR. Blocks of stone with finished squared matching faces.

BARREL VAULT. A VAULT, or SHELL, having, in cross-section, the profile of a curved ARCH but extended laterally with the same cross-section throughout [5.5 (h, i)].

BEAM. A STRUCTURAL ELEMENT capable, like an ARCH or CATENARY, of spanning a horizontal gap, but acting as a combination of the two. Usually straight horizontally, though it may vary in depth according to the expected variations in BENDING MOMENT [5.5(q)]. See also BOX BEAM, CANTILEVER.

BEARING WALL. A wall that serves as a vertical support for loads other than its self-weight.

BENDING. A type of DEFORMATION in which initially parallel cross-sections of a STRUCTURAL ELEMENT become inclined towards one another; and any STRUCTURAL ACTION that leads to this type of deformation [2.5 (c, d)].

BENDING MOMENT. An external MOMENT leading to the BENDING of a STRUCTURAL ELEMENT, and the COUPLE at any cross-section due to the internal TENSIONS and COMPRESSIONS that are called into play to resist the external moment.

BENT. A part of a larger COLUMN-and-BEAM FRAME carried by a single row of COLUMNS. Usually, in a framed building, one of a number of similar or identical bents constructed parallel to one another and linked by secondary BEAMS.

BOX BEAM, BOX GIRDER. A BEAM (or GIRDER) of hollow rectangular or other closed cross-section, usually with DIAPHRAGMS at intervals which divide it internally into a series of more or less enclosed boxes.

BOX FRAME. A FRAME with stiff box-like compartments, usually formed by interconnected BEARING WALLS and SLABS of reinforced concrete, though the term is sometimes also used for COLUMN-and-BEAM FRAMES stiffened by infilling SHEAR WALLS and SLABS.

BRACE, BRACING. A secondary STRUCTURAL ELEMENT or set of elements that serves to maintain the configuration of the primary elements of a structure in the face of lateral loads or disturbances and thereby stiffens and strengthens it.

BRACED FRAME. A FRAME, particularly one composed of COLUMNS and BEAMS, to which overall rigidity is imparted wholly or largely by BRACING rather than by RIGID JOINTS or full TRIANGULATION.

BRITTLE. Prone to sudden disruption with relatively little prior deformation as soon as the ultimate strength is reached. See, for instance, the stress-strain curve for cast iron in Figure 3.2.

BUCKLE. To bend suddenly out of the original line or surface under the action of a slowly increasing primary COMPRESSION that is not perfectly axial; or, more generally, to suffer a rapid loss of STIFFNESS under this compression, owing to increases in secondary BENDING MOMENTS due to its non-axiality.

BUOYANT FOUNDATION. A FOUNDATION involving substantial excavation of the original ground and the retention of a large enough void or basement below the original ground level to compensate for much of the weight of the structure erected above.

BUTTRESS. A vertical or inclined STRUCTURAL ELEMENT designed to resist an outward THRUST or provide lateral or transverse STABILITY. It may be either free-standing or simply a projection from a WALL. See also FLYING BUTTRESS.

CABLE. A slender flexible STRUCTURAL ELEMENT made of a number of steel wires or fibres of other materials. These are laid alongside one another to build up the desired cross section. They may be twisted together in STRANDS or laid parallel to one another and bound by circumferential wrapping or encased in a matrix. Greater flexibility is obtained by twisting, at the cost of greater extension under load. In the locked coil cable, this extension is reduced by shaping the individual wires to pack as closely together as possible in the unloaded state.

CABLE NETWORK. A network of hanging CABLES, analogous structurally to a similarly supported continuous MEMBRANE of the same surface geometry. It usually consists of two families of cables intersecting roughly at right angles.

CAISSON. A continuous WALL within which excavation may be made for a FOUNDATION in poor ground or under water. It differs from a COFFERDAM in being fabricated at or near ground level, often at another site, and then progressively sunk into the ground by various expedients as excavation proceeds, and in subsequently remaining in place as a part of the foundation that is constructed within it. For shallow excavation it may be open at the top. For greater depths it may be closed at the top and internally pressurised to exclude water and is then known as a pneumatic caisson.

CAISSON FOUNDATION. A FOUNDATION constructed using a CAISSON and usually of greater width or diameter than a CYLINDER FOUNDATION and often multi-cellular internally

CANTILEVER. A BEAM or SLAB supported at one end or edge only, projecting out from this support, and restrained from rotating about it. Or, as a verb, to project outwards in this manner.

CATENARY. Strictly, the curve assumed by a chain or cable uniformly loaded along its length and freely suspended from two horizontally separated points. But used here in the sense of *catena* to denote the hanging chain or cable itself as a STRUCTURAL ELEMENT capable (as the counterpart of the ARCH) of spanning a horizontal gap and carrying its own weight and other loads wholly by internal TENSION [5.5(f)].

CENTRING. The temporary supports on which an ARCH or VAULT is frequently constructed or on which the FORMWORK for it is carried. It is usually of timber or, today, steel. It may itself be propped up directly from the ground or span independently between the final SPRINGINGS of the arch or vault, when it is known as flying centring.

CHAIN ARCH. The name given here to a stone ARCH deliberately constructed in a manner that gives it the flexibility of a notionally inverted hanging chain or CATENARY.

CHORD. Of a TRUSS: one of the outer (usually top and bottom) members. It is analogous to the FLANGE of a BEAM.

CLOISTER VAULT. A VAULT approximating to a DOME, but polygonal rather than circular in plan. Also referred to as a domical vault.

COFFERDAM. A temporary wall of earth, timber, steel, or concrete, built *IN SITU* around an area to be excavated, and serving to exclude water, or to help to exclude it, as the excavation proceeds.

COLONNADE. A series of COLUMNS in line.

COLUMN. A vertical STRUCTURAL ELEMENT of relatively small cross-section capable of serving as a support and acting wholly or primarily in COMPRESSION [5.5(e)]. See also PIER.

COMPONENT. Of a STRUCTURAL ACTION: that lesser action that would be equivalent to it in another direction or another plane and would act solely in that other direction or plane. See also STRUCTURAL COMPONENT.

COMPRESSION. Contraction and the STRUCTURAL ACTION that leads directly to it [2.5(b)].

CONTINUOUS. Of a STRUCTURAL ELEMENT such as a BEAM, COLUMN, or SLAB that has more than the minimum number of supports or attachments to keep it in place: capable of transmitting BENDING MOMENTS from one side of an intermediate support or attachment to the other. Such moments are thereby shared by all SPANS or lengths and the STRUCTURAL ACTION is made STATICALLY INDETERMINATE by the continuities or, from another point of view, by the additional supports or attachments.

CORBEL. A block of stone or other material projecting from a PIER or WALL and acting as a short CANTILEVER.

COUPLE. 1. A pair of equal but oppositely directed FORCES tending to cause rotation.

2. A pair of RAFTERS meeting as an inverted V.

CRACK. To respond to TENSION by a local or more extensive cleavage, or the cleavage itself.

CROWN. Of an ARCH or VAULT: the highest part.

CRUCK. An inclined timber BEAM structurally analogous to a principal RAFTER but serving (always as one of a pair) as a principal framing element of one of the simplest types of hut, house, or barn. Unlike a rafter, it extends as a single piece throughout the height of the building, and usually has a slight upward curve.

CRUSH. To respond to a primary COMPRESSION by multiple disruption brought about by secondary TENSIONS at right angles to the compression.

CYLINDER FOUNDATION. A FOUNDATION similar to a PILED FOUNDATION but having a smaller number of vertical supports of larger diameter, up to or exceeding one metre.

DAMPER. A device intended to dissipate energy. See also TUNED-MASS DAMPER.

DAMPING. Dissipation of the energy of disturbing FORCES, particularly PERIODIC ones, by turning it into a form of energy that is not associated with structural DEFORMATION or displacement.

DEAD LOAD. SELF-WEIGHT plus the weight of other permanent construction carried by the STRUCTURAL ELEMENT or STRUCTURE under consideration.

DEFORMATION. Change in shape or dimensions as a result of or as part of a STRUCTURAL ACTION.

DENSITY. The MASS of a unit volume of a material. More loosely, the weight of a unit volume.

DIAPHRAGM. A transverse plate or similar member in a BOX BEAM, BOX GIRDER, or COLUMN of similar form introduced for the purpose of stiffening the sides where these are in COMPRESSION and, more generally, of maintaining the shape of the cross-section and the uniformity of the internal distribution of LOADS and STRESSES.

DODECAHEDRON. A plane-faced three-dimensional body having twelve faces. In the regular dodecahedron these faces are all regular pentagons, having five equal sides [1.5].

DOME. A spanning and space-enclosing STRUCTURAL ELEMENT circular in plan and commonly hemispherical or nearly so in total form. To distinguish the thin shell-like form from the more massive typical masonry form, the shell-like form is referred to here as a domical shell.

DOMICAL VAULT. See CLOISTER VAULT.

DOUBLY CURVED. Of a line: not traceable on a plane but only on a curved surface.

Of a surface: not capable, without stretching, of being unrolled to lie flat.

DRUM. A cylindrical WALL, or one polygonal in plan, interposed between a DOME or CLOISTER VAULT and a system of supporting PENDENTIVES, SQUINCHES, or ARCHES, or supports of some other kind.

DUCTILE. Capable of carrying loads at or near the ULTIMATE STRENGTH over a wide range of DEFORMATION rather than failing in a BRITTLE manner. (See, for instance, the stress strain curve for mild steel in figure 3.2). Hence also ductility.

DYNAMIC LOAD. A LOAD that is applied or changes sufficiently rapidly to bring into play significant INERTIAL RESISTANCES and thus to have an effect markedly different from that of an otherwise identical load applied more slowly and acting more continuously.

ELASTIC. Capable of returning to the original shape and dimensions when a LOAD is removed. A property well exemplified by a spring provided that it is not overloaded. Hence also ELASTICITY.

ELASTIC DEFORMATION. That part of a DEFORMATION that disappears when the LOAD causing it is removed.

ELEMENT. See STRUCTURAL ELEMENT.

ENERGY. Capacity for doing mechanical work such as moving or deforming a STRUCTURAL ELEMENT against internal or external resistances. Energy is always conserved but may either be stored within the element and its supports in a form capable of returning them to their original state or dissipated in another form such as heat. In the former case it is known as elastic strain energy when it is associated with DEFORMATION, and as potential energy when it is associated with bodily displacement.

ENTABLATURE. The complete set of horizontal members surmounting a COLONNADE in a Classical temple and related derivative forms.

EQUILIBRIUM. A state of balance of all the LOADS acting on a STRUCTURE or any part of it. It is described as static if no DYNAMIC or INERTIAL loads are involved, and is then a state of rest. Otherwise it is dynamic and entails continually changing deformations and displacements. When unqualified it should usually be read as denoting the static state. It may also be either STABLE or UNSTABLE.

EXTRADOS. The upper (usually convex) surface of an ARCH.

FALSE ARCH. A STRUCTURAL ELEMENT spanning a horizontal gap and having the upwardly curved or pointed SOFFIT typical of most ARCHes, but constructed by progressive CANTILEVERing or CORBELling from the two sides with horizontal joints [6.3, 6.4]. Similarly, false vault.

FAN VAULT. A particular form of RIBBED VAULT in which numerous RIBS of identical profile radiate from each support.

FATIGUE. Failure by the slow development of initially minute CRACKS as a result of a very large number of repeated applications of LOADS that, singly, would be almost innocuous.

FATIGUE STRENGTH. The maximum LOAD that can be sustained without failure when repeatedly applied a given number of times. It is usually also necessary to specify the minimum load in each loading cycle.

FLANGE. Of a beam or column of stemmed or box-shaped cross-section: a projection from an extremity of the stem or from a side at right angles to the major BENDING ACTION, and acting in almost pure TENSION or COMPRESSION to resist this.

FLYING BUTTRESS. A free-standing inclined STRUT or prop (often arched) acting as a BUTTRESS and commonly spanning from, or slightly above, the SPRINGINGS of a VAULT to near the top of a free-standing outer PIER.

FLYING CENTRING. See CENTRING.

FOLDED PLATE. A STRUCTURAL ELEMENT akin to a BARREL VAULT supported to span longitudinally [5.5(h)], but with a V-shaped or folded (rather than curved) cross-section.

FORCE. A linear STRUCTURAL ACTION (usually the notional concentrated equivalent of a more dispersed one) that tends to initiate motion in, or change the state of motion of, any body on which it acts, or that results directly from an imposed change in the state of rest or motion of the body.

FORMWORK. The temporary mould of timber or metal boards or sheets that gives wet concrete its desired form and temporary local support until it has hardened sufficiently to no longer need this.

FOUNDATION. That part of a STRUCTURE that meets the ground and through which all LOADS are transferred to it.

FRAME. The most general term for an assembly of interconnected linear or planar STRUCTURAL ELEMENTs designed to serve a primary load-bearing function. In a building it is usually clothed or infilled to some extent by secondary elements that play little or no part in the primary STRUCTURAL ACTION but may give useful added stiffness. For particular types of frame see BOX FRAME, BRACED FRAME, PORTAL FRAME, RIGID JOINTED FRAME, SPACE FRAME, TRUSS.

FULCRUM. A support about which free rotation is possible.

FUNICULAR. Composed of ropes or cables carrying suspended weights or having a configuration similar to that which would be assumed by such ropes or cables.

GEODESIC. A name given to a type of framed DOME described in Chapter 9. The name derives from a word denoting the shortest possible line between two points on a surface, which, on a sphere, is part of a great circle.

GIRDER. A term used almost interchangeably with BEAM, but usually restricted to the larger sizes and, in steel, to those built up from smaller sections. See also BOX GIRDER.

GRID. An open STRUCTURAL FORM acting broadly in the manner of a slab, but consisting only of two, or sometimes three, intersecting and interconnected parallel sets of beams or plane trusses [5.11 (j, k. l)].

GRILLAGE FOUNDATION. A FOUNDATION in which heavier and more concentrated load than can be sustained by a simple SPREAD FOOTING is spread more widely by a grillage of superimposed layers of beams, with the beams of successive layers laid at right angles to one another.

GROIN. The curving line of intersection of the SOFFITS of two adjacent sections of a GROINED VAULT.

GROINED VAULT. A VAULT formed, notionally, by the intersection of two or more BARREL VAULTS and the omission of all those parts that would lie below part of another vault.

GROUT. To fill joints or other voids by means of a fluid mortar, either simply poured in under gravity or forced in under pressure.

HANGER. A TIE serving for vertical suspension.

HAUNCH. Of an ARCH: that part extending some distance above the springing at each side.

HINGE. See PINNED JOINT.

HYPERBOLIC PARABOLOID. A surface of ANTICLASTIC curvature that may be generated by moving one straight line along two others placed skew to one another and some distance apart, or a SHELL having this surface geometry. The name derives from the fact that vertical cross-sections of maximum curvature are upward- or downward-curved parabolas and horizontal sections are paired branches of hyperbolas.

ICOSAHEDRON. A plane-faced three-dimensional body having twenty faces. In the regular icosahedron, these faces are all equilateral triangles.

IMPOSED LOAD. The LOAD imposed on a STRUCTURAL ELEMENT or STRUCTURE by its users and environment. It is sometimes taken as excluding loads due to the wind, but is here taken as including these. Hence it is the total ACTIVE LOAD less the DEAD LOAD. In distinction to the latter, it may also be referred to as live load.

INERTIAL LOAD. A LOAD that results from setting a body rapidly in motion or from any rapid change in its state of motion. It is proportional to the acceleration and the MASS being accelerated.

INERTIAL RESISTANCE. An INERTIAL LOAD regarded as a resistance to an enforced acceleration.

IN SITU. In the final location, referring, for instance, to the casting of concrete STRUCTURAL ELEMENTS in their final location and to concrete so cast.

INSTABILITY. A tendency to collapse, particularly as a result of a slight disturbance.

INTRADOS. The lower (usually concave) surface or SOFFIT of an ARCH.

ISOSTATIC. A line indicating the direction of one of the PRINCIPAL STRESSES at every point through which it passes.

JACK (TO). To apply FORCES by means of jacks – either screw jacks or more frequently hydraulic jacks able to apply larger forces under close control.

JOIST. A timber or steel or precast-concrete BEAM of a floor or ceiling.

JOGGLES. Interlocking projections and recessions on adjacent STRUCTURAL COMPONENTS such as the VOUSSOIRS of an ARCH.

KEYSTONE. The uppermost or closing VOUSSOIR in a VOUSSOIR ARCH or a VAULT RIB.

LINTEL. A common name for a single block of stone spanning horizontally over an opening, or a simple BEAM of some other kind similarly used. Hence the term post-and-lintel referring to the COLUMN-and-BEAM structure seen in the Classical temple.

LIVE LOAD. See IMPOSED LOAD.

LOAD. Often synonymous with FORCE, but can denote also the local intensity of a distributed force such as a PRESSURE or a combination of forces such as a BENDING MOMENT.

LOCKED COIL CABLE. See CABLE.

MASS. 'The quantity of matter in a body.' It is the measure that determines the acceleration that a given FORCE will produce and the INERTIAL RESISTANCE that will automatically oppose a given acceleration. For most practical purposes it may be taken as proportional to the weight.

MEMBRANE. A thin sheet-like STRUCTURAL ELEMENT, either flat [5.5(b)], singly curved [5.5(g)], or doubly curved [5.5 (k, n)], and acting wholly in TENSION.

MOMENT. Of a couple: the product of the magnitude of the FORCE and the MOMENT ARM.

Of a single force about an axis: the product of the COMPONENT of the force acting at right angles to the axis and the MOMENT ARM. See also BENDING MOMENT.

MOMENT ARM. Of a couple: the perpendicular distance between the two constituent forces.

Of a single force about an axis: the perpendicular distance from the axis to the line of action of the force.

MONOLITHIC. Composed of a single block of stone.

NATURAL FREQUENCY. The number of cycles of a free oscillation or vibration in unit time (usually per second).

NATURAL PERIOD OF VIBRATION. The time taken for one complete cycle of a free oscillation or vibration.

OPUS QUADRATUM. Roman masonry composed of squared blocks of stone.

PARALLELOGRAM OF FORCES. A geometrical construction in which two FORCES acting in different directions in the same plane are represented in both magnitude and direction by two sides of a parallelogram and their RESULTANT is then similarly represented by the diagonal passing through their point of intersection. See also TRIANGLE OF FORCES.

PEDIMENT. The name given to the ENTABLATURE at the ends of a Classical temple when the top member (the cornice) is inclined upwards from each end to follow the pitch of the roof, and to a similar architectural feature in later buildings.

PENDENTIVE. A STRUCTURAL ELEMENT serving as a transition between two WALLS or ARCHes meeting at an angle and part of the circular base or springing line of a dome rising above them. Its inner surface is, or approximates to, a triangular portion of a large hemisphere whose base would circumscribe the feet of all the arches or the meeting points of all the walls and whose diameter at the level of the dome is equal to or slightly less than that of the dome itself [7.10]. When the dome is not a distinct element, but merely the upward continuation of the larger hemisphere to which the pendentives belong, these are known as merging pendentives [7.9].

PERIODIC. Of LOADS, displacements, etc.: varying with time in a rhythmically repetitive manner.

PERMANENT SET. That part of the deformation under load that remains when the load is removed. It is equal to the amount of PLASTIC deformation that has taken place. See Figure 3.2.

PIER. A vertical STRUCTURAL ELEMENT like a COLUMN, but greater in relative cross section.

PILASTER. A shallow PIER-like projection from a WALL, usually built integrally with the wall.

PILED FOUNDATION. A FOUNDATION in which the loads of the structure above are carried down to a depth where they can be safely borne by means of long posts or STRUTS of timber, steel, or reinforced concrete known as piles. These are either driven in by repeated blows or are cast IN SITU in prebored holes.

PINNED JOINT. A joint between STRUCTURAL ELEMENTS or COMPONENTS allowing them to rotate freely relative to one another. Strictly speaking, this calls for a frictionless pin to make the joint, but other kinds of joint, including many riveted ones, allow enough relative rotation with little resistance to be regarded as pinned. An equivalent joint in reinforced concrete is referred to as a hinge.

PLASTIC, PLASTICITY. Descriptive of the DEFORMATIONAL behaviour of a DUCTILE material after the limit of ELASTIC behaviour or ELASTICITY has been passed. In perfectly plastic behaviour (as illustrated by the horizontal parts of the stress-strain curve for mild steel in Figure 3.2) deformation continues to increase with no increase in load, and the increase remains as a PERMANENT SET when the load is removed.

PLASTIC DEFORMATION. Continued DEFORMATION of a DUCTILE material under a constant or near-constant load close to the ultimate strength [3.2].

PNEUMATIC CAISSON. See CAISSON.

POLYGONAL MASONRY. Masonry composed of blocks of stone dressed to fit one another with irregular polygonal exposed faces rather than rectangular ones as in ASHLAR masonry.

PORTAL FRAME. The simplest RIGID-JOINTED COLUMN-AND-BEAM FRAME, consisting of two columns spanned by a single beam rigidly joined to their heads [5.11(e)].

POST-AND-LINTEL. See LINTEL.

PRESTRESS. To apply loads by controlled external means to a STATICALLY INDETERMINATE STRUCTURE or STRUCTURAL ELEMENT during construction for the purpose of modifying beneficially the state of internal STRESS. It is tantamount to introducing deliberately undersized or oversized elements or members for the same purpose.

PRESSURE. A distributed FORCE such as that exerted by the wind, or the local or average intensity of such a force.

PRINCIPAL STRESS. The maximum or minimum direct STRESS in TENSION or COMPRESSION acting at any point within a STRUCTURAL ELEMENT, regarding one as positive and the other as negative. Also, in a three-dimensional situation, the direct STRESS acting at right angles to the plane in which the maximum and minimum stresses act. The maximum and minimum stresses themselves always act at right angles to one another. So in the three dimensional situation all three stresses are mutually perpendicular, though one or two of them may be zero.

PURLIN. A horizontal BEAM spanning between the principal RAFTERS in a roof to carry the roof covering, usually via secondary rafters.

QUADRANT VAULT. A laterally extended VAULT with the cross-sectional profile of half a semicircle. Resembling therefore a semicircular BARREL VAULT sliced along its CROWN and with one half removed.

RAFTER. One of the inclined BEAMS of a pitched roof that meet along the ridge and directly or indirectly carry the roof covering. Principal rafters receive no intermediate support. Secondary rafters receive intermediate support from PURLINS carried by the principal rafters.

REACTION (or REACTIVE LOAD). The load or set of loads brought into play to counterbalance the ACTION or ACTIVE LOAD on a STRUCTURAL ELEMENT or STRUCTURE.

REDUNDANT ELEMENT. Any STRUCTURAL ELEMENT whose introduction into a structure makes its STRUCTURAL ACTION STATICALLY INDETERMINATE. It is usually possible to make alternative choices of the element or elements that are to be regarded as redundant.

REINFORCEMENT. A general term denoting some kind of addition to a STRUCTURAL ELEMENT, usually in another material, for the purpose of improving its STRENGTH or STIFFNESS. It now usually refers to bars, cables, or prefabricated meshes of steel added to elements of concrete and bonded or anchored to the concrete so as to act intimately with it.

RESONANCE. A state in which a PERIODICALLY varying disturbance continually reinforces the natural oscillations or vibrations of a structure or STRUCTURAL ELEMENT, and thereby builds up an AMPLITUDE considerably greater than would otherwise result.

RESULTANT. Of a STRUCTURAL ACTION: that single action (FORCE or MOMENT) that would have the same effect.

RIB. A linear projection from the surface of a STRUCTURAL ELEMENT, particularly from the lower surface of a VAULT, SHELL or SLAB

RIBBED VAULT. A VAULT in which ARCH-like RIBS traverse the lower surface along the lines of any GROINS and sometimes elsewhere also. According to the details of construction, the ribs may or may not carry the intervening WEBS. See also FAN VAULT.

RIGID JOINT. A joint between STRUCTURAL ELEMENTS or COMPONENTS allowing no relative rotation where they meet. Most joints in reinforced concrete and prestressed concrete when made IN SITU, and most bolted and welded joints in steel, may be so regarded.

RIGID-JOINTED FRAME. A FRAME, particularly one composed of COLUMNS and BEAMS, to which overall rigidity is imparted wholly or largely by RIGID JOINTS.

RISE. Of an ARCH, SHELL, or VAULT: the vertical distance between the SOFFIT at the CROWN and the level of the SPRINGINGS.

ROLLER BEARING. A type of support incorporating hardened steel rollers set between hardened steel plates to allow free movement in one direction.

SECTION. A name given to a rolled or extruded form of iron, steel, or other metal, with a constant shape of cross section such as an I, T, or U.

SEGMENTAL. Of an ARCH: having the profile of a circular arc substantially less than a full semicircle.

SELF-WEIGHT. The LOAD on a STRUCTURAL ELEMENT or STRUCTURE due to the action of gravity on the element or structure itself.

SEVERY. See WEB of a RIBBED VAULT.

SHEAR. A slipping or racking type of DEFORMATION which is equivalent to a contraction in one direction and an equal extension at right angles to it, and the type of STRUCTURAL ACTION that leads directly to it [2.5(f)].

SHEAR CONNECTOR. A mechanical dowel-like interconnection to prevent or limit relative slip at the interface between two STRUCTURAL ELEMENTS or parts of an ELEMENT, usually between a FLANGE of a steel BEAM and a reinforced-concrete SLAB.

SHEAR LAG. Reductions in the direct STRESSES (TENSION and COMPRESSION) developed in the FLANGES of a BEAM and in analogous situations as a result of SHEAR deformations within the flanges or their equivalents.

SHEAR WALL. A STRUCTURAL ELEMENT geometrically indistinguishable from any other WALL, but loaded primarily, or to an important extent, in SHEAR rather than vertically [5.5(c)].

SHELL. A relatively thin STRUCTURAL ELEMENT with a singly or doubly curved surface capable of spanning horizontal gaps and simultaneously serving as a roof or ceiling and acting wholly or primarily in direct COMPRESSION or a combination of direct COMPRESSION and TENSION [5.5 (h, i, l, m, o, p)]. Any bending should be confined to the vicinities of the supports and of any concentrated loads.

SIMPLY SUPPORTED. Of a beam: supported at or near its two ends in such a way that it is free at both of them to rotate in the plane of the loads and free at one of them to expand or contract longitudinally.

SINGLY CURVED. Of a line: traceable on a single plane. Of a surface: capable, without stretching, of being unrolled to lie flat, i.e. having one direction in which the curvature is everywhere zero.

SKEWBACK. An inwardly inclined seating for the SPRINGING of an ARCH or VAULT to give it both vertical and inward horizontal support.

SLAB. A STRUCTURAL ELEMENT capable of spanning a horizontal gap in the manner of a BEAM, but extended laterally [5.5(r)]. It may also be capable, if suitably supported, of spanning simultaneously in a second direction [5.5 (s, t, u)].

SLIDING BEARING. A type of support usually incorporating one or more sheets of a material like neoprene rubber sandwiched between steel plates to allow almost free movement at right angles to the direction of support.

SOFFIT. The under surface of a structural element such as an ARCH or a BEAM, SLAB, OR VAULT.

SPACE FRAME. A three-dimensional FRAME, usually a triangulated one composed of STRUTS and TIES assembled in several families of intersecting plane TRUSSES which are so interconnected that they all share, in carrying any load [5.11 (c, f, i, j, 1)].

SPALL. Of part of the surface of a STRUCTURAL ELEMENT: to break away under the action of secondary TENSIONS at right angles to a primary COMPRESSION acting parallel to the surface.

SPAN. 1. The horizontal distance between the supports of a BEAM, SLAB, etc., or between the SPRINGINGS of an ARCH, VAULT, etc.
2. A single ARCH or that part of a BEAM, SLAB, etc. that spans between adjacent supports.

SPANDREL. The more or less triangular space above the HAUNCH of an ARCH and below the level of the CROWN, or, in a bridge, below any roadway or deck. Also, if this space is filled, the WALL-like filling.

SPREAD FOOTING. A FOUNDATION in which the intensity of load imposed by a WALL or PIER on the ground is reduced by giving it an increased width below ground level.

SPRINGING. Of an ARCH or VAULT: the position at which the arch or vault begins to curve inwards from a support.

SQUINCH. An alternative STRUCTURAL ELEMENT to the PENDENTIVE in which the angle between two WALLS or ARCHES is bridged by a smaller ARCH set diagonally to help provide a more nearly circular base for the support of a DOME [7.11].

STABLE. 1. A state of EQUILIBRIUM that is not at the mercy of any slight disturbance but tends to be automatically re-established as soon as the disturbance passes, as with a hanging pendulum moved a little to one side and then released.
2. In a state of STABLE EQUILIBRIUM and therefore not prone to collapse unless loaded beyond the capacity of its constituent elements or subjected to a major disturbance.

STABILITY. Another name for the state of STABLE EQUILIBRIUM as defined above, though sometimes used more loosely to denote little more than a state of EQUILIBRIUM under STATIC loads.

STANCHION. A name frequently given to a prefabricated COLUMN of steel or timber.

STATICAL DETERMINACY. A characteristic of the STRUCTURAL ACTION of a structure or any part of it that stems from an ability to carry a given set of loads in only one way that is uniquely determined by the loads and its geometry. Hence also statically determinate. It should be noted that it is possible to scrutinise the structural action of most structures at different levels and that a structure that is statically determinate at one level of scrutiny may be statically indeterminate at a deeper level.

STATICAL INDETERMINACY. A characteristic of the STRUCTURAL ACTION of a structure or any part of it that stems from a potential ability to carry a given set of loads in a multiplicity of different ways, the action in a particular instance being dependent on relative STIFFNESSES and the pre-existing internal LOADS or STRESSES resulting from the construction process and prior loadings, etc., as well as on the present loads and the geometry. Hence also statically indeterminate.

STATIC LOAD. A LOAD that is applied sufficiently slowly and remains sufficiently constant not to bring into play significant INERTIAL RESISTANCES

STAY. A slender straight inclined support acting in TENSION.

STIFFENER. A RIB-like projection from a thin structural member loaded in COMPRESSION (either simple compression or that arising from BENDING or SHEAR), and particularly from a steel plate forming part of a large BEAM or COLUMN to increase the STIFFNESS in BENDING and thereby prevent a BUCKLING failure.

STIFFNESS. Resistance to DEFORMATION. More specifically the load that is required to produce a given deformation. For a STRUCTURE or STRUCTURAL ELEMENT it is measured in terms of a particular type and pattern of LOADING and a particular characteristic DEFORMATION. For a structural material it is measured in terms of a particular type of STRESS acting alone and the associated STRAIN.

STRAIN. A local proportionate DEFORMATION, or deformation per unit length. In TENSION or COMPRESSION it is measured in the direction of the extension or contraction. In SHEAR it is measured as the angular displacement of a line initially at right angles to the line or plane along which the shearing action takes place.

STRAND. A group of wires or fibres twisted together.

STRENGTH. Ability to carry load. More specifically the maximum load that can be carried before a chosen limit of behaviour is reached. For a STRUCTURE or STRUCTURAL ELEMENT it is measured in terms of a particular type and pattern of loading, and for a structural material in terms of a particular type of STRESS or combination of stresses. See also FATIGUE STRENGTH, ULTIMATE STRENGTH, YIELD STRENGTH.

STRESS. The local intensity of an internal FORCE measured as the force acting perpendicularly (for TENSION or COMPRESSION) or tangentially (for SHEAR) to a unit area of a given cross-sectional plane. Tensile and compressive stresses are also known as direct stresses. See also PRINCIPAL STRESS.

STRUCTURAL ACTION. A general term denoting the pattern or characteristics of some or all of the following: the FORCES acting on a structure or any part of it; the resultant internal FORCES and STRESSES; and the associated displacements, DEFORMATIONS, and STRAINS.

STRUCTURAL COMPONENT. A fully formed unit of construction, which may be as small as a brick or bolt, or as large as a prefabricated section of a large steel box beam or a precast-concrete wall or slab. Thus it may be a self-sufficient STRUCTURAL ELEMENT, or it may serve only as a part of such an element. It is identified essentially in terms of the construction or fabrication process rather than structural role.

STRUCTURAL ELEMENT. A basic unit of construction capable of carrying its SELF WEIGHT and other LOADS to its supports. See also ARCH, BEAM, BUTTRESS, CATENARY, COLUMN, DOME, FALSE ARCH, FOLDED PLATE, MEMBRANE, PENDENTIVE, PIER, SHEAR WALL, SHELL, SLAB, SQUINCH, STRUT, TIE, VAULT, WALL. FRAMES, GRIDS, and TRUSSES designed to act in analogous ways to some of these simpler elements may themselves also be regarded as elements. Alternatively their constituent BEAMS, COLUMNS, STRUTS, or TIES may be so regarded.

STRUCTURAL FORM. A term used here to denote both the external geometric configuration (the form *per se*) of a STRUCTURE or STRUCTURAL ELEMENT, its ability to carry LOADS and withstand DEFORMATION, and its intended role in so doing. It therefore also embraces those details of the hidden internal configuration, the materials used, and the manner of construction or fabrication that also partly determine the response to loading or deformation.

STRUCTURE. A system of STRUCTURAL ELEMENTS. Commonly, in a building or bridge, the complete system that plays the primary LOAD-bearing role as distinct from infillings, finishes, etc., that could be removed without significant loss of overall STIFFNESS, STRENGTH, or STABILITY.

STRUT. A STRUCTURAL ELEMENT of relatively small cross-section acting, like a COLUMN, wholly or primarily in COMPRESSION [5.5(e)] but, unlike the column, not necessarily vertical and usually forming part of a triangulated SPACE FRAME or TRUSS.

SURFACE OF REVOLUTION. A surface traced out by a line rotated about an axis in such a way that all points on the line remain at constant distances from the axis and therefore trace out circular arcs or full circles.

SWAY BRACE, SWAY BRACING. A BRACE or system of BRACING designed to limit lateral deformations of a FRAME or TRUSS.

SYNCLASTIC. Of a surface: having curvatures in the same sense (concave or convex) in all directions through any point.

TENDON. A bar, STRAND, or CABLE used as TENSILE REINFORCEMENT in prestressed concrete.

TENSEGRITY. A name usually reserved for a type of SPACE FRAME in which the STRUTS and TIES are so assembled that there are continuous paths for the transmission of TENSION but there is no such continuous path for the transmission of COMPRESSION, but sometimes used more loosely.

TENSION. Extension and the type of STRUCTURAL ACTION that leads directly to it [2.5a]. Hence also tensile.

TETRAHEDRON. The simplest plane-faced three-dimensional body, having four triangular faces. In the regular tetrahedron, these faces are all equilateral triangles.

THRUST. A COMPRESSIVE STRUCTURAL ACTION, particularly of the kind characteristic of an ARCH. More specifically the action exerted by an arch on its supports, the outward horizontal COMPONENT of this action, or an analogous action exerted by another STRUCTURAL ELEMENT.

THRUST LINE. The line of action of the RESULTANT COMPRESSION or THRUST within an ARCH or related STRUCTURAL ELEMENT.

TIE. A straight slender STRUCTURAL ELEMENT acting wholly or almost wholly in tension [5.5(a)], either as a HANGER or as part of a triangulated SPACE FRAME or TRUSS.

TORSION. Twisting and the type of STRUCTURAL ACTION that leads directly to it [2.5(e)].

TRIANGLE OF FORCES. A geometrical construction, equivalent to the PARALLELOGRAM OF FORCES, in which the two FORCES are similarly represented by two sides of a triangle and their RESULTANT is then represented by the third side.

TRIANGULATION. The achievement of rigidity, or at least the prevention of completely free relative movement, in an assembly of STRUTS and TIES, by arranging them in a continued series of triangles in one or more planes [5.11 (a, c, etc.)].

TRUSS. An assembly of interconnected linear STRUCT-URAL ELEMENTS designed to act broadly in the manner of a BEAM, and usually a triangulated assembly of STRUTS and TIES [5.11(g)].

TUNED-MASS DAMPER. A DAMPER actuated by a large mass that is so mounted on a structure that it is set in motion by the structure's to-and-fro movements and, being tuned to have the same NATURAL FRE-QUENCY, moves out-of-step with them.

ULTIMATE STRENGTH. The maximum attainable STRENGTH under STATIC LOAD irrespective of the associated DEFORMATION, though in practice often taken as the value of the load at which there would be a rapid loss of STIFFNESS.

UNSTABLE. Of a state of EQUILIBRIUM: liable (like that of a pendulum with a rigid arm standing pre-cariously upright instead of hanging) to be upset by a slight disturbance.

Of a STRUCTURE or STRUCTURAL ELEMENT: being in a state of UNSTABLE EQUILIBRIUM or on the verge of collapse.

VAULT. Used here to denote a STRUCTURAL ELEMENT akin to a SHELL, but usually thicker in relation to its SPAN and constructed of masonry incapable of resisting much TENSION and therefore acting more exclusively in direct ARCH-like COMPRESSION. See also BARREL VAULT, CLOISTER VAULT, DOME, GROIN-ED VAULT, QUADRANT VAULT, RIBBED VAULT.

VORTEX. An eddy or intense local spiral motion in a stream of air.

VOUSSOIR. A slightly wedge-shaped block of stone or equivalent structural component used for construct-ing an ARCH.

VOUSSOIR ARCH. An ARCH constructed of VOUSSOIRS.

WALL. A laterally extended vertical STRUCTURAL ELE-MENT with a thickness comparable with, or slightly less than, that of a COLUMN of the same height. As a primary element it serves usually as a vertical sup-port and acts wholly or primarily in COMPRESSION [5. 5(d)] but it may also or alternatively be loaded in SHEAR [5.5(c)]. See also BEARING WALL, SHEAR WALL.

WEB. Of a BEAM or COLUMN of stemmed or box-shaped cross-section: the stem or the sides of the box lying in the plane of the major BENDING ACTION.

Of a RIBBED VAULT: A section of the vault between adjacent RIBS. Also referred to as a severy.

WIND BRACE, WIND BRACING. A BRACE or system of BRACING designed to limit DEFORMATIONS of a FRAME or TRUSS under wind loads.

YIELD. To give way to load by PLASTIC DEFORMATION. Also the onset of this giving-way.

YIELD STRENGTH. The STRENGTH as limited by YIELD.

Notes and References

List of abbreviations

The following abbreviations are used below in referring to frequently cited English-language periodicals and to books cited more than once:

AB – *Art Bulletin*

AF – *Architectural Forum*

Alberti *De re aedificatoria* – Alberti, L. B., *De re aedificatoria*, Florence, 1485 (reprinted with Italian translation, Orlandi, G., Polifilo, Milan, 1966); English translations: Leoni, J., *Ten books on architecture by Leone Battista Alberti*, London, 1726, (facsimile edn, ed. Rykwert, J., Tiranti, 1955); and Rykwert, J, Leach, N. and Tavernor, R., *Leone Battista Alberti: On the art of building in ten books*, MIT Press, 1988

Allen *Shelters* – Allen, E., *Stone shelters*, MIT Press, 1969

Ammann et al. *Tacoma Narrows Bridge* – Ammann, O. H., Karman, T. von, and Woodruff, G. B., *The failure of the Tacoma Narrows Bridge*, Federal Works Agency, 1941

ARec. – *Architectural Record*

ARev. – *Architectural Review*

Arnold *Building in Egypt* – Arnold, D., *Building in Egypt*, Oxford University Press, 1991

Belidor *Science des ingénieurs* – Belidor, B. F. de, *La science des ingénieurs*, Paris, 1729

Benevenuto *Introduction* – Benevenuto, E., *An introduction to the history of structural mechanics*, Springer, 1991 (original Italian edn 1981)

Berger *Light structures* – Berger, H., *Light structures – structures of light: the art and engineering of tensile architecture*, Birkhäuser, Basel, 1996

Berger *The Palace of the Sun* – Berger, R. W., *The Palace of the Sun: the Louvre of Louis XIV*, Pennsylvania State University Press, 1993

Bill *Maillart* – Bill, M., *Robert Maillart*, Verlag für Architektur, Zürich, 1949

Blake *Ancient Roman construction* – Blake, M. E., *Ancient Roman construction in Italy from the Prehistoric period to Augustus*, Carnegie Institution of Washington Publication 570, 1947

Blake *Roman construction* – Blake, M. E., *Roman construction in Italy from Tiberius through the Flavians*, Carnegie Institution of Washington Publication 616, 1959

Boaga and Boni *Morandi* – Boaga G. and Boni, B., *The concrete architecture of Riccardo Morandi*, Tiranti, 1965

Boucher *Palladio* – Boucher, B., *Andrea Palladio*, Abbeville Press, 1994

Caraffa *Cupola* – Caraffa, G., *La cupola della sala decagona degli Horti Liciniani restauri 1942*, Vittorio Ferri, Rome, 1944

CE – *Civil Engineering* (American Society of Civil Engineers)

Chadwick *Paxton* – Chadwick, G. F., *The works of Sir Joseph Paxton*, Architectural Press, 1961

Choisy *Histoire* – Choisy, A., *Histoire de l'architecture*, Paris, 1899

Clarke and Engelbach *Egyptian masonry* – Clarke, S. and Engelbach, R., *Ancient Egyptian masonry*, Oxford University Press, 1930

Clark and Stephenson *Tubular bridges* – Clark, E. and Stephenson, R., *The Britannia and Conway Tubular Bridges*, London, 1850, 2 vols. text, 1 vol. plates

Collins *Concrete* – Collins, P., *Concrete: the vision of a new architecture*, Faber & Faber, 1959

Condit, *Nineteenth century* – Condit, C. W., *American building art: the nineteenth century*, Oxford University Press, 1960

Condit, *Twentieth century* – Condit, C. W., *American building art: the twentieth century*, Oxford University Press, 1961

Coulton *Greek architects* – Coulton, J. J., *Greek architects at work*, Elek, 1977

Davey *Materials* – Davey, N., *A history of building materials*, Phoenix House, 1961

De l'Orme *Nouvelles inventions* – De l'Orme, P., *Nouvelles inventions pour bien bastir et a petits fraiz*, Paris, 1561

Diamant *Industrialised building* – Diamant, R. M., *Industrialised building*, Iliffe, 1964-5, 2 vols

Di Stefano *Cupola* – Di Stefano, R., *La cupola di San Pietro*, Edizioni Scientifiche Italiane, Naples, 1963 (rev. edn 1980)

Fairbairn *Tubular bridges* – Fairbairn, Sir W., *An account of the construction of the Britannia and Conway Tubular Bridges*, London, 1849

Fairbairn *Application* – Fairbairn, Sir W., *On the application of cast and wrought iron to building purposes*, London, 1854

Fitchen *Construction* – Fitchen, J. F., *The construction of Gothic cathedrals*, Clarendon Press, 1961

Fitchen *Dutch barn* – Fitchen, J. F., *The New World Dutch barn: a study of its characteristics, its structural system, and its probable erection procedures*, Syracuse University Press, 1967

Frankl *Gothic architecture* – Frankl, P., *Gothic architecture*, Pelican History of Art, 1962

Freitag *Architectural engineering*, – Freitag, J. K., *Architectural engineering*, New York, 1895

Fugl-Meyer *Chinese bridges* – Fugl-Meyer, H., *Chinese bridges*, Kelly & Walsh, Shanghai, 1937

Gabriel *Beyond the cube* – Gabriel, J. F., ed., *Beyond the cube*, John Wiley, 1997

Galileo *Discorsi* – Galileo Galilei, *Discorsi e dimostrazioni matematiche intorno a due nuove scienze*, Leiden, 1638; English translation, *Dialogues concerning two new sciences*, trans. Henry Crew and Alfonso de Salvio, Macmillan, 1914

Gappoev et al. *Suchov* – Gappoev, M., Graefe, R. and Pertschi, O., eds, *V. G. Suchov Kunst der Konstruktion*, Institut für Auslands-beziehungen, Stuttgart, 1990

Gauthey *Dissertation* – Gauthey, E. M., *Dissertation sur les dégradations survenues aux piliers du dome du Panthéon française et sur les moyens d'y remédier*, Paris, 1799

Gazzola *Ponti* – Gazzola, P., *Ponti Romani*, Olschki, Florence, 1963

Giovannoni *Sala termale* – Giovannoni, G., *La sala termale della Villa Liciniana e le cupole Romane*, Stabilmento Tipo-Litografico del Genio Civile, Rome, 1904

Giuliani *Spatial structures* – Giuliani, G. C., ed., *Spatial structures: heritage, present and future*, International Association of Shell and Space Structures, Milan, 1995, 2 vols

Graefe *Geschichte* – Graefe, R., ed., *Zur Geschichte des Konstruierens*, Deutsche Verlags-Anstalt, Stuttgart, 1989

Guarini *Architettura* – Guarini, G., *Architettura civile*, Turin, 1737 (facsimile edition, Gregg, 1964)

Guasti *Cupola* – Guasti, C., *La cupola di Santa Maria del Fiore illustrata con i documenti dell'archivio dell'Opera Secolare*, Florence, 1857

Guasti *Santa Maria del Fiore* – Guasti, C., *Santa Maria del Fiore: la costruzione della chiesa e del campanile secondo i documenti tratti dall' archivo dell'opera secolare*, Florence, 1887

Haegermann et al. *Vom Caementum zum Spannbeton* – Haegermann, G., Huberti, G., Möll, H. and Deinhard, J-M., *Vom Caementum zum Spannbeton*, Bauverlag, Wiesbaden, 1964, 2 vols

Hamilton *Reinforced concrete* – Hamilton, S. B., *A note on the history of reinforced concrete in buildings*, National Building Studies Special Report 24, HMSO, 1956

Heyman *Coulomb's memoir* – Heyman, J., *Coulomb's memoir on statics*, Cambridge University Press, 1972

Hewett *Historic carpentry* – Hewett, C. A., *English historic carpentry*, Phillimore, 1980

Hodge *Greek roofs* – Hodge, A. T., *The woodwork of Greek roofs*, Cambridge University Press, 1960

Hopkins *Bridges* – Hopkins, H. J., *A span of bridges*, David & Charles, 1970

Horn and Born *Barns* – Horn, W. and Born, E., *The barns of the Abbey of Beaulieu and its granges at Great Coxwell and Beaulieu St Leonards*, University of California Press, 1965

Ito et al. *Cable-stayed bridges* – Ito, M., Fujino, Y., Miyata, T. and Narita, N., *Cable-stayed bridges: recent developments and their future*, Elsevier, Amsterdam, 1991

JACI – *Journal of the American Concrete Institute*

JICE – *Journal of the Institution of Civil Engineers*

JRIBA – *Journal of the Royal Institute of British Architects*

JSAH – *Journal of the Society of Architectural Historians*

Joedicke *Shell architecture* – Joedicke, J., *Schalenbau*, Karl Kramer, Stuttgart, 1962 (trans. J. C. Palmes, Tiranti, 1963)

Joedicke et al. *Nervi* – Joedicke, J. et al., *The works of Pier Luigi Nervi*, Architectural Press, 1957

Killer *Grubenmann* – Killer, J., *Die Werke der Baumeister Grubenmann*, 3rd edn, Birkhäuser, Basle, 1985

Koncz *Precast concrete* – Koncz, T. (trans. C. van Amerongen), *Manual of precast concrete construction*, Bauverlag, Wiesbaden, 1968-71, 3 vols

Krafft *Charpente* – Krafft, J. C., *Plans coupes et élévations de diverses productions de l'art de la charpente exécutées tant en France que dans les pays étrangers*, Paris, 1805

Kranakis *Constructing a bridge* – Kranakis, E, *Constructing a bridge*, MIT Press, 1997

La Hire *Traité* – La Hire, P. de, *Traité de méchanique*, Paris, 1695

Lawrence *Greek architecture* – Lawrence, A. W., *Greek architecture*, Pelican History of Art, 1957 (5th edn, rev. R. A. Tomlinson, 1996)

Lauer *Pyramide* – Lauer, J-P., *La pyramide à degrés, l'architecture*, Institut Français de l'Archéologie Orientale, Service des Antiquités de l'Égypte, Cairo, 1936

Le Seur et al. *Parere* – Le Seur, T., Jacquier, F. and Boscovich, R. G., *Parere di tre mattematici sopra i danni, che si sono trovati nella cupola di S. Pietro*, Rome, 1743

Leonhardt *Bridges* – Leonhardt, F., *Brücken Bridges*, Architectural Press, 1982

Leupold *Theatrum pontificiale* – Leupold, J., *Theatrum pontificiale oder Schau-Platz der Brücken und Brücken baues*, Leipzig, 1726

Licht *Rotunda* – Licht, K. de F., *The Rotunda in Rome*, Jutland Archaeological Society Publications 8, Copenhagen, 1968

Lugli *Tecnica edilizia* – Lugli, G., *La tecnica edilizia romana con particolare riguardo a Roma e Lazio*, Giovanni Bardi, Rome, 1957

MacDonald *Architecture of the Roman empire* – MacDonald, W. L., *The architecture of the Roman empire: an introductory study*, Yale Publications in the History of Art 17, 1965

Mainstone *Hagia Sophia* – Mainstone, R. J., *Hagia Sophia: Architecture, structure and liturgy of Justinian's Great Church*, Thames and Hudson, 1988

Mainstone *Structure in architecture* – Mainstone, R. J., *Structure in architecture: history, design and innovation*, Variorum Collected Studies Series, forthcoming, 1999

Makowski *Steel space structures* – Makowski, Z. S., *Steel space structures*, Michael Joseph, 1965.

Min.Proc.ICE – *Minutes of Proceedings of the Institution of Civil Engineers*

Needham *Civil engineering and nautics* – Needham, J., *Science and civilisation in China, vol. 4, pt III: Civil engineering and nautics*, Cambridge University Press, 1971

Nervi *Structures* – Nervi, P. L., *Structures* (trans. G. and M. Salvadori), E. W. Dodge, 1956

Nervi *New structures* – Nervi, P. L., *New structures*, trans. G. Nicoletti, Architectural Press, 1963

Ordonez *Freyssinet* – Ordonez, J. A. F., *Eugène Freyssinet*, 2C Editions, Barcelona, 1979

Otto *Das hangende Dach* – Otto, F., *Das hangende Dach*, Ullstein, Berlin, 1954

Otto and Schleyer *Tensile structures* – Otto, F. and Schleyer, F-K., *Tensile structures: cables, nets and membranes*, trans. D. Ben-Yaakov and T. Pelz, MIT Press, 1969

Palladio *Quattro libri* – Palladio, A., *I quattro libri dell'architettura*, Venice, 1570 (reproduced in facsimile, Hoelpi, Milan, 1945 and later reprints). Principal English translations: Leoni, J., *The architecture of Andrea Palladio*, London, 1715-1720; Ware, I., *The four books of Andrea Palladio's Architecture*, London, 1738 (reproduced in facsimile, Dover, 1965); and Tavenor, R. and Schofield, R., *Andrea Palladio: The four books on architecture*, MIT Press, 1997

Parsons *Engineering in the Renaissance* – Parsons, W. B., *Engineers and engineering in the Renaissance*, Williams and Wilkins, 1939 (reprinted MIT Press, 1968)

Patte *Mémoire* – Patte, P., *Mémoire sur la construction de la coupole projettée pour couronner la nouvelle église de Sainte Geneviève à Paris*, Amsterdam and Paris, 1770

Petzet *Sainte-Geneviève* – Petzet, M., *Soufflots Sainte-Geneviève und der französische Kirchenbau des 18 Jahrhunderts*, Walter de Gruyter, Berlin, 1961

Plot *Oxfordshire* – Plot, R., *The natural history of Oxfordshire*, Oxford, 1672

Poleni *Memorie* – Poleni, G., *Memorie istoriche della gran cupola del Tempio Vaticano*, Padua, 1748

Pope *Survey* – Pope, A.U., ed., *A survey of Persian art*, Oxford University Press, vol. 2, 1939

Pope *Persian architecture* – Pope, A. U., *Persian architecture*, Thames & Hudson, 1965

Pope *Treatise* – Pope, T., *A treatise on bridge architecture in which the superior advantages of the flying pendent lever bridge are fully proved*, New York, 1811

Portoghesi *Guarini* – Portoghesi, P., *Guarino Guarini*, Electra, Milan, 1956

Proc.ASCE – *Proceedings of the American Society of Civil Engineers*

Proc.ICE – *Proceedings of the Institution of Civil Engineers*

Procopius *Buildings* – Procopius, *Buildings*, ed. and trans., H. B. Dewing, Loeb Classical Library, William Heinemann, 1940

Radding and Clark *Medieval architecture* – Radding, C. M. and Clark, W. W., *Medieval architecture, medieval learning*, Yale University Press, 1992

Report into the application of iron – *Report of the Commissioners appointed to inquire into the application of iron to railway structures*, London, 1849

Report on Ronan Point – Ministry of Housing and Local Government, *Report of the inquiry into the collapse of flats at Ronan Point, Canning Town*, London, 1968

Robbin *New architecture* – Robbin, R, *Engineering a new architecture*, Yale University Press, 1996

Rohlfs *Primitive Kuppelbauten* – Rohlfs, G., *Primitive Kuppelbauten in Europa*, Bayerischen Akademie der Wissenschaften, 1957

Rondelet, *Traité 1814* – Rondelet, J., *Traité théorique et pratique de l'art de bâtir*, 1st edn Paris, 1814

Rondelet, *Traité 1838* – Rondelet, J., *Traité théorique et pratique de l'art de bâtir*, 8th edn Paris, 1838

Rondelet, *Mémoire* – Rondelet, J., *Mémoire historique sur le dôme du Panthéon français*, Paris, 1797

Ruddock *Arch bridges* – Ruddock, T., *Arch bridges and their builders 1735-1835*, Cambridge University Press, 1979

SE – *The Structural Engineer*

Salvadori and Heller *Structure* – Salvadori, M. and Heller, R., *Structure in architecture*, Prentice-Hall, 1963 (rev. edns 1975, 1986).

Sanayei *Restructuring America* – Sanayei, M., ed., *Restructuring America and beyond*, American Society of Civil Engineers, 1995

Schodek *Structures* – Schodek, D. L., *Structures.*, Prentice-Hall, 1980

Séjourné *Grandes voûtes* – Séjourné, P., *Grandes voûtes*, Tardy-Pigelet et Fils, Bourges, 1913-16, 6 vols

Sgrilli *Descrizione* – Sgrilli, S. B., *Descrizione e studi dell'insigne fabbrica di S. Maria del Fiore*, Florence, 1733

Shirley-Smith *Bridges* – Shirley Smith, H., *The world's great bridges*, 2nd rev. edn, Phoenix House, 1964

Smeaton *Edystone* – Smeaton, J., *A narrative of the building and a description of the construction of the Edystone Lighthouse with stone*, 2nd edn, London, 1793

Tang Huangcheng *Chinese bridges* – Tang Huangcheng, *Ancient Chinese bridges*, Cultural Relic Press, Beijing, 1987 (in Chinese and English).

Telford *Life* – Telford, T., *Life of Thomas Telford written by himself*, ed. J. Rickman, London, 1838, 1 vol. text, 1 vol. plates

Thompson *On growth and form* – Thompson, D'A. W., *On growth and form*, 2nd edn rev., Cambridge University Press, 1942

Torroja, *Philosophy of structures* – Torroja, E., *Philosophy of structures*, University of California Press, 1958

Torroja, *Structures* – *The structures of Eduardo Torroja*, F. W. Dodge, 1958

Trans.ASCE – *Transactions of the American Society of Civil Engineers*

Trans.NS – *Transactions of the Newcomen Society*

Verantius *Machinae novae* – Verantius, F., *Machinae novae*, Venice, n.d. (c.1615)

Vitruvius *De architectura* – Vitruvius, *De architectura*, ed. and trans. Granger, F., Loeb Classical Library, Heinemann, 1931. English translation only, Morgan, M. H., *The ten books on architecture*, Harvard University Press, 1914 (reprinted Dover, 1960)

Viollet-le-Duc *Dictionnaire raisonné* – Viollet-le-Duc, E. E., *Dictionnaire raisonné de l architecture française* du XIe au XVIe siècle, Paris, 1854-68

Ward-Perkins *Roman Imperial Architecture* – Ward-Perkins, J. B., *Roman Imperial Architecture*, Pelican History of Art, 2nd edn 1981 (first published as parts 2-4 of Boëthius, A. and Ward-Perkins, J. B., *Etruscan and Roman Architecture*, Pelican History of Art, 1970)

Wilber *Architecture of Islamic Iran* – Wilber, D. N., *The architecture of Islamic Iran: the Il Khanid period*, Princeton University Press, 1955

Wittkower *Architectural principles* – Wittkower, R., *Architectural principles in the age of humanism*, 3rd edn rev., Tiranti, 1962

Wittkower *Art and architecture in Italy* – Wittkower, R., *Art and architecture in Italy 1600-1750*, Pelican History of Art, 1958 (3rd edn rev., 1973)

Wittfoht *Bridges* – Wittfoht, H. (trans. Kluttz, E.), *Building bridges*, Beton-Verlag, Düsseldorf, 1984

Woodward *St. Louis Bridge* – Woodward, C. M., *A history of the St. Louis Bridge*, 1881

Wren *Parentalia* – Wren, S., *Parentalia or memoirs of the family of the Wrens*, London, 1750 (facsimile edn of the extra-illustrated "heirloom" copy, Gregg, 1965)

Part 1 Introductory

Chapter 1 Introduction

1. Thompson *On growth and form*, gives the classic account, from the present point of view, of natural forms. In Chapter 16, 'On form and mechanical efficiency', pp. 958-1025, he draws a number of illuminating parallels between these forms and man-made ones. See also Mainstone, R. J., 'On construction and form', *Program* (Columbia University School of Architecture), no. 3, 1964, pp. 51-70; Otto, F., *Natürliche Konstruktionen*, Deutsche Verlags-Anstalt, Stuttgart, 1982.

2. Smeaton *Edystone*, pp. 42-44. and pl. 13.

3. Matthew vi, 27.

4. Thompson *On growth and form*, pp. 1018-1019.

5. The most relevant studies of the symbolism of the dome are Lehmann, K., 'The dome of heaven', *AB*, vol. 27, 1945. pp. 1-27; and Smith, E. B., *The dome: a study in the history of ideas*, Princeton University Press, 1950, though the argument of the latter tends towards circularity. Mathews, T. F., 'Cracks in Lehmann's "Dome of heaven"', *Source notes in the history of art*, vol.1, 1982, pp. 12-16 is more sceptical. Mango, C. and Parker, J., 'A twelfth-century description of St. Sophia', *Dumbarton Oaks Papers*, vol. 14, 1960, pp. 233-245, discusses a specific instance.

6. Demus, O., *Byzantine mosaic decoration*, Kegan Paul Trench Trubner, 1947.

7. Jerphanion, G. de, *Les églises rupestres de Cappadoce*, Librairie Orientaliste Paul Geuthner, Paris, 1925-1942, 4 vols text, 3 vols plates, was the first major study of these churches. There have been many subsequent publications.

8. Wittkower *Architectural principles*.

9. Andrews, P.A., 'Zwiebelkuppel und Gitterzelt', in Graefe *Geschichte*, pp. 48-69.

10. Panofsky, E., *Abbot Suger on the Abbey Church of St. Denis and its treasures*, Princeton University Press, 1946; Simson, O. von, *The Gothic cathedral*, Routledge & Kegan Paul, 1956; Radding and Clark *Medieval architecture*.

Chapter 2 Structural actions

1. The discussion in this chapter, being purely qualitative though illustrated with visual evidence of some of the actions considered, differs considerably from the quantitative but more abstract discussions to be found in standard texts on structural mechanics and structural analysis and design, and in the vast specialized literature on these subjects. It seems unnecessary, for the present purpose, to refer in detail here to this literature. Of the few previous attempts at a more qualitative discussion, the best are probably Torroja *Philosophy of structures*, and Salvadori and Heller *Structure*. See also Schodek *Structures*.

2. Thompson *On growth and form*, p. 1019.

Chapter 3 Structural materials

1. A fuller discussion of the materials used up the rise of iron and steel will be found in Davey *Materials*. There is no comparable comprehensive account of the development of the materials important today but Gordon, J. E., *The new science of strong materials*, Penguin, 2nd edn 1976, is an excellent highly readable general account and Doran, D. K., *Construction materials reference book*, Butterworth-Heinemann, 1992, is a useful reference work dealing with the manufacture and properties of most of them.

2. Clarke and Engelbach *Egyptian masonry* and now also Arnold *Building in Egypt* give excellent accounts of Egyptian masonry techniques, including the quarrying, fitting, and dressing of stone, the use of cramps, and the making and use of bricks. For further discussion concentrating on practice in ancient Greece and on the quarrying, cutting and transport of marble for a particular structure see Korres, M., 'The geological factor in ancient Greek architecture', in Marinos, P. G. and G. C. Koukis eds, *Engineering geology of ancient works and historical sites*, Rotterdam, 1988; pp. 1779-1793; and idem, *From Pentelicon to the Parthenon*, Melissa, Athens, 1995.

3. See, for instance, Viollet-le-Duc *Dictionnaire raisonné*, article 'chaînage', vol. 2, pp. 396-404; Balanos, N., *Les monuments de l'Acropole: relèvement et conservation*, Charles Massin and Albert Lévy,

Paris, 1938, pl. 16 and 110-115; Livadefs, C. J., 'The structural iron of the Parthenon', *Journal of the Iron and Steel Institute*, vol. 182, 1956, pp. 49-66; Lugli *Tecnica edilizia*, vol 1, pp. 235-242 and vol. 2, pls 31 and 32; Orlandos, A. K., *Building Materials and methods of the ancient Greeks*, Athens (in Greek), 1959-1960, vol. 2, pp. 222-243; and Adam, J-P., *Roman building: materials and techniques*, Batsford, 1994, pp. 54-58.

4. Blake, *Ancient Roman construction*; idem, *Roman construction*; and Lugli *Tecnica edilizia* give comprehensive accounts of the development of Roman concrete as well as of Roman uses of cut stone. For later Roman developments see Ward-Perkins, J. B., 'Notes on the structure and building methods of Early Byzantine architecture', in Rice, D. T., ed., *The Great Palace of the Byzantine Emperors*, Second report, Walker Trust, 1958, pp. 52-104; and MacDonald, W., 'Some implications of later Roman construction', *JSAH*, vol. 17, 1958, pp. 2-8.

5. Davey *Materials*, pp. 97-111, gives a good account of this with references. See also Skempton, A. W., 'Portland cements 1843-1887', *Trans.NS*, vol. 35, 1962-63, pp. 117-152.

6. Deneux, H., 'L'évolution des charpentes du XIᵉ au XVIIIᵉ siècle', *L'architecte*, 1927, pp. 49-53, 57-60, 65-68, 73-75, and 81-89, gives an excellent summary of the development of French jointing techniques. See also Moles, A., *Histoire des charpentiers*, Librairie Gründ, Paris, 1949, for structural uses of timber generally; Hewett, C. A., *English historic carpentry*, Phillimore, 1980, and numerous earlier publications by this author for English medieval practices; Fitchen *Dutch barn* for later North American practice; and Engel, H., *The Japanese house*, Tuttle, 1964, for a different approach.

7. Linkwitz, K., 'Two examples of integrated formfinding and numerically controlled premanufacturing: the timber shells of Bad Neuenahr 1993 and Maulbronn 1995' in Giuliani *Spatial structures*, pp. 1397-1406; idem, 'Holzschalendächer Hölderlin-Haus-der-Anthroposophia, Maulbronn', *Bautechnik*, vol. 73, 1996, pp. 205-214.

8. Hamilton, S. B., 'The structural use of iron in antiquity', *Trans.NS*, vol. 31, 1957-59, pp. 29-47, discusses some of the evidence for early uses of iron, but omits mention of some of uses referred to by Vitruvius and of the extensive sixth-century use of iron tie bars in the church of Hagia Sophia, Istanbul. For the latter see Mainstone *Hagia Sophia*, pp. 70-71, 77, 82-83. See also the works cited in note 3 above; Reeves, J., Simpson, G. and Spenser, P., 'Iron reinforcement of the tower and spire of Salisbury Cathedral', *The Archaeological Journal*, vol. 149, 1992, pp. 380-406; and further references in Chapter 8. Wilcox, R. P., *Timber and iron reinforcement in early buildings*, Society of Antiquaries, London, 1981, is less helpful than its title promises.

9. Brown, J. M., 'W. B. Wilkinson (1819-1902) and his place in the history of reinforced concrete', *Trans.NS*, vol. 39, 1966-67, pp. 129-142.

10. The continued development is summarized in Hamilton *Reinforced concrete*; Collins *Concrete*; Condit, *Nineteenth century*, pp. 231-240; Haeger-

mann et al., *Vom Caementum zum Spannbeton*, vol.1; de Courcy, J. W., 'The emergence of reinforced concrete, 1750-1910', *SE*, vol. 65A, 1987, pp. 315-322; and the papers by Bussell, M. N. in 'Historic concrete', *Proc.ICE: Structures and Buildings*, vol. 116, 1996, pp. 295-334. Some mention should also be made of ferrocemento, developed chiefly by Nervi in the present century and using a much closer mesh of very small diameter reinforcement. For this see Nervi *Structures*, pp. 50-62.

11. Walley, F., 'The childhood of prestressing', *SE*, vol. 62A, 1984, pp. 5-10.

12. See, for instance, Richmond, B. and Head, P. R., 'Alternative materials in long-span bridge structures', in Institution of Structural Engineers, *1st Oleg Kerensky Memorial Conference*, 1988, session 5, pp. 6-12; and Liddell, W. I., 'Structural fabrics and foils', ibid., pp. 13-17.

Chapter 4 Construction and form

1. The earlier situation is discussed further, with references, in Mainstone, R. J., 'Structural theory and design before 1742, *ARev.*, vol. 143, 1968, pp. 303-310.

2. Institution of Structural Engineers, *Industrialized building and the structural engineer*, 1966; and idem, *Design for industrial production*, 1971, discuss the constraints from the point of view of later practice.

3. In addition to general works on the history of reinforced concrete cited in Chapter 3, see Peterson, J. L., 'History and development of precast concrete in the United States', *JACI*, vol. 25, 1954, pp. 477-496; and Koncz *Precast concrete*.

4. Newby, F., 'On cast iron and precast concrete', *Architectural Design*, vol. 30, 1960, 306-314.

5. Sefton, W., 'Lift slab design and construction', in Institution of Structural Engineers, *Proceedings of the fiftieth anniversary conference*, 1958, pp 275-282.

6. See, for instance, Adler, F., 'Jig for the construction of a 17-storey block of flats at Barras Heath, Coventry', *Proc.ICE*, vol. 27, 1964, pp. 433-463; and Wittfoht *Bridges*, pp. 232-235.

7. This was the procedure adopted for constructing the arched ribs of the Gothic-style vaults of the Cathedral of St John and St Paul, Washington, DC. To achieve precisely the desired profile calls either for temporary weighting of that part of the centering that is not yet loaded by the voussoirs finally placed, or for repeated adjustment of the wedges as the centering is deformed by the placing of the voussoirs. At the cathedral, the voussoirs were initially held apart only by small pieces of lead, the joints being grouted with very wet mortar only when the whole rib was complete (as I observed in 1966 by courtesy of Mr Howard B. Trevillian). For a much fuller discussion of probable Gothic procedures, see Fitchen *Construction*.

8. Perronet, J. R., 'Mémoire sur le cintrement et le décintrement des ponts, et sur les differens mouvemens que prennent les voûtes pendent leur construction', *Mémoires de l' Académie Royale des Sciences*, 1773, pp. 33-50.

9. Alberti *De re aedificatoria*, bk 3, ch. 14.

10. Torroja *Structures*, pp. 70-81. For the original Melan system of reinforcing concrete, see also Chapter 6.

11. Alberti *De re aedificatoria*, bk 3, ch. 14.

12. Eiffel, G., *Notice sur le Pont du Douro, a Porto (Pont Maria-Pia)*, Paris, 1879; idem, *Mémoire presenté a l'appui du projet définitif du viaduc de Garabit*, Paris, 1889; Woodward *St. Louis Bridge*.

13. Finsterwalder, U., and Schambeck, H., 'Von der Lahnbrücke Balduinstein bis zur Rheinbrücke Bendorf', *Der Bauingenieur*, vol. 40, 1965, pp. 85-91; Finsterwalder, U., 'Free-cantilever construction of prestressed concrete bridges and mushroom shaped bridges', in American Concrete Institute, *Concrete bridge design*, 1969, pp. 467-494; and Wittfoht *Bridges*, pp. 196-214, give general accounts of the development of the process as well as referring to the Bendorf Bridge. Kerensky, O. A. and Little, G., 'Medway Bridge: design', *Proc.ICE*, vol. 29, pp. 19-52; and Kier, M., Hansen, F., and Dunster, J. A., 'Medway Bridge: construction', ibid., pp. 53-100, describe in detail the Medway Bridge.

14. Virlogeux, M., 'Erection of cable-stayed bridges' in Ito et al. *Cable-stayed bridges*, pp. 77-106; idem, 'Normandie Bridge: design and construction', *Proc.ICE: Structures and Buildings*, vol. 99, 1991, 281-302.

15. Mainstone, R. J., *Tests on the new Government Offices, Whitehall Gardens*, National Building Studies Research Paper 28, HMSO, 1960.

16. Clark and Stephenson *Tubular bridges*, ch. 4 and Appendix C.

17. Freyssinet, E. 'Pre-stressed concrete: principles and applications', *JICE*, vol. 33, 1950, pp. 331-380; Ordonez *Freyssinet*, ch. 4 and Appendix C. Also see again Walley, F., 'The childhood of prestressing', *SE*, vol. 62A, 1984, pp. 5-10.

18. Chaudesaigues, M. J., 'La reconstruction en béton précontraint des ponts sur la Marne à Annet, Trilbardou, Esbly, Ussy et Changis-Saint-Jean', *Annales de l'Institut Technique du Bâtiment et des Travaux Publics*, new series, no. 228, 1952.

Chapter 5 Structure and form

1. Galileo *Discorsi* (Crew and Salvio p. 2).

2. Boada, I. P., *El templo de la Sagrada Familia*, Omega, Barcelona, 1952; Collins, G. R., *Antonio Gaudi*, George Braziller, 1960; idem, 'Antonio Gaudi: structure and form', *Perspecta* (Yale University School of Architecture), vol. 8, 1963, pp. 63-90.

3. No unique validity is claimed for the classification in this figure. Other classifications, explicit or implicit, may be found in Torroja *Philosophy of structures*; Lisborg, N., *Structural design*, Batsford, 1961; Siegel, C., *Structure and form in modern architecture*, (trans. T. E. Burton), Reinhold, 1962; Salvadori and Heller *Structure*; Engel, H., *Tragsysteme: structure systems* (in German and English), Deutsche Verlags-Anstalt, Stuttgart, 1967; and Schodek *Structures*.

4. It is even less possible here than in fig. 5.5 to do justice to the vast potential range of these forms. To gain a further idea of it, as already realized in

actual construction up to 1963, see Makowski *Steel space structures*.

5. Baker, J. E., Williams, E. L., and Lax, D., 'The design of framed buildings against high-explosive bombs', in Institution of Civil Engineers, *The civil engineer in war*, 1948, vol. 3, pp. 80-112.

6. *Report on Ronan Point*.

Part 2 Structural elements

Chapter 6 Arches and catenaries

1. Maxwell, G., *A reed shaken by the wind*, Longmans Green, 1957, reproduces (between pp. 162-163) an excellent series of photographs of the sequence of construction.

2. Garstang, J., *Tombs of the third Egyptian dynasty at Reqaqnah and Bet Khalláf: report of excavations at Reqaqnah 1901-2*, Constable, 1904. For another early example (at Khorsabad) see Place, V., *Ninive et l'Assyrie*, Paris, 1867-1870, vol 1, pp. 170-178 and vol. 2, pls 9-12.

3. Clarke, S., *Christian antiquities of the Nile Valley*, Oxford University Press, 1912, p. 29.

4. Lugli, G., 'Considerazioni sull'origine dell'arco a conci radiali', *Palladio*, new series, vol. 2, 1952, pp. 9-31; Boyd, T. D., 'The arch and the vault in Greek architecture', *American Journal of Archaeology*, vol. 82, 1978, 83-100. See also Blake, *Ancient Roman construction*; idem, *Roman construction.*, and Lugli *Tecnica edilizia*, for this and other Roman innovations discussed in this chapter.

5. Strictly speaking, the joggled voussoir, like some other predominantly Roman developments, was probably a Hellenistic Greek innovation in the first place since it was used already in Ptolemaic Egypt. See Clarke and Engelbach *Egyptian masonry*, p. 187.

6. Fugl-Meyer *Chinese bridges*, pp. 85-99. See also Needham *Civil engineering and nautics*, pp.168-170 and pl. 345, and Tang Huangcheng *Chinese bridges*, pp. 110-162.

7. Robert Hooke enunciated the idea in 1675 as the 'riddle of arch, *ut pendet continuum flexile, sic stabit inversum rigidum*', though he had already claimed, at the Royal Society, to have the answer in 1670. See also Gregory, D., 'Catenaria', *Philosophical Transactions of the Royal Society*, vol. 19, no. 231, 1697, pp. 637-652; and idem, 'Responsio ad animadversionem ad Davidis Gregorii Catenariam, Act. Eruditorum Lipsiae', ibid., vol. 21, no. 259, 1699, pp. 419-426 (both in Latin).

8. See, for instance, La Hire *Traité*, pp. 465-470; idem, 'Sur la construction des voûtes dans les édifices', *Mémoires de l'Académie Royale des Sciences*, vol. 69, 1713, pp. 69-77; and Coulomb, C. A. de, 'Essai sur une application des règles de maximis & minimis a quelques problèmes de statique, relatifs a l'architecture', *Mémoires de Mathématique et de Physique, présentés a l'Académie Royale des Sciences, par divers savans, et lûs dans les Assemblées*, vol. 7, 1776, pp. 343-382 (reprinted in Heyman *Coulomb's memoir*). Ruddock *Arch bridges* and Séjourné *Grandes voûtes* review actual developments in the design and construction of bridge arches.

9. Séjourné *Grandes voûtes*, vol. 2, pp. 67-82. Immediately afterwards, even this span was slightly exceeded at Plauen, Saxony (ibid., vol. 3, pp. 52-58).

10. The relief is best illustrated in Cichorius, C., *Die Reliefs der Traianssäule*, vol. 3, Berlin, 1900, scene 99, and its reproduction in Lepper, F.and Frere, S., *Trajan's Column*, Alan Sutton, 1988, pl. 72 with commentary, pp. 148-151. Gazzola *Ponti*, vol. 2, pp. 136-137, gives a plan of the site. See also Dio Cassius, *Roman History*, ed. E. Cary, Loeb Classical Library, William Heinemann, 1925, vol. 8, bk 68.

11. Baines, Sir F., *Westminster Hall: report to the first commissioner of H.M. Works, London, Cd. 7436*, HMSO, 1914, contains very precisely detailed drawings of the construction. See also Courtenay, L. T., 'The Westminster Hall roof: a new archaeological source', *Journal of the British Archaeological Association*, vol. 143, 1990, pp. 95-111. The representations in Viollet-le-Duc *Dictionnaire raisonné*, article '*charpente*', vol. 3, pp. 41-45, are less accurate and reliable.

12. De l'Orme *Nouvelles inventions*; Verantius *Machinae novae*, pl. 31.

13. Booth, L. G., 'The development of laminated timber arch structures in Bavaria, France and England in the early 19th century', *Journal of the Institute of Wood Science*, vol. 5, 1971, pp. 3-16; idem, 'Laminated timber arch railway bridges in England and Scotland', *Trans.NS*, vol. 44, 1971-72, pp. 1-16.

14. Reproduced in Maguire, R. and Matthews, P., 'The Ironbridge at Coalbrookdale', *Architectural Association Journal*, vol. 74, 1958, pp. 29-45. See also Cossons, N. and Trinder, B., *The iron bridge*, Moonraker, 1979.

15. Excellent fuller accounts of this development, copiously illustrated and with extensive bibliographies, are given in James, J. G., 'The cast iron bridges of Thomas Wilson', *Trans.NS*, vol. 50, 1978-79, pp. 55-72; idem, 'The evolution of iron bridge trusses to 1850' ibid., vol. 52, 1980-81, pp. 67-101; 'Russian iron bridges to 1850', ibid., vol. 54, 1982-83, pp. 79-104; idem, 'Some steps in the evolution of early iron arched bridge designs', ibid., vol. 59, 1987-88, pp. 153-186; and idem, 'Thomas Paine's iron bridge work', ibid., vol. 59, 1987-88, pp. 189-221.

16. Perronet, J. R., 'Mémoire sur la réduction de l'épaisseur des piles et sur la courbure qu'il convient de donner aux voûtes, le tout pour que l'eau puisse passer plus librement sous les ponts', *Mémoires de l'Académie Royale des Sciences*, 1777, pp. 553-564.

17. 'The St Louis Memorial Arch', *The Engineer*, vol. 221, 1966, pp. 373-375.

18. Humber, W., *A complete treatise on cast and wrought iron bridge construction, including iron foundations*, 2nd edn, 2 vols, London, 1864, pp. 231-234 and pls 78-80; and Shirley-Smith, Sir Hubert, 'Royal Albert Bridge, Saltash' in Pugsley, Sir Alfred, ed., *The works of Isambard Kingdom Brunel*, Thomas Telford, 1976, pp. 163-182.

19. For the Thur and Valtschiel arches see Bill *Maillart*, pp. 94-98 and 58-61. See also Maillart, R., 'Leichte Eisenbeton-Brücken in der Schweiz', *Der Bauingenieur*, vol. 12, 1931, pp. 165-171.

20. Freyssinet, E., 'Hangars à dirigeables en ciment armé en construction à l'aéroport de Ville-neuve Orly', *Le Génie Civil*, vol. 83, 1923, pp. 265-273, 291-297, and 313-319; and 'Le nouveau pont en béton armé, sur le Seine', ibid., pp. 417-421.

21. Graefe, R., *Vela erunt*, Von Zabern, Mainz, 1979, 1 vol. text (German with English summary) and 1 vol. plates, presents all the surviving evidence, both in Rome and elsewhere, and discusses possible reconstructions.

22. Verantius *Machinae novae*, pl. 34; Needham *Civil engineering and nautics*, pp. 193-207.

23. Séguin, M., *Des ponts en fil de fer*, Paris, 1824; Vicat, L. J., 'Ponts en fil de fer sur le Rhône', *Annales des Ponts et Chaussées, Mémoires et Documents*, vol. 1, 1831, pp. 93-144.

24. For the modern spinning process, essentially that developed by the Roeblings but more mechanized, see, for instance, John A. Roebling's Sons Company, *Suspension bridges: a century of progress*, 1934; and Shirley-Smith *Bridges*, pp. 91, 185-188.

Chapter 7 Vaults, domes, and curved membranes

1. Brodrick, A. H., 'Grass roots', *ARev.*, vol. 115, 1954, pp. 101-111; and Rapoport, A., *House form and culture*, Prentice Hall, 1969, discuss contemporary hut forms that have probably developed fairly directly from this presumed earliest form.

2. Mallowan, M. E. L., and Rose, J. C., 'Excavations at Tall Arpachiyah 1933', *Iraq*, vol. 2, pt 1, 1935, pp. 1-178, especially pp. 25-34; Dikaios, P., *Khirokitia: final report of the excavation of a Neolithic settlement in Cyprus on behalf of the Department of Antiquities 1936-1946*, Oxford University Press, 1953.

3. Frankfort, H., Jacobsen, T., and Preusser, C., *Tell Asmar and Khafaje – the first season's work in Eshnunna 1930-31*, Chicago University Oriental Institute Communication 13, 1932, fig. 11. For false domes in stone see Rohlfs *Primitive Kuppelbauten*; Lilliu, G., 'L'architettura Nuragia', *Atti del XIII Congresso di Storia dell'Architettura*, Rome, 1966, vol. 1, pp. 2-92; and Allen *Shelters*.

4. Boak, A. E. R., and Peterson, E. E., *Karanis: topographical and architectural report of excavations during the seasons 1924-28*, University of Michigan Humanistic Series 25, 1931.

5. Faegre, T., *Tents*, John Murray, 1979, describes and illustrates a wide range of the tents built by present nomads, some of which probably derive from much earlier forms.

6. Palladio *Quattro libri*, bk 4, ch. 23, p. 92. An almost identical dome is shown on the Temple of Vesta by the Tiber in Rome.

7. Vitruvius *De architectura*, bk 5, ch. 10.

8. Maiuri, A., 'Il restauro di una sala termale a Baia', *Bolletino d'Arte*, vol. 10, 1930, pp. 241-253.

9. For fuller accounts, see MacDonald *Architecture of the Roman empire* and Ward-Perkins *Roman Imperial Architecture*. See also Hansen, E., La 'Piazza d'Oro' e la sua cupola', *Analecta Romana Instituti Danici I Supplementum*, 1960; and MacDonald, W. L. and Pinto, J., *Hadrian's villa and its legacy*, Yale University Press, 1995.

10. Giovannoni, G., 'La cupola della Domus Aurea Neroniana in Roma', *Atti del I Congresso Nazionale di Storia dell'Architettura 1936*, 1938, pp. 3-6; and MacDonald *Architecture of the Roman empire*, pp. 38-41.

11. Best described in Licht *Rotunda*. Piranesi, F., *Raccolta de'tempi antichi*, Rome, 1790, vol. 2, pl. 28, was wrong in showing a complex system of embedded ribs rising right to the eye of the dome. Bouteloup, P., 'Les coupoles en maçonnerie non armée', *Annales des Ponts et Chaussées, Mémoires et Documents*, vol. 128, 1958, pp. 429-503, gives a useful tabular summary of the dimensions of the principal domes constructed up to the end of the nineteenth century.

12. Terenzio, A., 'La restauration du Pantheon de Rome', in *La conservation des monuments d'art et d'histoire*, l'Institut de Cooperation Intellectuelle, Paris, 1933, pp. 280-285, illustrates the complete extent of the cracking as observed during the most recent restoration in about 1930, pl. 26.

13. Of the numerous analyses that have been made of the structural behaviour, see, for instance, Mark, R. and Hutchinson, P., 'On the structure of the Roman Pantheon', *AB*, vol. 68, 1986, pp. 24-34, and my comment on this, Mainstone R. J., 'Letters', loc.cit., pp. 673-674.

14. Giovannoni *Sala termale*; Caraffa *Cupola*. Choisy, A., *L'art de bâtir chez les Romains*, Paris, 1872, illustrates numerous examples of such ribs, but his drawings exaggerate the precision and regularity of their construction and make them appear more independent of the main concrete mass than they really were. Rasch, J. J., 'Zur Konstruktion spätantiker Kuppeln vom 3. bis 6. Jahrhundert', *Jahrbuch des Deutschen Archäeologischen Instituts*, vol. 106, 1991, pp. 311-383 and pls 77-84 presents the results of recent photogrammetric studies of all surviving similarly ribbed domes and much additional data.

15. Verzone, P., 'Le cupole di tubi fittili nel v e vi secolo in Italia', *Atti del I Congresso Nazionale di Storia dell'Architettura 1936*, Florence, 1938, pp. 7-11; De Angelis d'Ossat, G., 'Nuovi dati sulle volte costruite con vasi fittili', *Palladio*, vol. 5, 1941, pp. 241-251; Kostof, S. K., *The Orthodox Baptistery of Ravenna*, Yale Publications in the History of Art 18, 1956; Arslan, E. A., 'Osservazioni sull'impiego e la diffusione delle volte sottili in tubi fittili', *Bollettino d'Arte*, series 5, year 50, 1965, pp. 45-52; Storz, S., 'Zur Funktion von keramischen Wölbröhren in römischen und frühchristlichen Gewölbebau', *Architettura*, vol. 2, 1977, pp. 89-105; Roberti, G. M., Lombardini, N. and Falter, H., 'Late Roman domes in clay tubes: historical and numerical study of San Vitale in Ravenna' in Giuliani *Spatial structures*, pp. 1237-1244.

16. This is a very simplified discussion of the alternatives. For a fuller discussion introducing some further distinctions see Mainstone, R. J., 'Squinches and pendentives: comments on some problems of definition', *Art and Archaeology Research Papers*, no. 4, December 1973, pp. 131-137, noting a missing line – 'geometric form but with the individual' – before the words 'blocks or bricks' four lines from the end of p. 131.

17. Many of the holes referred to in the dome of Hagia Sophia have long been visible, having been used to support lighting hoops. Recent cleaning of the mosaics has disclosed others at a lower level which can have had no use other than the support of circumferential centering. Timber struts across the window openings, of which vestiges remain, will have assisted in maintaining circumferential continuity in the earlier stages of construction. The dome of the Mausoleum of Diocletian is illustrated in Adam, R., *Ruins of the palace of the emperor Diocletian at Spalatro in Dalmatia*, London, 1764, pl. 34, and in numerous later publications including Mainstone, R. J., 'Brunelleschi's dome of S. Maria del Fiore and some related structures', *Trans.NS*, vol. 42, 1969-70, pp. 107-26, pl. 17. The herringbone bond is well depicted in a sketch by Antonio da Sangallo the Younger, Uffizi no. 900 (reproduced in Mainstone, R. J., ibid., fig. 2, p. 114), and may be seen today in a number of fifteenth- and sixteenth-century Florentine domes including, in addition to that of the cathedral, the crossing domes and aisle vaults of the churches of S. Lorenzo and S. Spirito, and the domical vault of a large octagonal vestibule in the Forezza da Basso designed by Sangallo. Other examples of both alternatives are well illustrated in Sanpaolesi, P., 'Strutture a cupola autoportanti', *Palladio*, vol. 21, 1971, pp. 3-64. (It seems to me, however, that Sanpaolesi goes too far in this paper in attributing both constructional and statical advantages to the usual Roman practice of casting concrete domes in superimposed horizontal layers. From the constructional point of view, this practice would, just as in the construction of any other form, have minimized the need for centering, given a certain thickness of layer and certain rates of construction and of hardening of the concrete. But these latter factors were the more significant ones in the determination of how far the potentially self-supporting character of each completed ring could, in practice, be exploited. The statical advantage claimed is that of reducing or eliminating outward thrusts on the supports. It should be clear from the discussion in Chapter 2 that the only way in which this can be achieved is by giving the lower part of the dome the ability to develop circumferential tensions adequate to contain within itself the radial thrusts inevitably developed in the upper part. The inclination of the layers of concrete – or rings of masonry – makes no difference to these radial thrusts.)

18. For the Iranian domes see Stronach, D. and Young, T. C., 'Three Seljuk tomb towers', *Iran*, vol. 4, 1966, pp. 1-20. The timber dome of the Dome of the Rock is illustrated and described in Vogüé, C. J. M. de, *Le Temple de Jérusalem, monographie du Haram-ech-Chérif*, Paris, 1864, pp. 91-93 and pl. 19; Richmond, E. T., *The Dome of the Rock in Jerusalem: a description of its structure and decoration*, Clarendon Press, 1924; and Creswell, K.A.C., *Early Muslim architecture*, 2nd edn, Clarendon Press, 1969, frontispiece and pp. 65-131. For origins see again Andrews, P.A., 'Zwiebelkuppel und Gitterzelt', in Graefe *Geschichte*, pp. 48-69.

19. Godard, A., 'The Mausoleum of Öljeitü at Sultaniya' in Pope *Survey*, pp. 1103-1118 ; Wilber *Architecture of Islamic Iran*, especially pp. 61-67;

Sanpaolesi, P., 'La cupola di Santa Maria del Fiore ed il Mausoleo di Soltanieh', *Mitteilungen des Kunsthistorischen Institutes in Florenz*, vol. 16, 1972, pp. 221-260 (note that some of the references here to my paper referred to in note 23 below appear to be based on misreadings of the text); and Mainstone, R. J., 'Brunelleschi's dome', *ARev.*, vol. 162, 1977, pp. 156-166, especially p. 165.

20. Two copies exist, one published in Guasti *Cupola*, doc. 51, pp. 28-30; and the other (more definitive) by Doren, A., 'Zum Bau der Florentiner Domkuppel', *Repertorium für Kunstwissenschaft*, vol. 21, 1898, pp. 249-262.

21. See Sanpaolesi, P., *La cupola di Santa Maria del Fiore*, (Reale Istituto d'Archeologia e Storia dell'Arte, Opera d'Arte 11), Rome, 1941 (reprinted Edam, Florence, 1977); Prager, F. D., and Scaglia, G., *Brunelleschi: studies of his technology and inventions*, MIT Press, 1970; Mainstone, R. J., 'Brunellschi's dome' (note 19); Saalman, H., *Filippo Brunelleschi: the cupola of Santa Maria del Fiore*, Zwemmer, 1980. The most comprehensive survey is that by G. B. Nelli in Sgrilli *Descrizione*.

22. Alberti *De re aedificatoria*, bk 3, ch. 14.

23. Mainstone, R. J., 'Brunelleschi's dome of S. Maria del Fiore and some related structures', *Trans.NS*, vol. 42, 1969-70, pp. 107-126; and idem, 'Structural analysis, structural insights and historical interpretation', *JSAH*, vol. 56, 1997, pp. 316-340, fig. 5.

24. See, for instance, Fontana, C., *Templum Vaticanum et ipsius origo*, Rome, 1694, especially pp. 315-323; Millon, H. A. and Smyth, C. H., *Michelangelo architect: the facade of San Lorenzo and the drum and dome of St. Peter's*, Olivetti, Milan, 1988; and Mainstone, R. J., 'The dome of St Peter's: structural aspects of its design and construction and inquiries into its stability', in idem, *Structure in architecture*.

25. See chiefly Baldinucci, F., *Vita del Cavaliere Gio. Lorenzo Bernino*, Florence, 1682, pp. 82-102; Le Seur et al. *Parere*; Poleni *Memorie*, cols 141-218 and 402-424 and pls 1-19 and K; Di Stefano **Cupola**, pp. 107-129 (rev. edn, pp. 81-95); and Di Stefano, R., *Luigi Vanvitelli ingegnere e restauratore*, Edizioni Scientifiche Italiane, n.d, pp.217-222 and pls 414-434.

26. Collins, G. R., 'The transfer of thin masonry vaulting from Spain to America', *JSAH*, vol. 27, 1968, pp. 176-201.

27. 'Erecting a large dome without falsework', *The Engineering Record*, vol. 60, 1909, pp. 508-510.

28. Godard, A., 'Voûtes Iraniennes', *Athar-é Iran* (Annales du service archéologique de l'Iran), vol. 4, pt 2, 1949, pp. 187-345, especially pp. 259-325

29. There are good illustrations of the *tas-de-charge* in Viollet-le-Duc *Dictionnaire raisonné*, article 'construction', especially figs 48ter and 49bis, vol. 4, pp. 93, 95. See also this whole article, pp. 1-279, the articles 'tas de charge' and 'voûte', vol. 9, pp. 7-12 and 464-550, and Fitchen *Construction* for more general discussions of the construction of such vaults.

30. A few bays farther east at Ourscamp, the latter has happened (illustrated in Mainstone, R. J., 'The springs of structural invention', *JRIBA*, vol. 70, 1963, pp. 57-7l, fig. 42; and in idem, 'Intuition and the springs of structural invention', *VIA* (Graduate School of Fine Arts, University of Pennsylvania),

vol. 2, 1973, pp. 46-67, fig. 40). See also Porter, A. K., *The construction of Lombard and Gothic vaults*, Yale University Press, 1911 (illustrating, in figs 32 and 35, similar behaviour at the Abbey of Longpont); Gilman, R., 'The theory of Gothic architecture and the effects of shell fire at Rheims and Soissons', *American Journal of Archaeology*, vol. 24, 2nd series, 1920, pp. 37-72; Sabouret, V., 'Les voûtes d'arêtes nervurées', *Le Génie Civil*, vol. 92, 1928, pp. 205-209; and Aubert, M., 'Les plus anciennes croisées d'ogives, leur rôle dans la construction', *Bulletin Monumental*, vol. 93, 1934, pp. 5-67 and 137-237. Most recent analyses of the behaviour using photoelastic and finite-element models ignore the consequences of heterogeneous construction, detailing of the interconnections, and sequence of construction. In this connection see again Mainstone, R. J., 'Structural analysis, structural insights and historical interpretation', *JSAH*, vol. 56, 1997, pp. 316-340.

31. The earlier development is reviewed in Ward, C., *Mediaeval church vaulting*, Princeton Monographs in Art and Archaeology V, 1915. For the later developments see, for instance, Frankl *Gothic architecture*; Schulze, K. W., *Die Gewölbesysteme im spätgotischen Kirchenbau in Schwaben von 1450 bis 1520*, Tübinger Forschungen zur Archäologie und Kunstgeschichte 16, Reutlingen, 1940; Pevsner, N., 'Bristol-Troyes-Gloucester', *ARev.*, vol. 113, 1953, pp. 88-98; and Cavallari-Murat, A., 'Static intuition and formal imagination in the space lattices of ribbed Gothic vaults', Student Publications of the School of Design (North Carolina State College), vol. 2, no. 2, 1963, pp. 2-39 (valuable more for its classification of patterns of ribbing than for its interpretations).

32. MS. B, Institut de France, fol. 10v. (reproduced in Reale Commisione Vinciana, *Il codice B (2173) nell'Istituto di Francia, I manoscritti e i disegni di Leonardo da Vinci V*, Rome, 1941, p. 18).

33. Heyman, J., 'Spires and fan vaults', *International Journal of Solids and Structures*, vol. 3, 1967, pp. 243-258. See also Mackenzie, F., *Observations on the construction of the roof of King's College Chapel, Cambridge*, London, 1840: Willis. R., 'On the construction of the vaults of the Middle Ages', *Transactions of the Royal Institute of British Architects*, vol. 1,1842, pp. 1-69; Howard, F. E., 'Fan vaults', *Archaeological Journal*, vol. 68, 1911, pp. 1-42; and Leedy, W., *Fan Vaulting*, Scholar, 1980.

34. Bauersfeld, W., 'Development of the Zeiss Dywidag process', lecture given in 1942 in Berlin, reprinted in Joedicke *Shell architecture*, pp. 281-283. One earlier 'experiment' was the central closure of the ribbed inner dome of a church in the Black Forest described in Kleinlogel, A., 'Die Eisenbetokuppel in Sanct Blasien', *Beton u. Eisen*, vol. 16, 1912, pp. 345-350. This closure was a part-spherical shell with a radius of curvature of 23.1 m, a span of 15.4 m, and only 120 mm thick.

35. Torroja *Structures*, pp. 23-28.

36. Laffaille, B., 'Mémoire sur l'étude générale des surfaces gauches minces', *Publications of the International Association for Bridge and Structural Engineering*, vol. 3, 1935, pp. 295-332; Aimond, F.,

'Étude statique des voiles minces en paraboloide hyperbolique travaillant sans flexion', ibid., vol. 4, 1936, pp. 1-112.

37. 'Saarinen challenges the rectangle', *AF*, vol. 98, no. 1, 1953, pp. 126-133; 'Grand Central of the air', ibid., vol. 104, no. 5, 1956, pp. 106-115; Becker, W. C. E., 'Intersecting ribs carry concrete roof shell', *CE*, vol. 25, 1955, pp. 430-433; Faber, C., *Candela: the shell builder*, Reinhold, New York, 1963, pp. 76-87. For wider reviews see Boyd, R., 'Engineering of excitement', *ARev.*, vol. 124, 1958, pp. 294-308 and again Joedicke *Shell architecture*.

38. Nervi *New structures*, pp. 32-39; Esquillan, N., 'The shell vault of the Exposition Palace, Paris', *Proc.ASCE*, vol. 86, no. ST1, 1960, pp. 41-70.

39. Isler, H., 'Evolution of shell structures' in Giuliani *Spatial structures*, pp. 271-292; Robbin *New architecture*, pp. 26-32.

40. Happold, E. and Liddell, W. I., 'Timber lattice roof for the Mannheim Bundesgartenschau', *SE*, vol. 53, 1975, pp. 99-135, gives a much fuller account.

41. Robbin *New architecture*, pp. 33-37.

42. For the first see Berger *Light structures*, pp. 29-30; for the second see Hähl, H., 'Der stahlkonstruktion für den US-Pavilon auf der Weltausstellung in Brüssel 1958', *Der Stahlbau*, vol. 27, 1958, pp. 117-121.

43 Little, J. B., 'Unique roof construction at Dulles Airport', *JACI*, vol. 60, 1963, pp. 835-850.

44. Otto *Das hangende Dach*; and Otto and Schleyer *Tensile structures*, which also reviews the whole development up to about 1966. For an earlier review see Boyd, R., 'Under tension', *ARev.*, vol. 124, 1963, pp. 324-334.

45. 'Parabolic pavilion', *AF*, vol. 97, no. 4, 1952, pp. 134-139; 'Parabolic cable roof', ibid., vol. 98, no. 6, 1953, pp. 170-171.

46. Berger *Light structures*, pp. 142-157.

47. Otto, F., and Trostel, R., *Tensile structures: pneumatic structures*, trans. D. Ben-Yaakov and T. Pelz, MIT Press, 1967, considers the possibilities and early achievements in much more detail.

48. Villecco, M., 'The infinitely expandable future of air structures', *AF*, vol. 133. no.2, 1970, 40-43.

Chapter 8 Beams and slabs

1. Clarke and Engelbach *Egyptian masonry*, p. 7; Lauer *Pyramide*, p. 203.

2. Quoted from the translation in Bowie, T., *The sketchbook of Villard de Honnecourt*, 2nd edn rev., George Wittenborn, 1959, p. 14

3. For Wren's design see Plot *Oxfordshire*, p. 272. For Independence Hall see Peterson, C. E., 'Independence Hall: its fabric restored' in idem, ed., *Building early America*, Chiltern Book Co., 1976, pp. 279-297. For some other examples see Yeomans, D., 'The Serlio floor and is derivatives', *Architectural Research Quarterly*, vol. 2, 1997, 74-83.

4. Alberti *De re aedificatoria*, bk 3, ch. 12; Francesco di Giorgio Martini, *Trattato di architettura ingegneria e arte militare*, Codice Torinese Saluzziano 148, fol. 22v. (facsimile edn, ed. Matteo, C., Polifilo, Milan, 1967, vol.1, pl. 40).

5. Fletcher, H. M., 'Sir Christopher Wren's carpentry', *JRIBA*, 3rd series, vol. 30, 1923, pp. 388-391.

6. Graves, H. G., 'Further notes on the early use of iron in India', *Journal of the Iron and Steel Institute*, vol. 85, 1912, pp. 187-202.

7. Dinsmoor, W. B., 'Structural iron in Greek architecture', *American Journal of Archaeology*, 2nd series, vol. 26, 1922, pp. 148-158.

8. The architrave blocks of the Temple of Zeus at Agrigento – Dinsmoor's chief example – are all now on the ground. Close inspection in 1969 of as many as were accessible failed to disclose even significant rust stains, and certainly none of the splitting that the rusting of solid embedded iron beams would have caused. And neither here nor elsewhere could the supposed beams have functioned as efficient reinforcements: they were incorrectly placed and inadequately anchored at the ends to act as ties. See also the comments by Hamilton, S. B., 'The structural use of iron in antiquity', *Trans. NS*, vol. 31, 1957-59, pp. 29-47; and by Jewett, R. A., 'Controversy', *Technology and Culture*, vol. 9, 1968, pp. 419-426. Choisy *Histoire,* vol. 1, p. 279, suggested that the cuttings at Agrigento were filled with '*un paquet de fers plates posés de champ*'.

9. The development of iron beams is outlined in Hamilton, S. B., 'The use of cast iron in building', *Trans.NS*, vol. 21, 1940-41, pp. 139-155; and, more definitively, in Skempton , A. W., 'The origin of iron beams', *Actes du VIIIe Congrès Internationale d'Histoire des Sciences*, Florence, 1956, pp. 1029-1039; and Jewett, R. A., 'Structural antecedents of the I-beam, 1800-1850', *Technology and Culture*, vol. 8, 1967, pp. 346-362. Among the contemporary published sources see particularly Fairbairn *Application*, and the works by Tredgold and Hodgkinson cited in the next two notes. See also Swailes, T., '19th century cast-iron beams: their design, manufacture and reliability', *Proc.ICE: Civil Engineering*, vol. 114, 1996, pp.25-35; and for 'fireproof' floors see note 25 below.

10. Tredgold, T., *A practical essay on the strength of cast iron*, London, 1822.

11. Hodgkinson, E., 'Theoretical and experimental researches to ascertain the strength and best form of iron beams', *Memoirs of the Manchester Literary and Philosophical Society*, 2nd series, vol. 5, 1831, pp. 407-544.

12. Two such collapses were of a beam in a cotton mill in Manchester described in Fairbairn, W., 'On some defects in the principle and construction of fire proof buildings', *Proc.ICE*, vol. 6, 1847, pp. 213-224, and of a span of Robert Stephenson's Dee Bridge shortly afterwards. The latter led to the setting up of a Royal Commission whose report, *Report into the application of iron*, sets out in detail the views of many leading engineers of the time. See also Sutherland, R. J. M., 'The introduction of structural wrought iron', *Trans.NS*, vol. 36, 1963-64, 67-84.

13. Barlow, S. and G Foster, 'Universal beams and their application', *SE*, vol. 35, 1957, 425-440.

14. Fairbairn *Tubular bridges;* Clark and Stephenson *Tubular bridges*, vol. 1, pp. 83-460 and vol. 2, pp. 552-589.

15. The development is summarized in Maggard, S. P., Sunbury, R. D. and Galambos, T. V., 'Trends in the design of steel box girder bridges', *Proc.ASCE*, vol. 93, no. ST3, 1967, pp. 165-180. Thul, H., 'Stählerne Strassenbrücken in der Bundesrepublik', *Der Bauingenieur*, vol. 41, 1966, pp. 169-189, illustrates some typical cross-sections together with others of alternative forms incorporating deep I-section or trussed beams. After several collapses of box spans during construction in the late 1960s and early 1970s, doubts about the adequacy of the original safety margin against buckling of the thin plates led to more stringent rules to guard against this risk.

16. Roberts, Sir G., 'Severn Bridge: design and contract arrangements', *Proc.ICE*, vol. 41, 1968, pp. 1-48; Hyatt, K. E., 'Severn Bridge: fabrication and erection', ibid., pp. 69-104.

17. Cordemoy, J. L. de, *Nouveau traité de toute l'architecture ou l'art de bastir utile aux entrepreneurs et aux ouvriers*, 2nd edn, Paris, 1714: Laugier, M. A., *Essai sur l'architecture*, Paris, 1753. See also Middleton, R. D., 'The Abbé de Cordemoy and the Graeco-Gothic ideal: a prelude to Romantic Classicism', *Journal of the Warburg and Courtauld Institutes*, vol. 25, 1962, pp. 278-320 and vol. 26, 1963, pp. 90-123; and Hermann, W., *Laugier and eighteenth century French theory*, Zwemmer, 1962.

18. For reinforced flat arches (*platebandes*) generally see Patte, P., *Mémoires sur les objets les plus importans de l'architecture*, Paris, 1769, pp. 259-318; and Rondelet *Traité 1814*, vol. 2, bk 3, pp. 94-101. For fuller discussion of those over the east colonnnade of the Louvre see Berger, R. W. and Mainstone, R. J., 'Materials and structure' in Berger *The Palace of the Sun*, pp. 65-74.

19. For the early tests see Brunel, M., 'Particulars of some experiments on the mode of binding brick construction', *Transactions of the Institute of British Architects*, vol.1 1836, pp. 40-43.

20. Pope *Treatise*, p. 279 and pl. 3.

21. Brown, J. M., 'W. B. Wilkinson (1819-1902) and his place in the history of reinforced concrete', *Trans.NS*, vol. 39, 1966-67, pp. 129-142; Cassie, W. F., 'Early reinforced concrete in Newcastle upon Tyne', *Magazine of Concrete Research*, vol. 6, 1955, pp. 25-30.

22. Hamilton *Reinforced concrete*. Also see again the other works cited in Chapter 3, note 10 for other related developments, particularly in North America and Germany.

23. Nervi *New structures*, pp. 108-117.

24. For the first, see 'Metropolitan gateway', *Progressive Architecture*, vol. 44, no. 4, 1963, pp. 156-160; for the second, Torroja *Structures*, pp. 2-18.

25. Hamilton, S. B., *A short history of structural fire protection of buildings*, National Building Studies Special Report 27, HMSO, 1958, gives a good general account. See also Eck, C. L. G., *Traité de construction en poteries et fer, à l'usage des batiments civils, industriels et militaires*, Paris, 1836; Barrett, J., 'On the construction of fire-proof buildings: *Min.Proc. ICE*, vol. 12, 1853, pp. 244-272; and Fairbairn, *Application*.

26. Sattler, K., 'Composite construction in theory and practice', *SE*, vol. 39, 1961, pp. 124-144; Johnson, R. P., *Composite structures of steel and concrete*, vol. 1, 2nd edn, Blackwell, 1994.

27. Nervi *Structures*, pp. 101-103; Joedicke et al. *Nervi*, pp. 88-91. The same was done by Felix Samuely and Partners in 1978 for roof intended to carry helicopters. For a further example see Dickson, M. G. T. and Gregson, S. J., 'Schlumberger Cambridge Research Ltd Phase 2', *Proc.ICE: Structures and Buildings*, vol. 110, 1995, pp. 288-295.

28. The American development is best described in Condit, *Twentieth century*, pp. 166-172. Talbot, A. N., and Slater, W. A., 'Tests of reinforced concrete buildings under load', *University of Illinois Engineering Experimental Station Bulletin*, no. 64, 1913, and two subsequent reports in the same series, record interesting tests on several early floors of this kind.

29. Maillart, R., 'Zur Entwicklung der unterzuglosen Decke in der Schweiz und in Amerika', *Schweizerische Bauzeitung*, vol. 87, 1926, pp. 263-265 (reprinted, slightly abridged, and translated in Bill *Maillart*, pp. 154-162) describes the development, including tests, and illustrates some early applications.

30. The main development of the modern form from the original battledeck took place in Germany immediately after the Second World war. Seegers, K. H., 'Neuere Flachblechfahrbahnen, insbesondere bei Strassenbrücken', *Der Bauingenieur*, vol. 39, 1964, pp. 173-19, and Kunert, K., 'Orthotropic steel decks in Germany' in British Constructional Steelwork Association, *Proceedings of the conference on steel bridges*, London, 1968, pp. 151-159, describe and illustrate representative examples.

Chapter 9 Trusses, portal frames, and space frames

1. Hodge *Greek roofs*. I do not accept a more recent contention, solely on the basis that similar roofs could easily be constructed today and are known to have been built later, that true trusses were indeed built from the fifth or fourth century BC. For this see Izenour, G. C., *Roofed theatres of classical antiquity*, Yale University Press, 1992.

2. Vitruvius *De architectura*, bk 4, ch. 2: '*sub tectis, si maiora spatia sunt, et transtra* [the inclined rafters] *et capreoli* [the bottom ties]'.

3. Forsyth, G. H., 'The Monastery of St. Catherine at Mount Sinai: the church and fortress of Justinian', *Dumbarton Oaks Papers*, vol. 22, 1968, pp. 3-19.

4. The lengths of timber bearing against the undersides of the rafters and interposed between them and the inclined struts would have obviated the need to weaken the rafters by cutting notches or mortices in them to receive the struts. These timbers were merely nailed to the rafters, so that they could have been slid up them in the final stage of erection until the whole assembly was tight. At such a remote and desolate site (believed to be that of Moses' Burning Bush), all the timber must have been brought from a considerable distance. Most likely, therefore, it was fully pre-cut ready for assembly; and the procedure suggested would have had the further advantage of

obviating also the need for any great accuracy in cutting. I am indebted to Professor Forsyth not only for the photograph reproduced in fig. 9.1, but also for placing at my disposal his further data about the roof.

5. Licht *Rotunda* reproduces a considerable number of these drawings on pp. 52-58.

6. Scamozzi, O. B., *Le fabriche e i disegni di Andrea Palladio*, Venice, 1776-1783.

7. For French medieval roofs see again Deneux, H., 'L'évolution des charpentes du XIᵉ au XVIIIᵉ siècle', *L'architecte*, 1927, pp. 49-53, 57-60, 65-68, 73-75, and 81-89; also Viollet-le-Duc *Dictionnaire raisonné*, article 'charpente', vol. 3, pp. 1-58. For English medieval roofs, see Howard, F. E., 'On the construction of mediaeval roofs', *Archaeological Journal*, vol. 71, 1914, pp. 293-352; Smith, J. T., 'Medieval roofs: a classification', *Archaeological Journal*, vol. 115, 1958, pp. 111-149; Cordingly, R. A., 'British historical roof-types and their members: a classification', *Ancient Monuments Society's Transactions*, new series, vol. 9, 1961, pp. 73-118; and Hewett *Historic carpentry*.

8. The arched rib was described in Chapter 6. For the roof as a whole see again the works by Baines, Sir P. and Courtenay, L. T., cited there in note 11.

9. For the analyses represented at (c), (d), and (e) see, respectively, Pippard, A. J. S. and Glanville, W. H., *Primary stresses in timber roofs (with special reference to curved bracing members)*, Building Research Technical Paper 2, HMSO, 1926; Heyman, J., 'Westminster Hall roof', *Proc.ICE*, vol. 37, 1967, pp. 137-162; and Mainstone, R. J., 'Discussion' on the previous paper, ibid., vol. 38, 1967, pp. 788-792. Subsequent analyses have given results broadly in line with (e). See, in particular, L. T. Courtenay and R. Mark, 'The Westminster Hall roof: a historiographic and structural study', *JSAH*, vol. 46, 1987, pp. 374-393.

10. For these roofs see Plot *Oxfordshire*, pp. 273-274; Killer *Grubenmann*, fig. 61; and Nicholson, P., *Carpenter's and joiner's assistant*, 2nd edn, London, 1805, p. 59. For a wider review see Yeomans, D., *The trussed roof: its history and development*, Scholar, 1992

11. Although no examples have survived, there are surviving representations of such parapets on bridges, including a relief of a bridge over the Danube on Trajan's column in Rome and a mosaic of a pontoon bridge in Ostia. See the works cited in Chapter 6, note 10 for the first and fig. 9.5 in the first edition of this book for the second.

12. For example in *MS. B*, Institut de France, fol. 28v. and, especially, in *Codice Atlantico*, fol. 266r.

13. Palladio *Quattro libri*, bk 3, chs 7 and 8, pp. 15,18. See also Boucher *Palladio*, pp. 209-211, and Mainstone, R. J., 'Structural analysis, structural insights and historical interpretation', *JSAH*, vol. 56, 1997, pp. 335-336.

14. Killer *Grubenmann*, pp. 21-62.

15. For fuller accounts of North American developments, see Brock, F. B., 'Truss bridges: an illustrated historical description of all expired patents on truss bridges, which under law are now public property and free to be used by any one', *Engineering News*, vol. 9, 1882, pp. 371-372 and

subsequent issues running into vol. 10, 1883; Cooper, T., 'American railroad bridges', *Trans.ASCE*, vol. 21, 1889, pp. 1-52; Fletcher, R. and Snow, J. P., 'A history of the development of wooden bridges', ibid., vol. 99, 1934, pp. 314-354; Edwards, L. N., *A record of the history and evolution of early American bridges*, Maine University Press, 1959; and Condit, *Nineteenth century*, pp. 75-162. For the European antecedents and parallel developments, see also Leupold *Theatrum pontificiale*; Krafft *Charpente*; Federov, S. G., 'Matthew Clark and the origins of Russian structural engineering 1810-1840s: an introductory biography', *Construction History*, vol. 8, 1992, pp. 69-88; and several papers, all with extensive further references, by James, J. G., especially 'The evolution of iron bridge trusses to 1850', *Trans.NS*, vol. 52, 1980-81, pp. 67-101; idem, 'The evolution of wooden bridge trusses to 1850', *Journal of the Institute of Wood Science*, vol. 16, 1982, pp. 116-135 and 168-193; and idem, 'The origins and worldwide spread of Warren truss bridges in the mid-nineteenth century: pt 1. Origins and early examples in the UK', *History of Technology*, vol. 11, 1986, pp. 65-123.

16. Whipple. S., *A work on bridge building*, New York, 1847.

17. Cubitt, J., 'A description of the Newark Dyke Bridge, on the Great Northern Railway', *Min.Proc.ICE*, vol. 12, 1853, pp. 601-612. See also Doyne, W. T. and Blood, W. B., 'An investigation of strains upon the diagonals of lattice beams, with the resulting formulae', ibid., vol. 11, 1852, pp. 1-14 (referring to the Warren truss and not the Town lattice).

18. Vogel, R. M., 'The engineering contributions of Wendel Bollman', *United States National Museum Bulletin*, no. 240, 1964, pp. 77-104.

19. Cilley, F. H., 'The exact design of statically indeterminate frameworks. An exposition of its possibility, but futility', *Trans.ASCE*, vol. 43, 1900, pp. 353-407, sets out the arguments in favour of statical determinacy as seen at the turn of the century. ('Discussion', ibid., pp. 408-443.). For the Eads Bridge joints see Woodward *St. Louis Bridge*, pls 24 and 25.

20. For the Sydney Harbour Bridge, see Freeman, R. et al., 'Sydney Harbour Bridge', *Min.Proc.ICE*, vol. 238, 1935, pp. 153-475. For the Bayonne Bridge, see Ammann, O. H., *First progress report on Kill Van Kull Bridge* and *Second progress report on Kill Van Kull Bridge*, 1930 and 1931.

21. Fitchen *Dutch barn* gives a full description of one application of this type of frame.

22. Hewett, C. A., 'The barns at Cressing Temple, Essex, and their significance in the history of English carpentry', *JSAH*, vol. 26, 1967, pp. 48-70.

23. The roof illustrated in fig 9.12 dates only from 1837-1841 and is a framed counterpart of the triple dome of St Paul's in London. The lower ribs were infilled with hollow pots in the completed dome. (See Ricard de Montferrand, A., *Église Cathédrale de Saint Isaac*, Paris, 1845, and Federov, S. G., 'Early iron domed roofs in Russian church architecture: 1800-1840', *Construction History*, vol. 12, 1996, pp. 41-66.) The first large iron-framed dome

352

(demolished in 1855) was constructed over the central court of the Corn Exchange, Paris, in 1809-1811, and had single sets of radial and circumferential ribs with a span of 39 m (see fig. 13.5, p. 241 and Rondelet *Traité 1838*, bk 7, pl. 164). Prior to this, many large ribbed timber domes had been constructed, the earliest one still surviving in part being the eleventh-century rebuilding of that of the Dome of the Rock in Jerusalem, with a span of 20 m (see Chapter 7, note 18). For Renaissance and later ribbed timber domes, see, for instance, De l'Orme *Nouvelles inventions*; and Krafft *Charpente*. A combined use of timber and iron is referred to in Vitruvius *De architectura*, bk 5, ch. 10, and this, rather than full iron or bronze framing, is probably what was referred to by Aelius Spartianus, *Antoninus Caracalla*, ch. 9, at the Baths of Caracalla:

> *quarum cellam soliarem architecti negant posse ulla imitatione qua facta est fieri. Nam et ex aere vel cupro cancelli subter-positi esse dicuntur, quibus camaratio tota concredita est . . .*

(*The Scriptores Historiae Augustae* ed. and trans. D. Magie, Loeb Classical Library, William Heinemann, 1924, vol. 2, pp. 24-25). In this connection, Middleton, J. H., *The remains of ancient Rome*, London, 1892, pp. 163-164, refers to the discovery, in late nineteenth-century excavations in what is now known as the *natatio* of the baths, of:

> an immense quantity, amounting to many tons, of fragments of iron girders . . . These were compound girders, formed of two bars riveted together thus ╫ and then cased in bronze.

24. Föppl, A., *Das Fachwerk im Raume*, Leipzig, 1892.

25. Wachsmann, K., *The turning point of building*, Reinhold, 1961, pp. 29-34

26. Motro, R., 'Tensegrity systems and geodesic domes', *International Journal of Space Structures*, vol. 5, 1990, 341-351; Hanaor, A., 'Tensegrity: theory and application' in Gabriel *Beyond the cube*, pp.385-408. But note that the term 'tensegrity' as used by Fuller denoted only a self-contained structure stiffened by internal stresses. Robbin *New architecture*, pp. 26-32, uses the term both in this sense and to denote cable roofs stiffened by tensioning the cables against continuous external compression rings. Examples of these, designed by Geiger, Levy, and Zetlin, have been referred to already in Chapter 7.

27. Joedicke et al. *Nervi*, pp. 36-43.

28. See now chiefly Robbin *New architecture*, pp 81-99, several other chapters of Gabriel *Beyond the cube*, and other publications of the Structural Morphology Group of the International Association for Shell and Space Structures. A classic earlier text is Pacioli, L., *De divina proportione*, MS., 1498, first printed Venice, 1509 (facsimile edn of the Ambrosiana Codex, Milan, 1956). Though not concerned with the present structural possibilities, this deals not only with the regular and some quasi-regular solid bodies, but also with the corresponding skeletal frameworks, both of which are illustrated, for each form discussed, in a beautiful series of drawings by Leonardo.

29. There is now a very extensive literature on Fuller's designs and ideas, not all of it helpful from the present point of view. See, for instance, the chapter on geodesic structures in Marks, R. W., *The dymaxion world of Buckminster Fuller*, Reinhold, 1960, pp. 55-64. Fuller's geodesic domes are set in the context of the numerous alternative triangulated framing systems in Makowski, Z. S., 'Braced domes, their history, modern trends and recent developments', *Architectural Science Review*, vol. 5, 1962, pp. 62-79.

30. Uyanik, M. E., 'The design of a geodesic dome', *Student Publication of the School of Design* (North Carolina State College), vol. 9, no. 2, 1960, pp. 46-56, describes the design of the Union Tank Dome, Baton Rouge, Louisiana − a similar structure to that illustrated in fig. 9.15.

31. For the first see Makowski, Z. S., 'Stressed skin space grids', *Architectural Design*, vol. 31, 1961, 323-237, and for the second see 'Graceful solution to controversy', *AF*, vol. 135, no.4, 1971, pp. 36-39. For more general reference see again Makowski *Steel space structures*, and Borrego, J., *Space grid structures*, MIT Press, 1968

32. Robbin *New architecture*, pp. 38-57.

Chapter 10 Supports, walls, and foundations

1. See again the works cited in Chapter 7, note 1.

2. See again the works cited in Chapter 8, note 1.

3. Lugli *Tecnica edilizia*; Blake, *Ancient Roman construction*.

4. For both Romanesque and Gothic pier and wall construction, see especially Viollet-le-Duc *Dictionnaire raisonné*, article 'construction', vol. 4, pp. 1-279.

5. Hamilton *Reinforced concrete*, Condit, *Nineteenth century*, and Haegermann et al., *Vom Caementum zum Spannbeton*, vol.1, describe the principal earlier stages in the development.

6. Bondale, D. S. and Clark, P. J., 'Composite construction in the Almondsbury Interchange', in British Constructional Steelwork Association, *Proceedings of the conference on structural steelwork*, 1966, pp. 91-100; Kerensky, 0. A. and Dallard, N. J., 'The four-level interchange between M4 and M6 motorways at Almondsbury', *Proc.ICE*, vol. 40, 1968, pp. 295-320.

7. There is a very extensive literature on these developments in the European architectural and building journals throughout the 1950s and 1960s under such headings as 'concrete large-panel construction' and 'concrete system building'. See also Koncz *Precast concrete*.

8. See, for instance, Horn, W., 'The great Tithe Barn of Cholsey, Berkshire', *Journal of the Society of Architectural Historians*, vol. 22, 1963, pp. 13-23; Horn and Born *Barns*.

9. Johnson, H. R., and Skempton, A. W., 'William Strutt's cotton mills, 1793-1812', *Trans.NS*, vol. 30, 1955-57, pp. 179-205. The development of cast-iron columns has been less well documented than that of cast-iron beams, but see also Hamilton, S. B., 'The use of cast iron in building', *Trans.NS*, vol. 21, 1940-41, pp. 139-155 and Elliott, C. D., *Technics and architecture*, MIT Press, 1992, ch. 4.

10. Detailed drawings appear in fig. 68 of

Hamilton, 'The use of cast iron in building' (previous note). Some ten years after the photograph reproduced here was taken, the whole structure (including a brick-vaulted lower storey) was saved from destruction by conversion to a shopping centre with a regrettable loss of its earlier spaciousness and atmosphere. When first visited with Dr Hamilton in the 1950s, the upper floor had become a bonded warehouse for port and sherry. With the sun just breaking through and dispersing an overnight fog at its open far end, the seemingly endless rows of branching columns disappearing into its light created an unforgettable impression.

11. Skempton, A. W., 'The Boat Store, Sheerness (1858-60) and its place in structural history', *Trans.NS*, vol. 33, 1959-60, pp. 57-78. For an indication also of the wide range of built-up sections in use a few decades later (as referred to below), see Birkmire, W H., *Skeleton construction in buildings*, 2nd edn, John Wiley, 1894, especially pp. 21-23.

12. Roberts, Sir G., 'Forth Road Bridge: design', *Proc.ICE*, vol. 32, 1965, pp. 333-405.

13. For the Stuttgart columns see Pehnt, W. and von Gerkan, W., *Flughafen Stuttgart, Stahl und Form*, Institut für Internationale Architektur-Dokumentation, Munich, 1991 (in German with English summary). For examples of branched timber columns see Wenzel, F., Frese, B. and Barthel, R., 'The timber ribbed shell roof in Bad Dürrheim. Design and construction of the roof above the new health spa', *Structural Engineering Review*, vol.1, 1988, pp. 75-81.

14. Many such ties are to be found in the sixth century parts of the Church of St Sophia, Istanbul, their contemporaneity with the structure itself being attested by the manner in which they are accommodated by it and by the original surface finishes.

15. Morandi, R., 'Tied bridges in prestressed concrete', *Civil Engineering and Public Works Review*, vol. 59, 1964, pp. 705-713 and 861-2; idem, 'Some types of tied bridges in prestressed concrete', in American Concrete Institute, *Concrete bridge design*, 1969, pp. 447-465.

16. Ohashi, M., 'Cables for cable-stayed bridges', in Ito et al. *Cable-stayed bridges*, 125-150, is a good wide-ranging discussion.

17. For fuller discussions see, for instance, Little, A. L., 'The development of foundation engineering', in Institution of Structural Engineers, *Proceedings of the fiftieth anniversary conference*, 1958, pp. 294-300; Terzaghi, K., and Peck, R. B., *Soil mechanics in engineering practice*, 2nd edn, John Wiley, 1967; and Tomlinson, M. J., *Foundation design and construction*, 5th edn, Longman, 1986. See also Skempton, A. W., 'Landmarks in early soil mechanics', in *7th European Conference in Soil Mechanics and Foundation Engineering*, 1979, vol.5, pp. 1-26.

18. Matthew 7, 25.

19. Navier, C. L. M., *Rapport à Monsieur Becquey . . . et mémoire sur les ponts suspendus*, 2nd edn, Paris, 1830, (*augmentée d'une notice sur le pont des Invalides*); Kranakis *Constructing a bridge*, pp.165-195.

20. Viollet-le-Duc *Dictionnaire raisonné*, article 'construction', vol. 4, pp 175-176. See also the brief

report of later excavations in the same bay of the north aisle of the choir in Durand, G., *Monographie de l'église Notre-Dame, cathedrale d'Amiens*, Paris, vol. 1, 1901, pp. 202-204. Durand's plan on p.203 shows the footings as continuous from pier to pier and standing on a deep bed of mortared rubble. They thus appear to have had something in common with the foundations at York referred to below.

21. Smeaton *Edystone*, pls 9 and 10; and Mainstone, R. J., 'The Eddystone Lighthouse' in Skempton, A. W., ed., *John Smeaton FRS*, Thomas Telford, 1981, pp. 83-102.

22. Piranesi, G. B., *Le antichita romane*, Rome, 1756, vol. 4. The least convincing are the longitudinal sections of the Pons Aelius and Hadrian's Mausoleum (pl. 7) and of the Pons Fabricius (pl. 19). The first shows the two structures standing on a single foundation that is continuous even under the three river spans of the bridge and is composed of huge ashlar blocks cramped together and reinforced with sections of embedded arch. It is 25 m deep under the bridge and much deeper under the mausoleum. Even more surprisingly, the second shows each of the two arches of the bridge continued below it to form a complete circle 24.5 m in diameter. Under the river bed, these circles are linked to other arches and incorporated in a similarly continuous foundation from abutment to abutment. Such foundations can have had no counterpart in actual construction even in Piranesi's day. His source for what is shown in pl. 7 must have been less massive continuous footings on land, such as those referred to on page 189 and in the following two notes. As for the continuous circles of masonry, Needham *Civil engineering and nautics*, pp. 170-171, makes tantalisingly brief reference to Chinese bridge arches continued in the same way below river level. A more likely source would have been the practice of constructing shallow inverted arches below ground level to connect individual footings for piers or columns. Many building manuals from Vitruvius *De architectura*, vol. 3, ch. 3, onwards refer to it, Alberti *De re aedificatoria*, bk 3, ch. 5, and Belidor *Science des ingénieurs*, bk 3, ch. 9, being more explicit. Good examples, with deep arches between the feet of the footings, were exposed during remedial works on Wren's Library at Trinity College Library in Cambridge (Insall, D., *The care of old buildings today*, Architectural Press, 1972, p.177-178). Similar arches were also used after Piranesi's day at All Souls' Church, Langham Place, London, (Beckmann, P. and Dekany, A., 'All Souls Langham Place and the Waldegrave Hall' *The Arup Journal*, vol.12, no.1, 1977, pp. 2-8).

23. For the latter, see Harrison, R. M., *Excavations at Saraçhane in Istanbul*, vol. 1, Princeton University Press, 1986; and idem, *A temple for Byzantium*, Harvey Miller, 1989 (a more popular account with a questionable reconstruction of the lost superstructure).

24. Dowrick, D. J., and Beckmann, P., 'York Minster structural restoration', *Proc.ICE*, vol. 49 supplement, 1971, pp. 93-156.

25. Crook, J. M., 'Sir Robert Smirke: a pioneer of concrete construction', *Trans.NS*, vol. 38, 1965-66, pp. 5-22. An excellent review of the whole recent

history of the use of concrete in foundations with full bibliographic references is given in Chrimes, M. M., 'Concrete foundations and substructures: a historical review', *Proc.ICE: Structures and Buildings*, vol. 116, 1996, 344-372.

26. Vitruvius *De architectura*, bk 2, ch. 9; and ibid., bk 3, ch. 4.

27. The relevant documents are translated and discussed in Parsons *Engineering in the Renaissance*, pp. 515-29.

28. See, for instance, Vitruvius *De architectura*, bk 5, ch. 12; Knight, W., 'Observations on the mode of construction of the present Old London Bridge as discovered in 1826-27, *Archaeologia*, vol. 23, 1831, pp. 117-119; Parsons *Engineering in the Renaissance*, pp. 544-545; Leupold *Theatrum pontificiale*; and Jensen, M., *Civil engineering around 1700 with special reference to equipment and methods*, Danish Technical Press, Copenhagen, 1969.

29. This is well described and illustrated in 'Pont de Saumur', in de Cessart, L. A., *Description des travaux hydrauliques*, vol. 1, section 1, Paris, 1806.

30. For the Eads Bridge caissons see Woodward *St. Louis Bridge*, pp. 44-47, 57-65, 201-237 and pls 7, 8, 13, 14. For bridge caissons more generally see Shirley-Smith *Bridges*, pp.77-80.

31. Peck, R. B., 'History of building foundations in Chicago', *University of Illinois Engineering Experimental Station Bulletin*, no. 373, 1948, describes the development there up to the early twentieth century.

32. The foundation illustrated is described in Wise, C. M., Bridges, H. W., and Walsh, S. R., 'The new Commerzbank headquarters, Frankfurt, Germany', *SE*, vol. 74, 1996, pp.111-122.

33. An earlier semi-buoyant foundation, with a continuous raft of inverted brick barrel vaults to spread the load, is described in Skempton, A. W., 'The Albion Mill foundations', *Géotechnique*, vol. 21, 1971, pp. 203-210, though it is not clear to what extent its buoyancy was taken into account in design.

34. For fuller reviews of modern developments in theory and practice, see again the works cited in note 17 above; Terzaghi, K., 'The science of foundations – its present and its future', *Proc.ASCE*, vol. 53, 1927, pp. 2263-2294; and Skempton, A. W., 'Foundations for high buildings', *Proc.ICE*, vol. 4, pt 3, 1955, pp. 246-269.

Part 3 Complete structures

Chapter 11 Early forms

1. This prehistory is summarized, with references, in Bradford, J., 'Building in wattle, wood, and turf', in Singer, C. J. et al., eds, *A history of technology*, vol. 1, Clarendon Press, 1955, pp. 299-326; Lloyd, S., 'Building in brick and stone', ibid., pp. 456-494; and Piggott, S., ed., *The dawn of civilisation*, Thames & Hudson, London, 1961. We are not, of course, concerned here with the myth of the 'primitive hut'; still less with the associated myth of the 'noble savage'. The form central to the first of these myths – a framed structure of four corner posts, four connecting beams set in the forked heads of the posts,

and gables raised above two of these by pairs of rafters connected by a ridge beam – appears to have been first described in Filarete, A., *Trattato di architettura*, c. 1460, fol. 5v. and 54v. (facsimile edn, *Filarete's treatise on architecture*, ed. and trans. J. R. Spencer, Yale Publications in the History of Art 16, 1965). It is absent from the more matter-of-fact account in Vitruvius *De architectura*, bk 2, preface, and is almost certainly far from primitive in the sense here intended.

2. Clarke and Engelbach *Egyptian masonry*, pp. 5-11; Lauer *Pyramide*.

3. This is illustrated in Arnold *Building in Egypt*, fig. 4.125, and was built somewhat later than the completion of Stonehenge in its final form, probably under Mediterranean influence. Structurally Stonehenge [3.4] was very similar except that the stability of its uprights owed more to burial of their feet in the ground than to the provision of broad square-dressed bases.

4. For Egyptian temples generally, see Smith, W. S., *The art and architecture of ancient Egypt*, Pelican History of Art, 1958 (2nd edn 1981); and Lange, K. and Hirmer, M., *Egypt: architecture, sculpture, painting*, trans. R. H. Boothroyd, 3rd edn, Phaidon, 1961.

5. For Greek temples generally, see Dinsmoor, W. B., *The architecture of ancient Greece*, 3rd edn rev., Batsford, 1950; and Lawrence *Greek architecture*.

6. This is open to doubt, however, because of the complex nature of behaviour in earthquakes. It is now clear that the multi-drum column of the typical temple is remarkably able to withstand quite severe shocks when free-standing. This is because it responds primarily by rocking rather than by bending and, since natural frequencies in rocking decrease with increasing amplitude, there is little risk of the resonance that can otherwise lead to serious damage or collapse. It is also immune from the risk that otherwise stems from the possible different responses of adjacent parts of the structure that are in contact.

7. For details of construction see Coulton *Greek architects*, pp.140-160, and the superbly illustrated *Studies for the restoration of the Parthenon*, especially vols 1 (Korres, M. and Bouras Ch.), 2a (Korres, M., Toganides, N. and Zambas, K.), 2b (plates), 3a (Koufopoulos, P.), 4 (Korres, M.), 5 (Toganides, N.), Committee for the preservation of the Acropolis monuments, Athens, 1983, 1989, 1994; Tanoulas, T., Ioannidou, M., and Moraitou, A., *Study for the restoration of the Propylaea*, Committee for the preservation of the Acropolis monuments, Athens, 1994; and Svolopoulos, D., *Temple of Apollo Epikourios. Architectural study*, 1 vol. text, 1 vol. plates, Committee for the preservation of the Temple of Apollo Epikourios, Athens, 1995 (all texts of these *Studies* in Greek with English summaries).

8. Hodge *Greek roofs* is an excellent presentation and discussion of the surviving evidence for temple roofs in general. See also Coulton *Greek architects*, pp. 74-85.

9. Fuller descriptions are given by Schmidt, E. F., *Persepolis I: structures, reliefs, inscriptions*, University of Chicago Oriental Institute Publications 68, 1953; Godard, A., *The art of Iran*, trans. M. Heron,

355

Allen & Unwin, 1965; and Pope *Persian architecture*.

10. See again Rohlfs *Primitive Kuppelbauten*; Allen *Shelters*; and Lilliu, G., 'L'architettura Nuragia', *Atti del XIII Congresso di Storia dell'Architettura*, Rome, 1966, vol. 1, pp. 2-92 for the early stone false dome generally; Daniel, G., *The megalith builders of western Europe*, Pelican Books, 1963 for tholos and related tombs outside Greece; and Lawrence *Greek architecture*, pp. 57-64, for the Mycenean tholos tomb.

11. See again the works cited in note 2; also Edwards, I. E. S., *The pyramids of Egypt*, rev. edn, Pelican Books, 1961; Fakhry, A., *The pyramids*, University of Chicago Press, 1961; and Arnold *Building in Egypt*, pp. 159-182.

12. Kubler, G., *The art and architecture of ancient America*, 3rd edn, Penguin, 1984; Coe, W. R., *Tikal*, University of Pennsylvania, 1978.

13. For Greek defence towers see Winter, F. E., *Greek fortifications*, Routledge & Kegan Paul, 1971, pp. 152-204; and Adam, J-P., *L'architecture militaire Grecque*, Picard, Paris, 1982, pp. 46-76 and 140-161.

14. Asin Palacios, M., 'Una descripción nueva del Faro de Alejandria', *Al-Andalus* (Revista de las Escuelas de Estudios Arabes de Madrid y Granada), vol. 1, 1933, pp. 241-192, transcribes and discusses the principal literary evidence; Otero, L. M., 'Interpretación gráfica de la descripción de Ibn al-Sayj', ibid., pp. 293-300, reconstructs the probable form. A summary, 'The Pharos of Alexandria', appeared in *Proceedings of the British Academy*, vol. 19, 1933, pp. 277-292.

Chapter 12 Pre-nineteenth-century wide-span buildings

1. Chiefly the fitting of a dome to a square base.

2. Vitruvius *De architectura*, bk 5, ch. 1. The translation by Morgan gives a reconstruction drawing on p. 135.

3. For the Basilica Ulpia see now the comprehensive account and superb graphic reconstructions in Packer, J. E., *The Forum of Trajan in Rome*, 2 vols text plus portfolio, University of California Press, 1997. For post-Republican Roman buildings generally, see MacDonald *Architecture of the Roman empire*; and Ward-Perkins *Roman Imperial Architecture*. Nash, E., *Pictorial dictionary of Ancient Rome*, Zwemmer, 1961 (2 vols), is more comprehensive within its field and valuable for its bibliographies on individual structures and photographs of their present states.

4. More precise plots of possible thrust-lines in representative Roman vaulted structures (including several of those discussed here) may be found in Giovannoni, G., *La tecnica della costruzione presso i Romani*, Societa Editrice d'Arte Illustrata, Rome, 1925. For reasons already touched on in Chapter 2 in discussions of statical indeterminacy, they can never be more than 'possible' thrust-lines. Unless the structure is on the verge of collapse, others will also be possible as sketched for the arch in fig. 2.11 (c), so it seems unnecessary to go beyond more general indications here. The fact that a structure that cannot develop significant tension remains standing is, of course, the best possible evidence that thrust lines in equilibrium with the loads do exist wholly within its masonry.

5. See again Giovannoni *Sala termale*, G.; Caraffa *Cupola*; and Rasch, J. J., 'Zur Konstruktion spätantiker Kuppeln vom 3. bis 6. Jahrhundert', *Jahrbuch des Deutschen Archäologischen Instituts*, vol. 106, 1991, pp. 311-383 and pls 77-84.

6. A date a hundred years later has also been suggested. For this and later Byzantine buildings generally, see Ward-Perkins, J. B., 'The Italian element in late Roman and early Mediaeval architecture', *Proceedings of the British Academy*, vol. 33, 1947, pp. 163-194; and Krautheimer, R., *Early Christian and Byzantine architecture*, Pelican History of Art, 1965 (4th edn 1987); and Mango, C., *Byzantine architecture*, New York, 1976. A detailed description of the Church of SS. Sergius and Bacchus referred to below is given in Sanpaolesi, P., 'La chiesa dei SS. Sergio e Bacco a Costantinopoli', *Rivista dell'Istituto Nazionale d'Archeologia e Storia dell'Arte*, new series, vol. 10, 1961, pp. 116-180. For San Vitale see Binda, L., Lombardini, N. and Guzzetti, F., 'San Vitale in Ravenna' in Sonderforschungsbereiches 315, *Historische Bauwerke*, University of Karlsruhe, 1996, pp. 113-124.

7. Procopius *Buildings*, bk 1, ch. 4, 9-16, and bk 5, ch. 1, 6, gives contemporary descriptions of the two churches. Hormann, H. et al., 'Die Johanneskirche', *Forschungen in Ephesos* (Österreichischen Archäologischen Institut), vol. 4, 1951, gives a detailed survey of the remains of the Church of St John. For St Mark's see the Visentini engravings in Zatta, A., *L'augusta basilica dell'evangelista San Marco*, Venice, 1761 (facsimile edn, Gregg, 1964), and Mainstone, R. J., 'The first and second churches of San Marco reconsidered', *The Antiquaries Journal*, vol. 71, 1991, 123-137. A photogrammetric survey of the present structure awaits publication.

8. It is impossible to describe this structure in as much detail as is really desirable. A fuller description and further illustrations and bibliography may be found in Mainstone *Hagia Sophia* and other briefer accounts in idem, 'Justinian's church of St Sophia, Istanbul: recent studies of its construction and first partial reconstruction', *Architectural History*, vol. 12, 1969, pp. 39-49, and idem, 'Hagia Sophia: Justinian's Church of Divine Wisdom, later the Mosque of Ayasofya, in Istanbul', *SE*, vol. 68, 1990, pp. 65-71. Van Nice, R. L., *Saint Sophia in Istanbul, an architectural survey*, Dumbarton Oaks, two instalments 1966 and 1986, is an indispensable, uniquely detailed, set of survey drawings. Among earlier accounts, see particularly Choisy, A., *L'art de bâtir chez les Byzantines*, Paris, 1883, pp. 135-141; Lethaby, W. R. and Swainson, H., *The church of Sancta Sophia, Constantinople*, London, 1894; and Swift, E. H., *Hagia Sophia*, Columbia University Press, 1940.

9. Procopius *Buildings*, bk 1, ch. 1, 46.

10. See, for instance, Creswell, K. A. C., *The early Muslim architecture of Egypt*, 2 vols, Oxford University Press, 1952; Wilber *Architecture of Islamic Iran*; Pope *Persian architecture*; Hoag, J. D.,

356

Islamic architecture, Faber & Faber, 1987; and Golombek, L. and Wilber, D., *The Timurid architecture of Iran and Turan*, 1 vol. text and 1 vol. plates, Princeton University Press, 1988.

11. Best described, together with other similar structures, in Brumfield, W.C., *A History of Russian architecture*, Cambridge University Press, 1993, especially pp. 114-129. For the introduction of masonry construction in Russia see Rappoport, P. A., *Building the churches of Kievan Russia*, Variorum, 1995.

12. For these mosques, see chiefly Charles, M. A., 'Hagia Sophia and the great Imperial mosques', *AB*, vol. 12, 1930, pp. 321-344; Goodwin, G., *A history of Ottoman architecture*, Thames and Hudson, 1971; and Kuran, A., *Sinan*, Institute of Turkish Studies, Washington DC, 1987.

13. Mainstone, R. J., 'Sinan's Suleymaniye Mosque and Hagia Sophia', *Public Assembly structures from antiquity to the present*, Mimar Sinan University, Istanbul, 1993, 53-62.

14. See again Sgrilli *Descrizione*. A composite plan and Nelli's most relevant cross section are reproduced in Mainstone, R. J., 'Brunelleschi's dome', *ARev.*, vol. 162, 1977, p. 158, figs 2 and 3.

15. See again Wittkower *Architectural principles*.

16. Ferrabosco, M. and Costaguti, G. B., *Archittetura della basilica di S. Pietro in Vaticano*, Rome, 1620, 2nd edn 1684, includes the most detailed and relevant plans and sections. See also Letarouilly, *Le Vatican et la Basilique de St. Pierre de Rome*, Paris, 1882, sections (c) and (d). Reduced facsimile edn, *Vatican 1*, Tiranti, 1953.

17. Bannister, T. C., 'The Constantinian Basilica of Saint Peter at Rome', *JSAH*, vol. 27, 1968, pp. 3-32; Krautheimer, Corbett, R. S. and Frazer, A. K., *Corpus Basilicarum Christianarum Romae*, vol. 5, Pontificio Istituto di Archeologia Cristiana, Vatican City, 1977 (devoted to the Lateran Church, St Paul's and St Peter's).

18. Alberti *De re aedificatoria*, bk 1, ch. 10.

19. Best described in Conant, K. J., *Carolingian and Romanesque architecture*, Pelican History of Art, 1959 (2nd edn 1978), and Kubach, H. E., *Romanesque architecture*, Faber & Faber, 1988. Choisy *Histoire*, pp. 194-232, gives a series of analytical axonometric views.

20. For Speyer see Kubach, H. E. and Haas, W., *Der Dom zu Speyer*, 3 vols including 1 vol. of excellent measured drawings, Deutscher Kunstverlag, Munich, 1972. For Durham, see Bilson, J., 'The beginnings of Gothic architecture: Part II: Norman vaulting in England', *JRIBA*, 3rd series, vol. 6, 1899, pp. 289-326; Thurlby, M., 'The Romanesque high vaults of Durham Cathedral' in Jackson, M., ed., *Engineering a cathedral*, Thomas Telford, 1993, pp. 63; and idem, 'The purpose of the rib in the Romanesque vaults of Durham Cathedral', ibid., pp. 64-76.

21. See, for instance, as well as the works on Gaudi cited in Chapter 5, note 2, the critical analysis and suggested redesign of St Ouen, Rouen, in Guadet, J., *Eléments et théorie de l'architecture*, Librairie de la Construction Moderne, Paris, 1901, vol. 3, pp. 342-348.

22. See again Hewett C. A., 'The barns at Cressing Temple, Essex, and their significance in the history of English carpentry', *JSAH*, vol. 26, 1967, pp. 48-70. Also Horn, W., 'On the origins of the mediaeval bay system', *JSAH*, vol. 17, 1958, pp. 2-23; idem, 'The great Tithe Barn of Cholsey, Berkshire', *JSAH*, vol. 22, 1963, pp. 13-23; and Horn and Born *Barns*.

23. Radding and Clark *Medieval architecture*, pp.67-76.

24. See, for instance, Conant, K. J., 'Observations on the vaulting problems of the period 1088-1211', *Gazette des Beaux-Arts*, 6th series, vol. 26, 1944, pp. 127-134; Viollet-le-Duc *Dictionnaire raisonné*, article 'arc-boutant', vol. 1, pp. 60-87, and article 'cathédrale', vol. 2, pp. 297-392; Aubert, M., 'Les plus anciennes croisées d'ogives, leur rôle dans la construction', *Bulletin Monumental*, vol. 93, 1934, pp. 5-67 and 137-237; Fitchen *Construction*; and Frankl *Gothic architecture*.

25. Lefèvre-Pontalis, M. E., 'L'origine des arcs-boutants', *Congrès archéologique de France*, vol. 82, 1920, pp. 367-396; Aubert, M., *Notre-Dame de Paris*, Laurens, Paris, 2nd edn, 1928, pp. 86-107; and James, J., 'Evidence for flying buttresses before 1180', *JSAH*, vol. 51, 1992, pp. 261-287, argue for and against priority at Notre Dame. See also Fitchen, J. F., 'A comment on the function of the upper flying buttress in French Gothic architecture', *Gazette des Beaux-Arts*, 6th series, vol. 45, 1955, pp. 69-90.

26. See, for instance, Conant, K. J., 'Observations' (note 24 above); idem, *Cluny: les églises et la maison du chef d'ordre*, Mediaeval Academy of America, 1968; and Bonelli, R., *Il duomo di Orvieto e l'architettura italiana del duocento*, Dell' Angelo, Citta di Castello, 1952 (the last describing a well documented example of later additions). Salisbury Cathedral furnishes good English examples.

27. For precise plots of possible thrust-lines, see Ungewitter, G. G., *Lehrbuch der gotischen Konstruktionen*, Tauchnitz, Leipzig, 1901, vol. 1, pp. 125-176; Rosenberg, G., 'The functional aspect of the Gothic style', *JRIBA*, vol. 43, 1936, pp. 273-290 and 364-371, fig. 29; and Dowrick, D. J., and Beckmann, P., 'York Minster structural restoration', *Proc.ICE*, vol. 49 supplement, 1971, pp. 93-156, fig. 20(a).

28. Goodyear, W. H., 'Vertical curves and other architectural refinements in the Gothic cathedrals and churches of northern France and in early Byzantine churches at Constantinople', *Memoirs of Art and Archaeology* (Brooklyn Institute of Arts and Sciences), vol. 1, no. 4, 1904, p. 60. Goodyear believed that this and other similar distortions were deliberate optical refinements like those to be seen in the Parthenon in Athens. On the basis of such a belief the very strong piers at the crossing of the Neo-Gothic cathedral of St Peter and St Paul in Washington DC were even constructed early in the present century with a built-in similar bow. But this mistaken interpretation was refuted, with particular reference to Amiens Cathedral, in Bilson, J., 'Amiens Cathedral and Mr Goodyear's "refinements". A criticism', *JRIBA*, 3rd series, vol. 13, 1906, pp. 397-417, and in subsequent exchanges with Goodyear, ibid., vol. 14, 1907, pp. 17-51 and pp. 84-91, and

vol. 16, 1909, pp. 715-740. Removal in the late 1960s by Lee Striker and Dogan Kuban of Ottoman facings from the piers of Goodyear's St Mary Diaconissa (now Kalenderhani Cami) in Istanbul confirmed beyond any doubt that the supposed 'refinements' there resulted entirely from structural distortions and repairs, while my own book cited in note 8 above and papers referred to therein have provided similar confirmation for Hagia Sophia.

29. Murray, S., *Beauvais Cathedral*, Princeton University Press, 1989, pp. 112-120. Earlier attempted diagnoses may be found in Heyman, J., 'Beauvais Cathedral', *Trans.NS*, vol. 40, 1967-68, pp. 15-35 (with a comment by the present writer) and Mark, R., *Experiments in Gothic structure*, MIT Press, 1982, pp. 69-71.

30. For instance in the nave of Florence Cathedral. See Guasti *Santa Maria del Fiore,* docs. 143-149, pp 168-174.

31. Reuther, H., *Balthasar Neumann*, Süddeutscher Verlag, Munich, 1983.

32. For the Salute, see Wittkower *Art and architecture in Italy*, pp. 191-195. For the Frauenkirche, see Hempel, E., *Baroque art and architecture in central Europe*, Pelican History of Art, 1965, pp. 196-198; the papers presented at a conference on the rebuilding after wartime destruction in 1945, *Wissenschaftliche Zeitschrift der Technischen Universität Dresden*, vol. 45, Sonderheft, 1996; Jäger, W., 'George Bährs Steinkuppel', *Die Dresdner Frauenkirche*, vol. 3, 1997, pp. 53-90; and Heinle, E. and Schlaich, J., *Kuppeln*, Deutsche Verlags-Anstalt, Stuttgart, 1996, p. 126. The last is useful also as a source of plans, sections and photographs of a much wider selection of domed structures than it has been possible to consider in this chapter and in Chapter 13.

33. For Wren, see the reports on Old St Paul's, Salisbury Cathedral and Westminster Abbey, in Wren *Parentalia*, pp. 271-278 and 295-308. For Guarini, see Guarini *Architettura*. For Soufflot, see the 'Mémoire sur l'architecture Gothique', presented to the Academy of Lyons in April 1741, printed in Petzet *Sainte-Geneviève*, pp. 135-142.

34. The most comprehensive published source is Botton, A. T. and Hendry, H. D., eds, *The Wren Society*, vols. 1, 1924; 2, 1925; 3, 1926; 8, 1936; 14, 1938; 15, 1937; and 16, 1939. Poley, A. F. E., *St Paul's Cathedral London*, privately published 1927, supplements it with a full series of measured drawings. The evolution of the design has been discussed in various publications including Sekler, E., *Wren and his place in European architecture*, Faber & Faber, 1956; and Fuerst, V., *The architecture of Sir Christopher Wren*, Lund Humphries, 1956. Now see chiefly Downes, K., *Sir Christopher Wren: the design of St Paul's Cathedral*, Trefoil, 1988.

35. The interrelation of the different elements of support is best illustrated by models and a series of beautiful analytical isometric drawings made in connection with restoration work between 1925 and 1930. These are illustrated in *The Wren Society* (see previous note) vol. 16, frontispiece and pls 5, 6 and 8. Some of them are also illustrated in Peach, C. S. and Allen, W. G., 'The preservation of St Paul's

Cathedral', *JRIBA*, vol. 37, 1930, pp. 655-676 (fig. 14) and in Hamilton, S. B., 'The place of Sir Christopher Wren in the history of structural engineering', *Trans.NS*, vol. 14, 1933-34, pp. 27-42 (pls 2 and 3).

36. Survey drawings including an excellent (but partly hypothetical) analytical axonometric were published in Denina, L. and Proto, A., 'La reale chiesa di San Lorenzo in Torino', *L'Architettura Italiana*, vol. 15, 1920, pp. 34-38 and pl. 20. The axonometric and some other drawings are also reproduced on a reduced scale in Portoghesi *Guarini*; and Meek, H. A., *Guarino Guarini*, Yale University Press, 1988.

37. Mainstone, R. J., 'Guarini and Leonardo', *ARev.*, vol. 147, 1970, p. 454, discusses this background and cites some of the most relevant Leonardo drawings (note that the two illustrations were reproduced in reverse order). See also Wittkower *Art and architecture in Italy*, p. 274; Turin Academy of Sciences, *Guarini e l'internazionalita del Barocco: Atti del convegno internazionale*, 2 vols, Turin, 1970; and, for the Guarini's designs of other related churches, Guarini, *Architettura*. Some of the latter are reproduced in Portoghesi *Guarini*.

38. Comprehensive illustrations of the structure as built are given in Rondelet, J., *Addition au mémoire historique sur le dome de l'Église Sainte-Geneviève*, Paris, 1814.

39. See again the works cited in note 17, Chapter 8.

40. The principal sources for the evolution of the design are Patte *Mémoire*; Rondelet *Mémoire*; and Gauthey *Dissertation*. See also Petzet *Sainte-Geneviève*; Braham, A, 'Drawings for Soufflot's Sainte-Geneviève' *Burlington Magazine*, vol. 112, 1971, pp. 582-591; and Canadian Centre for Architecture, *Le Panthéon: symbole des revolutions*, Picard, Paris, 1989.

Chapter 13 Ninetenth- and twentieth-century wide-span buildings

1. Notable examples were to be found in the Louvre (1779-1781) and the Theatre of the Palais Royal (1785-1790), both in Paris. They are illustrated in Rondelet *Traité 1838*, pls 155 and 154. See also Bannister, T. C., 'The first iron-framed buildings', *ARev.*, vol. 107, 1950, pp. 231-246.

2. For the Chatsworth Conservatory see Chadwick *Paxton*, pp.77-98 (referring also to related earlier greenhouses, both built and projected). For the Kew Palm House see Diestelkamp, E. J., 'Richard Turner and the Palm House at Kew Gardens' *Trans. NS*, vol. 54, 1982-83, pp. 1-26.

3. For the St Pancras roof, see 'Roof of the Saint Pancras Station – Midland Railway', *Engineering*, vol. 1, 1866, p. 12. For nineteenth-century trainsheds generally, see Meeks, C. L. V., *The railway station. An architectural history*, Yale University Press, 1957, (which includes a tabulation of dates and spans); and Condit, *Nineteenth century*, pp.197-222.

4. 'The Paris Exhibition', *Engineering*, vol. 47, 1889, pp. 415-503. See also Hautecoeur, L., 'La construction en fer au 19e siècle', *Bâtir*, no. 73, 1958, pp. 40-48, and no. 74, 1958, pp. 38-43 (reproducing

other engravings of the erection of the arches). For reviews of earlier and later arched and similar roofs not referred to here, see, respectively, Sutherland, R. J. M., 'Shipbuilding and the long span roof' *Trans.NS*, vol.60, 1988-89, pp. 107-126, and Wilkinson, C., *Supersheds*, Butterworth Architecture, 2nd edn 1996, especially pp. 12-24 dealing with airship hangars.

5. Klimke, H., Polónyi, S., Walochnik, W., and Ritchie, I., 'New Leipzig Fair – Design partnership Central Hall', *Conceptual design of structures*, International Association for Shell and Space Structures, Stuttgart, 1996, vol. 1, pp. 400-408 and 430-453.

6. Dilley, P., 'Kansai International Airport Terminal', *SE*, vol. 72, 1994, 293-297; idem and Taga, T., 'Kansai International Airport Terminal', *Civil Engineering*, vol. 108, 1995, pp. 2-11.

7. Davey, P., 'Waterloo International', *ARev.*, vol. 193, 1993, pp. 26-44.

8. For the Halle au Blé, see Wiebenson, D., 'The two domes of the Halle au Blé in Paris', *AB*, vol. 55, 1973, 262-279; and Deming, M. K., *La Halle au Blé de Paris*, Archives d'Architecture Moderne, Brussels, 1984, pp. 167-197. For the Albert Hall, see 'The Royal Albert Hall', *Engineering*, vol. 8, 1869, pp. 117-118 and 123; Scott, H., 'On the construction of the Albert Hall', *RIBA Sessional Papers*, 1872, pp. 83-100; and Sheppard, F. H. W., ed., *Survey of London*, vol. 38, 1975: *The museums area of South Kensington and Westminster*, pp. 177-195.

9. Freyssinet, E., 'Hangars à dirigeables en ciment armé en construction a l'aéroport de Villeneuve Orly', *Le Génie Civil*, vol. 83, 1923, pp. 265-273, 291-297, and 313-319.

10. Dantin, C., 'L'église en béton armé du Raincy', *Le Génie Civil*, vol. 83, 1923, pp. 1-4.

11. Nervi *New structures*, pp. 32-39 and 66-81; Joedicke et al. *Nervi*, pp. 59-68.

12. Esquillan, N., 'The shell vault of the Exposition Palace, Paris', *Proc.ASCE*, vol. 86, 1960, no. ST1, pp. 41-70.

13. Torroja *Philosophy of structures*, pp. 23-28. Notable British examples are described in Jordan, R. F., 'Brynmawr', *ARev.*, vol. 111, 1952, pp. 143-164; Arup, O. N., and Jenkins, R. S., 'The design of a reinforced concrete factory at Brynmawr', *Proc.ICE*, vol. 2, pt 3, 1953, pp. 345-397; and Ahm, P. and Perry, E. J., 'Design of the dome shell roof for Smithfield Poultry Market', ibid., vol. 30, 1965, pp. 79-108. At Smithfield, the ties were replaced by pretensioned edge members that distributed the vertical load uniformly to a number of columns spaced along each edge rather than concentrating it all on corner columns.

14. 'To cover this assembly bowl: a 400-ft prestressed saucer', *Engineering News Record*, vol. 166, 1961, 1 June, pp. 32-36; 'Deft concreting speeds work on huge dome', ibid., vol. 167, 1961, 5 October, pp. 30-32.

15. 'TWA's graceful new terminal', *AF*, vol. 108, no. 1, 1958, pp. 78-85.

16. 'Shaping a two-acre sculpture', *AF*, vol. 113, no. 2, 1960, pp. 118-123; TWA's concrete, wing-roofed terminal now ready for flight', *Engineering News Record*, vol. 168, 1962, 31 May, pp. 48-50.

17. Arup, O. N., and Zunz, G. J., 'Sydney Opera House', *SE*, vol. 47, 1969, pp. 99-132.

18. The building is more fully illustrated in 'Mies' enormous room', *AF*, vol. 105, no. 2, 1956, pp. 104-111.

19. Joyner, K. J., Makowski, Z. S., and Taylor, R. G., 'Structural aspects of the Boeing 747 hangar for BOAC at Heathrow Airport', *Proc.ICE*, vol. 47, 1970, pp. 483-513, describes one early example of the use of a steel space frame roof, here with deep trussed girders to bridge over the open front of the hangar.

20. 'Cathedral craft', *Concrete Quarterly*, no.100, 1974, pp. 22-32.

21. Nervi *New structures*, pp. 118-135.

22. Zunz, G. J., Manning, M. W. and Jofeh, C. G. H., 'The design of the structure for the new terminal building at Stansted Airport' *SE*, vol. 66, 1988, pp. 361-370; Davey, P., Davies, C. and Whitby, M., 'Stansted', *ARev.*, vol.189, 1991, pp. 42-76; and Zunz, J., Manning, M., Kaye, D. and Jofeh, C., 'Stansted Airport Terminal: the structure', *Arup Journal*, vol. 25, 1990, pp.7-15.

23. Boaga and Boni *Morandi*, pp. 52-65.

24. For Morandi's Via Olympica and Sulmona Bridges, see chiefly Boaga and Boni *Morandi*, pp. 72-85.

25. Banham, P. et al., 'Centre Pompidou, Paris' *ARev.*, vol. 161, 1977, pp. 270-294; Ahm, P. B., Clarke, F. G., Grut, E. L. and Rice, P., 'Design and construction of the Centre Nationale d'Art et de Culture Georges Pompidou' *Proc.ICE Part 1*, vol. 66, 1979, pp. 557-593.

26. The early development is well summarized with more copious references than could be given here in Graefe, R., 'Hängedächer des 19 Jahrhunderts', in Graefe *Geschichte*, pp. 168-187. The Lorient roof was described in Sganzin, J., and M. Reibell, *Programmes ou résumé des leçons d'un cours de construction*, 5th edn 1840, pl. 51, fig. 265.

27. Nervi *New structures*, pp. 164-167.

28. Boaga and Boni *Morandi*, pp. 182-191. At Frankfurt, two years before the Alitalia Hangar, two similar hangars of slightly lesser span were constructed back to back with similarly cantilevered roofs and with the stays continuous from one to the other so that each roof served as the counterweight to the other and there was no need for separate tie-downs. See Kirchner, G., 'Zum Entwurf des Schalendaches der Flugzeughalle III auf dem Flughafen Frankfurt/Main', *Beton- und Stahlbetonbau*, vol. 55, 1960, pp. 73-81.

29. Banham, P. et al., 'INMOS in Gwent', *ARev.*, vol. 172, 1982, pp. 26-33. For a wide review of other masted structures, see Harris, J. B. and Li, K. Pui-K., *Masted structures in architecture*, Butterworth Architecture, 1996.

30. See again Graefe, R., 'Hängedächer des 19 Jahrhunderts' (note 26 above). For an even more ambitious (and unrealizable) project illustrated in the *Chicago Tribune*, 9 March 1890, see Karlowicz, T. M., 'D. H. Burnham's role in the selection of architects for the World's Columbian Exposition', *JSAH*, vol. 29, 1970, pp. 247-254, fig. 1. For Suchov's pavilions see Graefe, R., 'Netzdächer, Hängedächer und

Gitterschalen' in Gappoev et al. *Suchov*, pp. 28-53.

31. Severud, F. N., 'Cable-suspended roof for Yale Hockey Rink', *CE*, vol. 28, 1958, pp. 666-669.

32. Otto *Das hangende Dach;* and Otto and Schleyer *Tensile structures.*

33. Leonhardt, F. and Schlaich, J., 'Structural design of roofs over the sports arenas for the 1972 Olympic Games: some problems of prestressed cable net structures', *SE*, vol. 50, 1972, pp. 113-119. Fuller details are given in idem, 'Vorgespannte Seilnetz-konstruktionen – Das Olympiadache in München' *Der Stahlbau*, vol. 41, 1972, pp. 257-266, 298-301, 367-378, and vol. 42, 1973, pp. 51-58, 80-86, 107-115, 176-185.

34. Berger *Light structures*, pp. 110-117.

Chapter 14 Bridges

1. The account presented in this chapter is necessarily much abbreviated. There are now many general histories including Whitney, C. S., *Bridges: a study in their art, science and evolution*, Rudge, 1929; Steinman, D. B. and Watson, S. R., *Bridges and their builders*, G. P. Putnam's Sons, 1941; and Shirley-Smith *Bridges*. Mock, E. B., *The architecture of bridges*, Museum of Modern Art, New York, 1949, is valuable chiefly for its illustrations; Hopkins *Bridges* for the nineteenth century; and Haegermann et al. *Vom Caementum zum Spannbeton*, vol.2, for bridges of reinforced concrete. The most valuable for the present purpose, and especially for the twentieth century, are Leonhardt *Bridges*; and Wittfoht *Bridges*.

2. For Roman bridges and aqueducts generally, see Gazzola *Ponti*; Van Deman, E. B., *The building of the Roman aqueducts*, Carnegie Institution of Washington Publications 423, 1934; Ashby, T., *The aqueducts of ancient Rome*, Clarendon Press, 1935; Casado, C. F., 'Historia del puente en España', *Informes de la construcción*, nos 76, 82, 97, etc. and 201, 205, etc.; O'Connor, C., *Roman bridges*, Cambridge University Press, 1993; and Çeçen, K., *The longest Roman water supply line*, Türkiye Sinai Kalkinma Bankasi, Istanbul, 1996. For the Pont du Gard, see Espérandieu, E., *Le pont du Gard*, 3rd edn rev., Henri Laurens, Paris, 1968.

3. Vitruvius *De architectura*, bk 5, ch. 12, describes Roman underwater foundations in late Republican times.

4. For medieval and Renaissance bridges generally, see Dartein, F. de, *Études sur les ponts en pierre remarquables par leur décoration anterieurs au XIXe siècle*, Ch. Beranger, Paris, 1909-1912; Séjourné *Grandes voûtes*, vol.1; Mesqui, J., *Le pont en France avant le temps des ingénieurs*, Picard, Paris, 1986; Pope *Survey*, pp. 1226-1251; and Parsons *Engineering in the Renaissance*, pp. 507-586.

5. For a Roman precedent in the Temple of Diana at Nîmes, see Chapter 12. This type of arch was commonest in France and England.

6. Illustrated in Mainstone, R. J., 'The springs of structural invention', *JRIBA*, vol. 70, 1963, fig. 40, and in idem, 'Intuition and the springs of structural invention', *VIA* (Graduate School of Fine Arts, University of Pennsylvania), vol. 2, 1973, fig. 36. These European examples have otherwise been completely ignored hitherto. For Chinese examples, see Fugl-Meyer *Chinese bridges*, pp. 85-99, and Tang Huangcheng *Chinese bridges*, pp. 110-162.

7. The bridges referred to are illustrated and described more fully in Baldaccini, R., *Monumenti Fiorentini. Il Ponte Vecchio*, Cya, Florence, 1947; Boyd, A., *Chinese architecture and town planning*, Tiranti, 1962, pp. 155-157 and pl. 24; Needham *Civil engineering and nautics*, pp. 175, 178 and pl. 348; and Tang Huangcheng *Chinese bridges*, pp.115-121.

8. As, for instance in Telford's Tongeland and Dean Bridges illustrated in Telford *Life*, pls 8 and 62.

9. Çeçen, K., *Sinan's water supply system in Istanbul*, ISKI, Istanbul, 1992, pp. 112-121 and plan 5.

10. The foundations are described, on the basis of sketches in a notebook kept by the engineer in charge of construction, in Parsons *Engineering in the Renaissance*, pp. 544-545.

11. Perronet, J. R., 'Mémoire sur la réduction de l'épaisseur des piles et sur la courbure qu'il convient de donner aux voûtes, le tout pour que l'eau puisse passer plus librement sous les ponts', *Mémoires de l'Académie Royale des Sciences*, 1777, pp. 553-564; idem, *Description des projets et de la construction des Ponts de Neuilly, de Mantes, d'Orléans et autres*, Paris, 1782-1783, 2 vols (new edn 1788, 1 vol. text, 1 vol. plates). For contemporary developments in Britain and Ireland, see Ruddock *Arch bridges* (with briefer reference also to iron arches).

12. Numerous examples are described and illustrated in Séjourné *Grandes voûtes*, vols 2-4. Vol. 5 gives details of design and construction procedures.

13. For the Valtschiel and Schwandbach Bridges, see Bill *Maillart*, pp. 58-61 and 90-93.

14. James, J. G., 'The cast iron bridge at Sunderland', *Occasional Papers in the History of Science and Technology* no.5, 1986, Newcastle upon Tyne Polytechnic; 'Thomas Paine's iron bridge work', *Trans.NS*, vol. 59, 1987-88, pp. 189-221.

15. Baker, Sir B., in 'Discussion' on the Hooghly 'Jubilee' Bridge, *Min.Proc.ICE*, vol. 92, 1888, pp. 116-117.

16. For these bridges see Telford *Life*, pls 47, 26, and 82; and James, J. G., 'Some steps in the evolution of early iron arched bridge designs', *Trans.NS*, vol. 59, 1987-88, pp. 153-186. His ambitious but unexecuted project for a single-span arch across the Thames on the site of Old London Bridge is described and illustrated in [Fourth] *Report from the select committee upon the improvement of the Port of London*, London, 1801, pp. 9-85; and in Telford, T. and Douglass, J., *An account of the improvements of the Port of London and more particularly of the intended iron bridge consisting of one arch of six hundred feet span*, London, 1801. But, though this was a more realistic project than that of Thomas Pope for a single timber span across the Hudson (Pope *Treatise*, pp.197-246 and pls 1-6), or that of Leonardo for a single masonry span over the Golden Horn (*MS. L*, Institut de France, fol. 66r., discussed in Stussi, F., 'Leonardo da Vincis Entwurf für eine Brücke über das Goldene Horn', *Schweizerische Bauzeitung*, vol. 71, 1953, pp. 113-116), it would be unwise to place much emphasis on it. The construction of adequate abutments would, if nothing

else, have proved a formidable task.

17. For the Maria Pia Bridge see Chapter 4. For the Garabit Bridge, see Eiffel, A. G., *Notice sur le viaduc de Garabit*, Paris, 1888; and idem, *Mémoire sur le Viaduc de Garabit*, Paris, 1889 (1 vol. text, 1 vol. plates).

18. For Maillart's Thur Bridge and Freyssinet's Sainte Pierre-du-Vauvray Bridge, see Bill *Maillart*, pp. 94-98; and Freyssinet, E., 'Le nouveau pont en béton armé, sur le Seine', *Le Génie Civil*, vol. 83, 1923, pp. 417-421. For examples of longer spans in steel and reinforced concrete, see the paper and reports referred to above in note 20 of chapter 9; Freyssinet, E., 'Le pont en béton armé Albert Louppe', *Le Génie Civil*, vol. 97, 1930, pp. 317-334; and Baxter, J. W., Gee, A. F. and James, H. B., 'Gladesville Bridge', *Proc.ICE*, vol. 30, 1965, pp. 489-530; and Haegermann et al. *Vom Caementum zum Spannbeton*, vol. 2, pp. 64-94.

19. As usual, Verantius illustrated the principle. See his *Machinae novae,* pls 32 and 33. Brunel's Saltash Bridge was also a tied arch, but with a catenary above deck level as the principal tie. For a few more recent examples, see Maunsell, G. A., and Pain, J. F., 'The Storstrøm Bridge', *JICE*, vol. 2, 1939, pp. 391-448; Stein, P. and Wild, H., 'Das Bodentragwerk der Fehmarnsundbrücke', *Der Stahlbau*, vol. 34, 1965, pp. 171-186; Hilton, N. and Hardenberg, G., 'Port Mann Bridge, Vancouver, Canada', *Proc.ICE*, vol. 29, 1964, pp. 677-712; and Smith, D. W., 'New Scotswood Bridge', ibid., vol. 42, 1969, pp. 217-249.

20. For other examples see, for instance, Caesar, *Gallic War*, ed. and trans. H. J. Edwards, Loeb Classical Library, William Heinemann, 1917, bk 4, 17, pp. 200-203 (with reconstructed plan and section facing p. 201), Fugl-Meyer *Chinese bridges*, pp. 57-81 and 133-137 (referring in particular to one stone bridge in Fukien with some spans exceeding 20 m), and Tang Huangcheng *Chinese bridges*, pp.23-63. Palladio's design appears in Palladio *Quattro libri*, bk 3, ch. 9, pp. 19-20, and is referred to in Boucher *Palladio*, pp. 211-213. The bridge was first built in 1569 and has three times been rebuilt to this design, most recently in 1948.

21. The principal source is Westhofen, W., 'The Forth Bridge' *Engineering*, vol. 49, 1890, pp. 213-283. See also Cooper, F. E., *Forth Bridge: Notes upon erection etc.*, 1896, MS, now in the Library of the Institution of Civil Engineers (a description of the erection by the resident engineer illustrated by careful drawings and numerous contemporary photographs); and Hammond, R., *The Forth Bridge and its builders*, Eyre & Spottiswoode, 1964.

22. Finsterwalder, U. and Schambeck, H., 'Von der Lahnbrücke Balduinstein bis zur Rheinbrücke Bendorf', *Der Bauingenieur*, vol. 40, 1965, pp. 85-91.

23. Described in 'Footbridge at Durham University', *ARev.*, vol. 135, 1964, pp. 264-266.

24. For Waterloo Bridge, see Buckton, E. J. and Cuerel, J., 'The new Waterloo Bridge', *JICE*, vol. 20, 1943, pp. 145-178; for the Zoo Bridge, Freudenberg, G. and Ratka, O., 'Die Zoobrücke über den Rhein in Köln', *Der Stahlbau*, vol. 35, 1966, pp. 225-235,

269-277, and 337-346.

25. There is now a vast literature in technical periodicals dealing with multi-span highway bridges. Leonhardt, F. and Andra, W., 'Stutzungs probleme der Hochstrassenbrücken', *Beton- und Stahlbetonbau*, vol. 55, 1960, pp. 121-132 (English version, 'Problems of supporting elevated road bridges', trans. C. van Amerongen, C & CA Library Translation, no. 95, 1962) discussed the problems of support. Wittfoht *Bridges*, is now the best source for developments in construction techniques. For a good selection of recent examples, mostly in continental Europe, see now Leonhardt *Bridges*, pp.109-208, and again Wittfoht. Fairhurst, W. A. and Beveridge, A., 'The superstructure of the Tay Road Bridge', *SE*, vol. 43, 1965, pp. 75-82, and Rawlinson, Sir J. and Stott, P. E., 'The Hammersmith Flyover', *Proc.ICE*, vol. 23, 1962, pp. 565-600, describe early British examples with, respectively, steel and reinforced concrete spans.

26. Ordonez *Freyssinet*, pp. 238, 247-251, 423-424, 430

27. Nissen, J., Falbe-Hansen, K. and Stears, H. S., 'The design of Kylesku Bridge', *SE*, vol. 63A, 1985, pp. 69-76; and Martin, J. M., 'The construction of Kylesku Bridge' *Proc.ICE*, vol. 80, 1986, pp. 317-342.

28. Peters, T. F., *Transitions in Engineering*, Birkhäuser, Basle, 1987, pp. 13-38, gives the best concise review of these early precursors of the modern suspension bridge and makes some reference also to their successors up to the early nineteenth century.

29. See, for example, Lee, D., 'Aspects of tension and compression for bridges' in Institution of Structural Engineers, *1st Oleg Kerensky Memorial Conference*, London, 1988, pp 27/2-32/2; Brüninghoff, H., 'The Essing Timber Bridge, Germany', *Structural Engineering International*, vol. 3, 1993, pp. 70-72; and Strasky, J., 'Precast stress-ribbon and suspension pedestrian bridges', in Giuliani *Spatial structures*, pp. 1003-1010.

30. Braun, F., and Moors, J., 'Wettbewerb zum Bau einer Rheinbrücke im Zuge der Inneren Kanalstrasse in Köln (Zoobrücke)', *Der Stahlbau*, vol. 32, 1963, pp. 251-253; and Finsterwalder, U., 'Prestressed concrete bridge construction: stress ribbon bridge', *JACI*, vol. 62, 1965, pp. 1042-1045.

31. This is illustrated in Sutherland, R. J. M., 'Some aspects of the history of the suspension bridge' in Institution of Structural Engineers, *1st Oleg Kerensky Memorial Conference*, London, 1988, pp. 15/1-23/1, which is now also the most useful single critical reference for other aspects of the early history. See also, for these other aspects, Kemp, E. L., 'Ellet's contribution to suspension bridge history', *Proc.ASCE*, vol. 99, 1973, pp. 331-351; idem, 'Links in a chain − the development of suspension bridges 1801-70, *SE*, vol. 57A, 1979, pp. 255-263; Day, T., 'Samuel Brown: his influence on the design of suspension bridges', *History of Technology*, vol. 8, 1983, pp. 61-90; and Picon, A., 'Navier and the introduction of suspension bridges in France', *Construction History*, vol. 4, 1988, pp. 21-34.

32. Verantius *Machinae novae*, pls 34 and 35.

33. Finley, J., 'A description of the patent chain bridge invented by James Finley', *The Port Folio*, new series, vol. 3, 1810, pp. 441-453; Kranakis *Constructing a bridge*, pp. 28-59. Kranakis also discussed the whole background to the development.

34. Pope *Treatise*, pp. 187-189.

35. Provis, W. A., *An historical and descriptive account of the suspension bridge constructed over the Menai Strait, in N. Wales with a brief notice of Conway Bridge*, London, 1828. See also Maunsell, G. A., 'Menai Bridge reconstruction', *JICE*, vol. 25, 1946, pp. 165-206; and Paxton, R. A., 'Menai Bridge and its influence on suspension bridge development' *Trans.NS*, vol. 49, 1977-78, pp. 87-110.

36. For Montrose Bridge, see Rendel, J. M., 'Memoir of the Montrose Suspension Bridge', *Proc. ICE*, vol. 1, 1841, pp. 122-129. For Roebling's bridges generally, see Condit, *Nineteenth century*, pp. 172-180; Hopkins *Bridges*, pp. 211-227; and Steinmann, D. B., *The builders of the bridge*, Harcourt Brace & Co., 1945. For the Cincinnati and Brooklyn Bridges in particular, see 'The Cincinnati Suspension Bridge', *Engineering*, vol. 4 1867, pp. 22-23, 49, 74-76, 98-99, 140-141, and pls facing pp. 346, 351; Seeley, H. R., Ammann, O. H., Gray, N. and Wessman, H. E., 'Technical survey – Brooklyn Bridge after sixty years', *Proc.ASCE*, vol. 72, 1946, pp. 3-68; and Trachtenberg, M., *Brooklyn Bridge: fact and symbol*, Oxford University Press, 1965. Vogel, R. M., 'Roebling's Delaware & Hudson Canal aqueducts', *Smithsonian Studies in History and Technology*, no. 10, 1971, describes Roebling's earliest surviving suspension structures.

37. Ammann, O. H. et al., 'George Washington Bridge', *Trans.ASCE*, vol. 97, 1933, pp. 1-377; Golden Gate Bridge and Highway District, *Golden Gate Bridge*, San Francisco, 1938; and Reidy, P. J., Ammann, O. H. et al., 'Verrazano-Narrows Bridge', *Proc.ASCE*, vol. 92, no. C02, 1966, pp. 1-192, describe these three bridges.

38. For the Menai Bridge, see Maunsell, G. A., 'Menai Bridge reconstruction' (note 35 above), pp. 192-193. For Brooklyn Bridge, see Seeley, H. R. et al., 'Technical survey' (note 36 above), pp. 29, 46. Measurements on the latter bridge in 1943 showed that most of the movements taking place as a result of relative changes in the tensions in the cables were rotations of the entire towers as rigid bodies about the more flexible timber grillages on which they were founded well below water level.

39. Ammann et al. *Tacoma Narrows Bridge*.

40. Roberts, Sir G., 'Forth Road Bridge: design', *Proc.ICE*, vol. 32, 65, pp. 333-405.

41. Roberts, Sir G., 'Severn Bridge: design and contract arrangements', *Proc.ICE*, vol. 41, 1968, pp. 1-48.

42. For the Great Belt see Ostenfeld, 'The Great Belt Bridge project' in Giuliani *Spatial structures*, pp. 947-960; and Gimsing, N. J., 'The Storebelt East Bridge', in Sanayei *Restructuring America*, pp. 253-263. For Messina see Leto, I. V., 'Preliminary design of the Messina Strait Bridge', *Proc.ICE: Civil Engineering*, vol. 102, 1994, pp.122-128. Some comparative elevations are given in Brown, D. J., *Bridges*, Beazley, 1993, pp.148 and 162-163

43. For the Shimotsui-Seto and Besan-Seto Bridges see Ishiyama, S. et al., 'Special topic: Honshu-Shikoku Bridge Project', *Civil Engineering in Japan*, vol. 27, 1988, pp. 19-69; and Honshu-Shikoku Bridge Authority, *Seto Ohasi Bridge*, Bridge and Offshore Engineering Association, Tokyo, 1988 (in Japanese with excellent illustrations and English translations of the main text). For the Akashi-Kaikyo Bridge see Saeki, S. et al., 'Technological aspects of the Akashi-Kaikyo Bridge', in Sanayei *Restructuring America*, pp. 237-252.

44. Kapsch, G., 'Die Strassenbrücke über den Rhein in Köln-Mülheim', *Die Bautechnik*, vol. 7, 1929, pp. 683-687, 775-808, 813-817, and 865-870. For a British example, see Buckton, E. J. and Fereday, H. J., 'The reconstruction of Chelsea Bridge', *JICE*, vol. 7, 1938, pp. 383-446.

45. Verantius *Machinae novae*, pl. 34. See again the works by Sutherland and Peters cited in note 31 for the early history.

46. Tamms, F. et al., *Nordbrücke Düsseldorf*, Landeshauptstadt, Düsseldorf, 1958.

47. Hess, H., 'Die Severinsbrücke Köln: Entwurf und Fertigung der Strombrücke', *Der Stahlbau*, vol. 29, 1960, pp. 225-261; Vogel, G., 'Die Montage des Stahlüberbaues der Severinsbrücke Köln', ibid., pp. 269-293; and Stadt Köln, *Kölner Rheinbrücken 1959-1966*, Wilhelm Ernst, Berlin, 1966, pp. 53-137.

48. Leonhardt, F. and Andrä, W., 'Fussgängersteg über die Schillerstrasse in Stuttgart', *Die Bautechnik*, vol 39, 1962, pp. 110-116.

49. Boaga and Boni *Morandi*, pp. 142-181; Simons, H., Wind, H., and Moser, W. H., *The bridge spanning Lake Maracaibo in Venezuela*, trans. H. Bucksch and H. Brückner, Bauverlag, Wiesbaden and Berlin, 1963.

50. Walther, R. et al., *Cable stayed bridges*, Thomas Telford, 1988; Leonhardt, F. and Zellner, W., 'Past present and future of cable-stayed bridges', in Ito et al. *Cable-stayed bridges*, pp. 1-33; and Gimsing, N. J., *Cable supported bridges*, John Wiley, 2nd edn 1997, deal with the requirements and possible ways of meeting them much more fully and the last of these works, in particular, cites numerous examples and contains extensive bibliography. Both Walther et al. and Wittfoht *Bridges*, pp. 96-100 and 295-297, give comparative similarly scaled elevations of numerous bridges.

51. For the Norderelbe Bridge, see Aschenberg, H., Freudenberg, G. and Havemann, H. K., 'Die Brücke über die Norderelbe im Zuge der Bundesautobahn Südliche Umgehung Hamburg', *Der Stahlbau*, vol. 32, 1963, pp. 240-248, 281-287, and 310-317. For the Brotonne Bridge see Brault, J.-L. and Mathivat, J., 'Le pont de Brotonne', *Travaux*, vol. 72, 1976, 22-43.

52. Lin, T. Y. and Redfield, C., 'Design of the Ruck-A-Chucky Bridge', *Concrete International*, vol. 1, 1980, pp. 12-16.

53. For the first, see Takenouchi, K., 'The aesthetics of cable-stayed bridges with spatial cable form', *Conceptual design of structures*, International Association for Shell and Space Structures, Stuttgart, 1996, vol. 2, pp. 671-678. For the second see Harvey, W. J., 'A reinforced plastic footbridge, Aberfeldy,

UK', *Structural Engineering International*, vol. 3, 1993, pp. 229-232.

54. Virolgeux, M., 'Normandie Bridge: design and construction', *Proc.ICE: Structures and Buildings*, vol. 99, 1991, 281-302; idem, 'The Normandie Bridge, towards very long cable-stayed spans' in Giuliani *Spatial structures*, pp. 961-970. This bridge will be slightly exceeded in span by the 890 m Tatara Bridge in Japan, due for completion in 1999.

Chapter 15 Multi-storey buildings and towers

1. If there are good bearing surfaces at the heads and feet of the columns, any tilting of them would slightly lift everything at a higher level.

2. Smeaton *Edystone*.

3. Beltrami, L., 'Fall of the Campanile of St Mark's, Venice', *JRIBA*, 3rd series, vol. 9, 1902, pp. 429-437.

4. Forsyth, W. A., 'The structure of Salisbury Cathedral tower and spire', *JRIBA*, vol. 53, 1946, pp. 85-87; Tatton-Brown, T., 'Building the tower and spire of Salisbury Cathedral', *Antiquity*, vol.65, 1991, pp.74-96.

5. The history of this period is well summarized in Jenkins, F. I., 'Some nineteenth-century towers', *JRIBA*, vol. 65, 1958, pp. 124-130.

6. For Trevithick's monument, see Trevithick, R., *Design for a gilded national monument of cast-iron*, London, 1832; for the second proposal, Burton, C., 'Proposal for the conversion of the Great Exhibition Building into a prospect tower 1,000 feet high', *The Builder*, vol. 10, 1852, pp. 280-281.

7. Bannister, T. C., 'Bogardus revisited. Part II: the iron towers', *JSAH*, vol. 16, 1957, pp. 11-19, describes and illustrates these two towers (which were demolished in 1907 and 1908) and summarizes the earlier development with full documentation.

8. 'The Paris Exhibition', *Engineering*, vol. 47, 1889, pp. 415-503; Eiffel, G., *La Tour de trois cents mètres*, Paris, 1900; idem, *La Tour Eiffel en 1900*, Paris, 1902; Vogel, R. M., 'Elevator systems of the Eiffel Tower', *United States National Museum Bulletin*, no. 228, 1961, pp. 1-40.

9. Petropavlovskaja, I. A., 'Hyperbolische Gitter-türme', in Gappoev et al. *Suchov*, pp. 78-91; idem, 'Der Sendeturm für die Radiostation Sabolovka in Moskau', ibid., pp. 92-103.

10. Faltus, F., 'Design, structural problems and construction of steel tower-shaped structures: general report', in *Proceedings of the symposium on tower-shaped steel and reinforced concrete structures, Bratislava 1966*, International Association for Shell Structures, Madrid, 1968, pp. 203-212, gives a general review up to 1966. Wise, C. M., 'Torre de Collserola, Barcelona', *Arup Journal*, vol. 27, 1992, pp. 3-7; and idem, 'Design of the Torre de Collserola, Barcelona' *SE*, vol. 71, 1993, pp. 353-359 describe a fine later example, 288 m high.

11. Leonhardt, F., 'The present position of rein-forced concrete tower design' in *Proceedings of the symposium on tower-shaped steel and reinforced concrete structures* (previous note), pp. 2-28; Rühle, H., 'TV towers and tower foundations' ibid., pp. 291-315; and, Leonhardt, F., 'Modern design of television towers', *Proc.ICE*, vol. 46, 1970, pp. 265-291, give general reviews.

12. Vitruvius *De architectura*, bk 2, ch. 8.

13. For Ostia, see Meiggs, R., *Roman Ostia*, Oxford University Press, 1960 (2nd edn 1973). For Roman tenements, see also Boethius, A., *The Golden House of Nero: some aspects of Roman architecture*, University of Michigan Press, 1960, pp. 129-185. For Trajan's Market, see MacDonald *Architecture of the Roman empire*, pp. 75-93; and Ward-Perkins *Roman Imperial Architecture*, pp. 86-89.

14. For the wall braced primarily with secondary framing members, see, for instance, Krafft *Charpente*, pt 1, pl. 2. For the balloon frame, see Bell, W. E., *Carpentry made easy*, Philadelphia, 1858; and Jensen, R., 'Board and batten siding and the balloon frame: their incompatibility in the nineteenth century', *JSAH*, vol. 30, 1971, pp. 40-50.

15. Bannister, T. C., 'The Roussillon vault, the apotheosis of a "folk" construction', *JSAH*, vol. 27, 1968, pp. 163-175. Collins, G. R., 'The transfer of thin masonry vaulting from Spain to America', ibid., pp. 176-201, discusses later uses.

16. For much fuller discussions of this phase of development, see Fairbairn, *Application*; Johnson, H. R., and Skempton, A. W., 'William Strutt's cotton mills, 1793-1812', *Trans.NS*, vol. 30, 1955-57, pp. 179-205; Bannister, T. C., 'The first iron-framed buildings', *ARev.*, vol. 107, 1950, pp. 231-246; and Skempton, A. W., and Johnson, H. R., 'The first iron frames', *ARev.*, vol. 131, 1962, pp. 175-186.

17. Barrett, J., 'On the construction of fire-proof buildings', *Min.Proc.ICE*, vol. 12, 1853, 244-272, (quoting a report by Sir Henry de la Beche and Thomas Cubitt) refers on pp. 246-247 to a progressive collapse partly attributable to this weakness that occurred in Oldham in 1844, and is referred to again in Fairbairn *Application*, Appendix 2, pp. 170-175.

18. Randall, F. A., *History of the development of building construction in Chicago*, University of Illinois Press, 1949; and Condit, C. W., *The Chicago school of architecture*, University of Chicago Press, 1964, discuss this phase.

19. For Badger and Bogardus see Bogardus, J. and Thomson, J. W., *Cast iron buildings: their construction and advantages*, New York, 1856; Badger, D. D., *Illustrations of iron architecture made by the architectural iron works of the City of New York*, New York, 1865 (facsimile edn of both works: *The origins of cast iron architecture in America*, intro. W. Knight Sturges, Da Capo, 1970); Bannister, T. C., 'Bogardus revisited. Part 1: the iron fronts', *JSAH*, vol. 15, 1956, pp. 12-22; and Friedman, D., *Historical Building Construction*, Norton, 1995

20. Wyatt, M. D., 'On the construction of the building for the exhibition of the works of industry of all nations in 1851', *Min.Proc.ICE*, vol. 10, 1850-51, pp. 127-191 (quotation from p. 151). See also Downes, C. and Cooper, C., *The building erected in Hyde Park for the Great Exhibition of the works of industry of all nations*, London, 1852; and, for the origins of the design, Chadwick *Paxton*, pp. 104-133.

21. Skempton, A. W., 'The Boat Store, Sheerness (1858-60) and its place in structural history', *Trans. NS*, vol. 33, 1959-60, pp. 57-78.

22. 'The St Ouen Docks, Paris', *The Builder*, vol. 23, 1865, pp. 296-299.

23. The most useful contemporary source is Freitag *Architectural engineering*. For general accounts see again the works cited in note 18 above.

24. Freitag *Architectural engineering* contains numerous detailed drawings of the structure and photographs of the building under construction, the photographs (figs 20 and 21) showing well the slenderness of the frame.

25. Condit, C. W., 'The first reinforced-concrete skyscraper', *Technology and Culture*, vol. 9, 1968, pp. 1-33.

26. For flat-slab structures see Chapter 8 and the works cited in notes 28 and 29 of that chapter. For Perret's buildings, see Collins *Concrete*.

27. Condit, *Twentieth century*, pp. 10-18 describes both buildings. For the Woolworth Building, see also Landau S. B. and Condit, C. W., *Rise of the New York Skyscraper*, Yale University Press, 1996, pp. 381-391; and for the Empire State Building, Rathbun, J. C., 'Wind forces on a tall building', *Trans.ASCE*, vol. 105, 1940, pp. 1-84.

28. For the CLASP system, see, for instance, Heathcote, F. W., 'Steel components for system building', in *Industrialized building and the structural engineer*, Institution of Structural Engineers, 1966, pp. 67-75. For York University, see 'University of York', *Architects' Journal*, vol. 142, 1965, pp. 1435-1458; and Brawne, M., 'University of York: first and second phases', *ARev.*, vol. 138, 1965, pp. 408-420.

29. 'The mobile home is the 20th century brick', *ARec.*, vol. 143, no. 4, 1968, pp. 137-143, illustrates one large-scale project of this kind.

30. Boyd, R., 'Habitat's cluster', *AF*, vol. 126, no. 5, 1967, pp. 36-41; 'Housing on Mackay Pier, Cité du Havre, Montreal', *Architects' Journal*, vol. 145, 1967, pp. 1059-1075; Fitzgerald, D. J., 'Construction of Habitat '67', *JACI*, vol. 65, 1968, pp. 801-810; Safdie M., 'Habitat: anatomy of a system', *JRIBA*, vol. 74, 1967, pp. 489-494; idem, *Beyond Habitat*, MIT Press, Cambridge, MA, 1970.

31. See, for instance Diamant *Industrialised building*, vol. 1, pp. 123-126.

32. This exploitation drew also on a much improved understanding of the structural behaviour, an understanding that was still growing. For this, see especially Baker, J. F., *The steel skeleton*, vol. 1: *Elastic behaviour and design*, Cambridge University Press, Cambridge, 1954; and Baker, J. F., Horne, M. R. and Heyman, J., *The steel skeleton*, vol. 2: *Plastic behaviour and design*, Cambridge University Press, 1956.

33. For the Lever Building, see 'Lever House, New York: glass and steel walls', *ARec.*, vol. 111, June, 1952, pp. 130-135.

34. 'IBM's exterior truss walls', *Progressive Architecture*, vol. 43, no. 9, 1962, pp. 162-167.

35. Brandenburger, J., Eatherley, M. and Raines, D., 'Bush Lane House', *The Arup Journal*, vol. 11, 1976, pp. 6-17.

36. For Highpoint One and its immediate successors, see Robertson, H. M., '"High Point", Highgate', *Architect and Building News*, vol. 145, 1936,

pp. 49-55; Arup, O. N., 'Box frame construction', *Architects' Journal*, vol. 101, 1945, pp. 439-440; and Lubetkin, B., 'Flats in Rosebery Avenue, Finsbury', *ARev.*, vol. 109, 1951, pp. 138-149. For the Essex University towers, see 'University of Essex − 2', *Architects' Journal*, vol. 144, 1966, pp. 1559-1573; and Sutherland, R. J. M., 'Design engineer's approach to masonry construction', in *Designing, engineering and constructing with masonry products*, ed. F. B. Johnson, Gulf Publishing Co., 1969, pp. 375-385. For concrete bearing-wall structures like those at Thamesmead, see Koncz *Precast concrete*, vol. 3; and Diamant *Industrialised building*.

37. Illustrated in 'Laboratory tower in cantilever construction, Racine, Wisconsin', *AF*, vol. 88, no. 1, 1948, pp. 114-115.

38. 'A landmark tower of New Haven's skyline', *ARec.*, vol. 148, no.8, 1970, 109-115.

39. Williams, G. M. J., and Rutter, P. A., 'The design of two buildings with suspended structures in high yield steel', *SE*, vol. 45, 1967, pp. 143-151. See also Schneider, M., 'Hochhauser mit hangenden Geschossen', *Der Stahlbau*, vol. 37, 1968, pp. 33-44 and 89-96; and Zunz, G. J., Heydenrych, R. A. and Michael, D., 'Standard Bank Centre, Johannesburg', *Proc.ICE*, vol. 48, 1971, pp. 195-222.

40. Gauntt, G. C. and Weinberg, B. E., 'Marina City', *CE*, vol. 32, no. 12, 1962, pp. 60-67; 'Marina City: outer-space image and inner-space reality', *AF*, vol. 122, no. 4, 1965, pp. 68-77

41. Frischmann, W. W., Lippard, D. C. and Steger, E. H., 'National Westminster Tower: design', *Proc. ICE*, vol. 74, 1983, pp. 387-434.

42. For fuller and rather more technical discussions of this and of the alternative possibilities considered in the following paragraphs, see le Messurier, W. J., 'The return of the bearing wall', *ARec.*, vol. 132, no. 1, 1962, pp. 168-171; 'Optimizing design in very tall buildings'. *ibid*, vol. 148, no. 2, 1970, pp. 133-136; Khan, F. R., 'Current trends in concrete high-rise buildings', in *Tall buildings*, ed. A. Coull and B. S. Smith, Pergamon, 1967, pp. 571-590; idem, 'Recent structural systems in steel for high-rise buildings', in British Constructional Steelwork Association, *Proceedings of the conference on steel in architecture*, 1970, pp. 55-65; Fintel, M. et al., 'Response of buildings to lateral forces', *JACI*, vol. 68, 1971, pp. 81-106; and the *Monographs on Tall Buildings and Urban Habitat*, 5 vols, various editors, American Society of Civil Engineers, 1978-1981.

43. Both buildings are described in the first of the papers by Khan, F. R., cited in the previous note.

44. 'The tallest steel bearing walls', *ARec.*, vol. 135, no. 5, 1964, pp. 194-196; Feld, L. S., 'Superstructure for 1,350-ft World Trade Center', *CE*, vol. 41, no. 6, 1971, pp. 66-70.

45. Iyengar, H. and Khan, F., 'Structural steel design of the Sears Tower', in Australian Institute of Steel Construction, *Conference on steel developments*, 1973; and Iyengar, H. S., 'Bundled-tube for Sears Tower', *CE*, vol. 42, no.11, 1972, pp. 71-75.

46. Khan, F. R., 'The John Hancock Center', *CE*, vol. 37, no. 10, 1967, pp. 38-42; and 'The tall one', *AF*, vol. 133, no. 1, 1970, pp. 36-45.

47. 'Engineering for architecture 1:The Citycorp

Centre, *ARec.*, vol. 160, mid August, 1976, pp. 66-71. For the tuned-mass dampers see the next note.

48. For fuller and more technical discussions see, for instance, Beedle, L. S., ed., *Second century of the skyscraper*, Van Nostrand Reinhold, 1988; Taranath, B. S., *Structural analysis and design of tall buildings*, McGraw-Hill, 1988; Schueller, W., *The vertical building structure*, Van Nostrand Reinhold, 1990; Iyengar, H., 'Tall structures' in Blanc, A., McEvoy, M. and Plank, R., eds, *Architecture and construction in steel*, Spon, 1993, pp. 214-233; and Kowalczyk, R. M., Sinn, R. and Kilmister, M. B., eds, *Structural systems for tall buildings*, McGraw-Hill, 1995. For damping generally, including provisions made in the Citycorp Centre and with numerous further references, see Wiesner, K. B., 'The role of damping systems' in the first of the above publications, pp. 789-802.

49. Fairweather, V., 'Record high-rise, record low steel, *CE*, vol. 56, no.8, 1986, pp. 42-45; Blake, P., 'Scaling new heights', *ARec.*, vol. 189, January, 1991, pp. 77-83.

50. Nötzold, F. and Kolmar, W., 'Messeturm Frankfurt am Main – Konstruktion und Berechnung' *Beton- und Stahlbetonbau*, vol. 86, 1991, pp. 79-82 and 120-123. See also Iyengar, H., 'Concrete core braced system for ultra-tall buildings', *Structural Engineering International*, vol. 3, 1992, pp. 168-169.

51. Mohamad, H. et al., 'The Petronas Towers – the tallest building in the world' in Council on Tall Buildings and Urban Habitat, *Habitat and the High-Rise: tradition and innovation*, 1995, pp. 321-357.

52. Wise. C. M. et al., 'The new Commerzbank headquarters, Frankfurt, Germany', *SE*, vol. 74, 1996, pp. 111-122; Bailey, P. et al., 'Commerzbank, Frankfurt', *The Arup Journal*, vol. 32, no. 2, 1997, pp. 3-12.

53. See, for instance, Habraken, N. J., *De dragers en de mensen: het einde van de massawoningbouw*, Scheltema & Holkema, Amsterdam, 1961 (English translation: *Supports: an alternative to mass housing*, B. Valkenburg, Architectural Press, 1972); 'Archigram', *Architectural Design*, vol. 35, 1965, pp. 559-573; Friedman, Y., 'Towards a coherent system of planning', in *Architects' Year Book XII, urban structure*, ed. David Lewis, 1968, pp. 52-64; and Dietrich, R. J., 'Metastadtprojekt I 1965-66', *Deutsche Bauzeitung – Die Bauzeitung*, vol. 103, 1969, pp. 18-21.

54. Zunz, G. J., Glover, M. J. and Fitzpatrick, A. J., 'The structure of the new headquarters for the Hongkong & Shanghai Banking Corporation, Hong Kong', *SE*, vol. 63A, 1985, pp. 255-284; idem, 'The Hongkong Bank: The new headquarters – The structure', *The Arup Journal*, vol. 20, no.4, 1985, pp. 2-26; Davey, P. et al., 'Hongkong and Shanghai Bank' *ARev.*, vol. 179, no.4, 1986, pp. 30-110.

Part 4 Design

Chapter 16 Design

1. For fuller discussions of these intuitions see Mainstone, R. J., 'The springs of structural invention', *JRIBA*, vol. 70, 1963, pp. 57-7l; idem, 'Intuition and the springs of structural invention', *VIA* (Graduate School of Fine Arts, University of Pennsylvania), vol. 2, 1973, pp. 46-67; and idem, 'The springs of invention revisited', *SE*, vol. 73, 1995, pp. 392-399. I am using the word 'intuition' in two slightly different senses. First, in the usual present sense of something directly known through experience rather than through reasoned deduction. What I have called the primary intuitions are initially of this kind, though they are capable of development by the application of reason to further observation and experiment. Second, in the extended sense of something known through such a development of the primary intuitions, but nevertheless held similarly at the back of the mind, similarly available as a guide and source of ideas without any conscious concentration on it, and distinguished in this way from the knowledge held in books. Both kinds of intuition, so defined, are essentially personal, like a physical accomplishment or skill in the fluent use of a language, even though a part of what is thus known may be put into words and thereby conveyed to others.

2. Among the expertises are (in chronological order) those relating to the cathedrals of Chartres (Frankl, P., *The Gothic: literary sources and interpretations through eight centuries*, Princeton University Press, 1960, pp. 844-845), Siena (Milanesi, G., *Documenti per la storia dell'arte Senese*, Siena, 1854, vol. 1, pp. 144-253 passim), Milan (*Annali della fabbrica del Duomo di Milano dall' origine fino al presente*, Milan, 1877, pp. 68-69 and 202-227 passim, reprinted and discussed in Ackermann, J. S., 'Gothic theory of architecture at the Cathedral of Milan', *AB*, vol. 31, 1949, pp. 84-111), Florence (Guasti *Santa Maria del Fiore*, docs 143-149, pp. 168-174 and Guasti *Cupola*, doc. 391, pp. 177-183), and Gerona (Frankl, P., op. cit., pp. 846-847), and to S. Petronio, Bologna (Gaye, G., *Carteggio inedito d'artisti dei secoli XIV, XV, XVL*, Florence, 1840, vol. 3, pp. 487-510, discussed in Booz, P., *Der Baumeister der Gotik*, Deutscher Kunstverlag, Berlin, 1956, pp. 54-61).

3. MS. Français 19093, Bibliothèque Nationale, Paris. I am greatly indebted to the Bibliothèque Nationale for the opportunity in 1966 to examine the Laon plan under glancing light and to Robert Branner who suggested to me the possible value of doing so.

4. Discussed further in Mainstone, R. J., 'Structural theory and design before 1742', *ARev.*, vol. 143, 1968, pp. 303-310. For the use in practice of the discipline up to early Renaissance times see, for instance, the works by Ackermann and Booz relating to Milan Cathedral and S. Petronio, Bologna, cited in note 2 above; Velte, M., *Die Anwendung der Quadratur und Triangulatur bei der Grund- und Aufrissgestaltung der gotischen Kirchen*, Birkhäuser, Basle, 1951; and Saalman, H., 'Early Renaissance architectural theory and practice in Antonio Filarete's Trattato di architettura', *AB*, vol. 41, 1959, pp. 89-106.

5. Archimedes, *On the equilibrium of planes*, English translation in *The works of Archimedes*, ed. T. L. Heath, Cambridge, 1897 (reprinted Dover, 1950), pp. 189-220.

6. Mainstone *Hagia Sophia*, pp.201-217.

7. Vitruvius *De architectura*.

8. Alberti *De re aedificatoria*, bk 3, chs 13 and 14.

9. Mainstone, R.J., 'Brunelleschi's dome', *ARev.*, vol.162, 1977, pp. 165-166, and idem, 'Le origine della concezione strutturale della cupola di Santa Maria del Fiore' [The sources of the structural idea for the dome of S. Maria del Fiore], *Filippo Brunelleschi: la sua opera e il suo tempo*, Centro Di, Florence, 1980, 883-892.

10. See, for instance, Sanabria, S. L., 'The mechanisation of design in the 16th century: The structural formulae of Rodgrigo Gil de Hontanón', *JSAH*, vol. 41, 1982, pp. 281-293.

11. Wren *Parentalia*, pp. 353-358.

12. Leonardo da Vinci, *Codice Atlantico*, fol. 310 r (reproduced as fig. 16.7 in the first edition of this book). For Leonardo's statical ideas more generally, see chiefly Richter, J. P., *The literary works of Leonardo da Vinci*, 3rd edn, Phaidon, 1970, vol. 2, pp. 59-78; and Uccelli, A., *I libri di meccanica*, Hoepli, Milan, 1940.

13. Mainstone, R. J. and Berger, R. W, 'Materials and structure', in Berger *The Palace of the Sun*, p. 73.

14. Summaries of this development in theory (with extensive references) are to be found in Heyman *Coulomb's memoir*, pp. 82-88 and 168-183; idem, *The masonry arch*, Ellis Horwood, 1982, pp. 42-59 and 112-114; and Benevenuto *Introduction*, Part 2, pp. 321-336. The principal sources are La Hire *Traité*; idem, 'Sur la construction des voûtes dans les édifices', *Mémoires de l'Académie Royale des Sciences*, vol. 69, 1713, pp. 69-77; Belidor *Science des ingénieurs*, bk 2, Paris, 1729; and Frézier, A. F., *La théorie et la pratique da la coupe des pierres . . . ou traité de stereotomie a l'usage de l'architecture*, vol. 3, Paris, 1739, pp. 345-423;. For the background to the Newtonian breakthrough see Clagett, M., *The science of mechanics in the middle ages*, University of Wisconsin Press, 1959, and Benevenuto *Introduction*, Part 1.

15. Le Seur et al. *Parere*.

16. Poleni *Memorie*, cols 42-50, 282-292, 323-330 and 402-424 and pls E and K;. Di Stefano *Cupola;* and Mainstone, R. J., 'The dome of St Peter's: structural aspects of its design and construction and inquiries into its stability', in idem, *Structure in architecture*.

17. Musschenbroek, P. Van, *Physicae experimentales: Introductio ad cohaerentiam corporum firmorum*, Leyden, 1729. For examples of somewhat later tests see Rondelet *Mémoire*; idem, *Traité 1814*, vol. 3, pp. 72-102; Barlow, P., *An essay on the strength and stress of timber*, London, 1817; and Tredgold, T., *A practical essay on the strength of cast iron*, London, 1822.

18. Patte *Mémoire*.

19. Gauthey, E. M., *Mémoire sur l'application des principes de la méchanique à la construction des voûtes et des dômes*, Dijon, 1771; and idem, *Dissertation*.

20. For Galileo, see *Discorsi* (Crew and Salvio, pp. 109-152). Hooke's law of a linear relationship between stress and strain for elastic materials was first published as an anagram in an appendix to Hooke, R., *Description of helioscopes and some other instruments*, London, 1676, p. 31; then in idem, *De potentia restitutiva,* London, 1678.

21. Pugsley, A. G., 'The history of structural testing', *SE*, vol. 22, 1944, pp. 492-505, deals solely with testing. Related developments in structural theory are reviewed up to their respective dates of publication in Todhunter, I., and Pearson, K., *A history of the theory of elasticity and of the strength of materials*, Cambridge, 1886-1893, 2 vols (reprinted Dover, 1960); Timoshenko, S. P., *History of strength of materials,* Mc Graw Hill, 1953; Charlton, T. M., *A history of theory of structures in the nineteenth century*, Cambridge University Press, 1982; and much more briefly in Hamilton, S. B., 'The historical development of structural theory', *Proc.ICE*, vol. 20, pt 3, 1952, pp. 374-402. Important early sources include Bernouilli, J., 'Véritable hypothèse de la resistance des solides avec la démonstration de la courbure des corps qui font ressort', *Mémoires de l'Académie Royale des Sciences*, vol. 62, 1706, pp. 176-186 and pl. 11; Euler, L., 'Additamentum I. De curvis elasticis', in *Methodus inveniendi lineas curvas . . .*, Lausanne and Geneva, 1744, pp. 245-310 (translation in Oldfather, W. A., Ellis, C. A. and Brown, D. M., 'Leonhard Euler's elastic curves translated and annotated', *Isis*, vol. 20, 1933, pp. 72-160); idem, 'Sur la force des colonnes', *Mémoires de l'Académie de Berlin*, vol. 13, 1759, pp. 252-282; Coulomb, C. A. de, 'Essai sur une application des règles de maximis & minimis à quelques problèmes de statique, relatifs à l'architecture', *Mémoires de l'Académie Royale des Sciences*, 1776, pp. 343-382; Boistard, L. C., 'Expériences sur la stabilité des voûtes' in Lesage, P. C., *Recueil de divers mémoires extraits de la Bibliothèque Impériale des Ponts et Chaussées, à l'usage de MM. les ingénieurs*, Paris, 1810, vol. 2, pp. 171-217; Navier, C. L. M., *Rapport et mémoire sur les ponts suspendus*, Paris, 1823; and idem, *Leçons sur l'application de la méchanique*, Paris, 1826.

22. Fairbairn *Tubular bridges;* Clark and Stephenson *Tubular bridges*, pp. 83-460 and 552-589. Some of the tests are also reported in *Report into the application of iron*. For the use of testing to assist in the design of early suspension bridges, see Smith, D., 'The use of models in nineteenth-century British suspension bridge design', *History of Technology*, vol. 2, 1977, pp.169-214.

23. Cross, H., 'Analysis of continuous frames by distributing fixed-end moments', *Proc.ASCE*, vol. 56, 1930, pp. 919-928.

24. See especially Ros, M., *Versuche und Erfahrungen an ausgeführten Eisenbeton Bauwerken in der Schweiz*, 1924-1937, Zürich, 1937, Eidgenössische Materialprüfungs und Versuchsanstalt für Industrie, Bauwesen und Gewerbe, Report 99; and Maillart, R., 'Zur Entwicklung der unterzuglosen Decke in der Schweiz und in Amerika', *Schweizerische Bauzeitung*, vol. 87, 1926, pp. 263-265 (abridged translation in Bill *Maillart*, pp. 154-162). Maillart, R., 'Aktuelle Fragen des Eisenbeton baues', *Schweizerische Bauzeitung*, vol. 111, 1938, pp. 1-5 (abridged translation in Bill *Maillart*, pp. 15-16) is a forceful statement of some of the shortcomings of early standard design procedures. See also Billington,

D. P., *Robert Maillart's bridges*, Princeton University Press, 1979.

25. Freyssinet, E. 'Pre-stressed concrete; principles and applications', *JICE*, vol. 33, 1950, pp. 331-380;. Ordonez *Freyssinet*, ch.4; and Harris, A. J., 'Freyssinet: the genius of prestressing', *SE*, vol. 75, 1997, pp. 201-206.

26. Major field tests on buildings began early in this century at the University of Illinois as mentioned in Chapter 8, note 28, and have continued on both sides of the Atlantic since then. The principal ones up to 1955 are referred to in Mainstone, R. J., *Tests of the New Government Offices, Whitehall Gardens*, National Building Studies Research Paper 28, HMSO, 1960. See also, for instance, Institution of Civil Engineers, *Conference on the correlation between calculated and observed stresses and displacements in structures*, 1955; and idem, *Stresses in service*, 1967. The test illustrated in fig. 16.14 is described in Mainstone, R. J., Menzies, J. B. and Weeks, G. A., 'Dynamic measurements on an experimental helicopter platform and on the Moat Street Flyover, Coventry', *Stresses in service*, pp. 99-111.

27. Reports on the three collapses illustrated are: Ammann et al. *Tacoma Narrows Bridge*; *Report on Ronan Point*; and Central Electricity Generating Board, *Report of the committee of inquiry into the collapse of cooling towers at Ferrybridge*, 1966. See also Institution of Civil Engineers, *Natural draught cooling towers – Ferrybridge and after*, 1967. Among numerous attempts to draw more general lessons from reviews of large numbers of failures the following are worth noting: McKaig, T. H., *Building failures*, McGraw-Hill, 1962; Feld, J., *Lessons from failures of concrete structures*, American Concrete Institute/ Iowa State University, 1964; Feld, J., *Construction failure*, John Wiley, 1968; Smith, D. W., 'Bridge failures' *Proc.ICE Part 1: Design and Construction*, vol. 60, 1976, pp. 367-382; Sibly, P. G. and Walker, A. C., 'Structural accidents and their causes', ibid., vol. 62, 1977, pp. 191-208; and Blockley, D. I. and Henderson, J. R., 'Structural failures and the growth of engineering knowledge', ibid., vol. 68, 1980, pp. 719-728.

28. Rowe, R. E. and Base, G. D., 'Model analysis and testing as a design tool', *Proc.ICE*, vol. 36, 1967, pp. 677-695, is a good review of tests of this kind.

29. Otto and Schleyer *Tensile structures*.

30. See again Happold, E. and Liddell, W. I., 'Timber lattice roof for the Mannheim Bundesgartenschau', *SE*, vol. 53, 1975, pp. 99-135, and Leonhardt, F. and Schlaich, J., 'Structural design of roofs over the sports arenas for the 1972 Olympic Games: some problems of prestressed cable net structures', *SE*, vol. 50, 1972, pp. 113-119. The development of these designs might be taken as twentieth-century counterparts of the development of the design of the Conway and Britannia bridges.

31. For a brief outline of the development of the method up to 1992 by one who played a major part in it see Zienkiewicz, O. C., 'The finite element method: its genesis and future', *SE*, 70, (1992): 355-360. Its use to analyse the structures typically employed in modern buildings is well covered in Macleod, I. A., *Analytical modelling of structural systems*, Ellis Horwood, London, 1990.

32. Zunz, G. J., Glover, M. J. and Fitzpatrick, A. J., 'The structure of the new headquarters for the Hongkong & Shanghai Banking Corporation, Hong Kong', *SE*, vol. 63A, 1985, pp. 266-268 and 273-276; idem, 'The Hongkong Bank: The new headquarters – The structure', *The Arup Journal*, vol. 20, no.4, 1985, pp. 10-11 and 16-18.

33. Barnes, M. R. and Wakefield. D. S., 'Form-finding, analysis and patterning of surface-stressed structures' in Institution of Structural Engineers, *1st Oleg Kerensky Memorial Conference*, 1988, pp. 8/4-15/4. See also Barnes, M. R., 'Computer aided design of the shade membrane roofs for EXPO 88', *Structural Engineering Review*, vol. 1, 1988, pp. 3-13; and idem, 'Form and stress engineering of tension structures', ibid., vol. 6, 1994, pp. 175-202. For form-finding for space frames, see, for instance, Nooshin H. and Hadker, D., 'Exploring configuration processing', in Parke, G. A. R. and Howard C. M., eds, *Space structures 4*, 1993, pp. 1033-1043; and Nooshin, H., Disney, P.L. and Champion, O.C., 'Computer-aided processing of polyhedric configurations' in Gabriel *Beyond the cube*, pp. 343-408.

34. Pugsley, A., *The safety of structures*, Arnold, 1966 is a good discussion of the most relevant aspects.

35. See, for instance, Arup's account, in *SE*, vol. 47, 1969, pp. 419-421, of working with Utzon in his introduction to the discussion of the paper, Arup, O. N. and Zunz, G. J., 'Sydney Opera House', ibid., pp. 99-132.

36. See again the papers referred to in note 1. The best account by a recent leading designer is Torroja *Philosophy of structures*. See also Salvadori and Heller *Structure*; Happold, E., Liddell, I. and Dickson, M., 'Design towards convergence', *Architectural Design*, vol. 46, 1976, pp. 430-435; Harris, A. J., 'Can design be taught', *Proc.ICE Part 1: Design and Construction*, vol. 68, 1980, pp. 409-416; the collection of papers by Arup in *The Arup Journal*, vol. 20, no. 1, 1985, pp. 1-47; Institution of Structural Engineers, *Aims of structural design*, 2nd edn, 1987; Rice, P., *An engineer imagines*, Artemis, 1994; and (for a more detached view) Addis, W., *The art of the structural engineer*, Artemis, 1994, pp. 9-21.

Indexes

References in bold type are to figures. References in the form 349(10.6) are to notes (here to note 6, Chapter 10, on page 349). They are given only when the note will not easily be found by first turning to the text pages.

Bridges and buildings

Individual structures are indexed under the appropriate place name except for bridges and aqueducts
outside towns, which are indexed under their own
name. Dates in parenthesis following the entry indicate the approximate dates A.D. of construction or
completion unless otherwise noted.

Architects, engineers, and writers

This index includes all names of designers mentioned in the main text and of others mentioned as having contributed to the development and transmission of ideas. Because of the necessary collaboration of larger numbers of designers or groups of designers on large recent structures, these designers are mostly mentioned only in the captions in order to avoid over-burdening a text that is primarily concerned with their works. Hence their names are not included here. Names mentioned only in the notes are also omitted.